# MCM C/Mixed Technologies and Thick Film Sensors

# NATO ASI Series

## Advanced Science Institutes Series

*A Series presenting the results of activities sponsored by the NATO Science Committee, which aims at the dissemination of advanced scientific and technological knowledge, with a view to strengthening links between scientific communities.*

The Series is published by an international board of publishers in conjunction with the NATO Scientific Affairs Division

| | |
|---|---|
| **A Life Sciences**<br>**B Physics** | Plenum Publishing Corporation<br>London and New York |
| **C Mathematical and Physical Sciences**<br>**D Behavioural and Social Sciences**<br>**E Applied Sciences** | Kluwer Academic Publishers<br>Dordrecht, Boston and London |
| **F Computer and Systems Sciences**<br>**G Ecological Sciences**<br>**H Cell Biology**<br>**I Global Environmental Change** | Springer-Verlag<br>Berlin, Heidelberg, New York, London,<br>Paris and Tokyo |

**PARTNERSHIP SUB-SERIES**

| | |
|---|---|
| **1. Disarmament Technologies** | Kluwer Academic Publishers |
| **2. Environment** | Springer-Verlag / Kluwer Academic Publishers |
| **3. High Technology** | Kluwer Academic Publishers |
| **4. Science and Technology Policy** | Kluwer Academic Publishers |
| **5. Computer Networking** | Kluwer Academic Publishers |

*The Partnership Sub-Series incorporates activities undertaken in collaboration with NATO's Cooperation Partners, the countries of the CIS and Central and Eastern Europe, in Priority Areas of concern to those countries.*

**NATO-PCO-DATA BASE**

The electronic index to the NATO ASI Series provides full bibliographical references (with keywords and/or abstracts) to more than 50000 contributions from international scientists published in all sections of the NATO ASI Series.
Access to the NATO-PCO-DATA BASE is possible in two ways:

– via online FILE 128 (NATO-PCO-DATA BASE) hosted by ESRIN, Via Galileo Galilei, I-00044 Frascati, Italy.

– via CD-ROM "NATO-PCO-DATA BASE" with user-friendly retrieval software in English, French and German (© WTV GmbH and DATAWARE Technologies Inc. 1989).

The CD-ROM can be ordered through any member of the Board of Publishers or through NATO-PCO, Overijse, Belgium.

**Series 3: High Technology – Vol. 2**

# MCM C/Mixed Technologies and Thick Film Sensors

edited by


W. Kinzy Jones

Department of Mechanical Engineering,
Florida International University,
Miami, Florida, U.S.A.

Karel Kurzweil

BULL, SA,
Les Clayes-sous-Bois, France

Gábor Harsányi

Department of Electronics Technology,
Technical University of Budapest,
Budapest, Hungary

and

Sylvia Mergui

Department of Electrical Engineering,
Florida International University,
Miami, Florida, U.S.A.



**Kluwer Academic Publishers**

Dordrecht / Boston / London

Published in cooperation with NATO Scientific Affairs Division

Proceedings of the NATO Advanced Research Workshop on
Advances in Multi-Chip Modules (MCM) and High Performance Electronic Materials
Islamorada, Florida, U.S.A.
May 23–25, 1994

A C.I.P. Catalogue record for this book is available from the Library of Congress

ISBN 0-7923-3460-4

Published by Kluwer Academic Publishers,
P.O. Box 17, 3300 AA Dordrecht, The Netherlands.

Kluwer Academic Publishers incorporates the publishing programmes of
D. Reidel, Martinus Nijhoff, Dr W. Junk and MTP Press.

Sold and distributed in the U.S.A. and Canada
by Kluwer Academic Publishers,
101 Philip Drive, Norwell, MA 02061, U.S.A.

In all other countries, sold and distributed
by Kluwer Academic Publishers Group,
P.O. Box 322, 3300 AH Dordrecht, The Netherlands.

*Printed on acid-free paper*



Printed in the Netherlands

# Table of Contents

## PREFACE

Advances in electronic devices with higher speeds and complexity have placed higher demands on the electronic packaging of systems. Multi-chip modules (MCM's), with high wiring density, controlled impedance interconnects and thermal management capabilities have recently evolved to address these new requirements.

Under the auspices of the NATO Science Committee, an Advanced Research Workshop (ARW) was held on MCM-C/Mixed Technologies (Florida International University, host at Islamorada, FL, May 23-25, 1994). This volume presents the majority of papers presented at this workshop.

The book addressed advances in multi-chip packaging, with primary emphasis on ceramic multi-chip technology (MCM-C), although other technologies were reviewed. The extent and advantages of MCM technology are reviewed relative to many application areas, including the full range from high volume, cost sensitive, to low volume, high reliability space applications.

Applications in satellite communications, telecommunications, and military applications are reviewed. MCM technologies and "known good die" availability allows the introduction of MCM's into communication products at a very attractive and competitive cost. The European Economic Community is sponsoring cooperative technical projects by European companies and academic institutions. APACHIP (Advanced Packaging for High Performance) has significantly contributed to the advancement of packaging technology in Europe, covering various aspects of high performance packaging at lower cost with improved thermal characteristics and ease of assembly. Considerable research and development has been undertaken on three dimensional (3-D) packaging. Mass memories represent a major application of 3-D, in particular for space applications. 3-D interconnections will surely allow moving into microsystems which are more autonomous, denser and smarter.

This book highlights recent advances in MCM-C technology. Developments in materials and processes which have lead to the increase density of interconnection in MCM-C are reviewed. Finer resolution thick film inks, high performance low temperature dielectric tapes, precision via generation by both mechanical and laser methods, and enhanced screen printing technologies has produced feature resolution to the 50 μm line/space level. New material developments, from cofire AlN to diamond has greatly expanded the thermal management capabilities of MCM-C technology. The work of the ARW, as demonstrated in the papers, addressed new materials development and characterization methods for electronic packaging.

Because of the compatibility of thick film sensors to MCM-C technology and the work by researchers in the Cooperating Partner (CP) countries, ceramic thick film based sensors were reviewed over a wide range of applications. Environmental gas sensors, pressure and temperature sensors, as well as the development of novel materials to support the sensor development, are presented.

The organizers of the ARW gratefully acknowledge the financial support by the NATO Scientific Committee, and ISHM, The Microelectronic Society. Special thanks are due to Dr. Karel Kurzweil, who organized the NATO European participants, and Dr. Gabor Harsanyi, who organized participants from cooperating partner (CP) countries. Special thanks also to Richard Breck and Janet Kingston of ISHM for their support and guidance in organizing the conference and providing support for the manuscript.

# 3-D INTERCONNECTION MICROSYSTEMS APPLICATIONS

CHRISTIAN M. VAL
*Thomson-CSF/DOI*
*160 Blvd. de Valmy, 92704 Colombes, France*

## 1. Introduction

The reliability of the 3-D interconnection has been proven by several companies [1,2] for applications designed around the stacking of memories. The particularity of the approach initiated in 1989 at Thomson is that it was based on the concept of dual-purpose technologies. In fact, the French Ministry of Defence recommended the development of techniques that could be used for consumer products as well as ruggedized products in order to attain the level of military standards. It is a well-known fact that the inverse, that is, the transition from a military product to a mass-produced civil product, is practically impossible on account of cost.

This dual-purpose approach forced us to devise a 3-D technique based on a very limited number of very well-known operations. The following criteria were retained:

a) Use of standard chips (no bumping or re-routing, ...).
b) Stacking of any type of chip (size, function, ...).
c) Capability of testing and burning-in chips prior to stacking.
d) Power dissipation inside the block to ensure very low thermal resistance.
e) Manufacturing cost compatible with use in consumer products.

The compliance to these criteria enabled us to define a 3-D technology in which non-identical chips can be stacked together, thereby opening a broader field of application :



*W. K. Jones et al. (eds.), MCM C/Mixed Technologies and Thick Film Sensors,* 1–18.

- Stacking of mixed chips to build Multichip Modules Vertical (MCM-V).*
- Stacking of heterogeneous components, resulting in microsystems.

## 2. Description of the 3-D Technique

It consists of six main steps listed in Table 1 below.

TABLE 1.

1. Chip on tape.
2. Electrical testing and burn-in.
3. Stacking chips on film and moulding.
4. Sawing the block.
5. Plating globally.
6. Trimming the cube.

### 2.1. CHIP ON TAPE

The automatic wiring of chips glued to a golden copper conductor tape is a well-known process (see Figure 1).

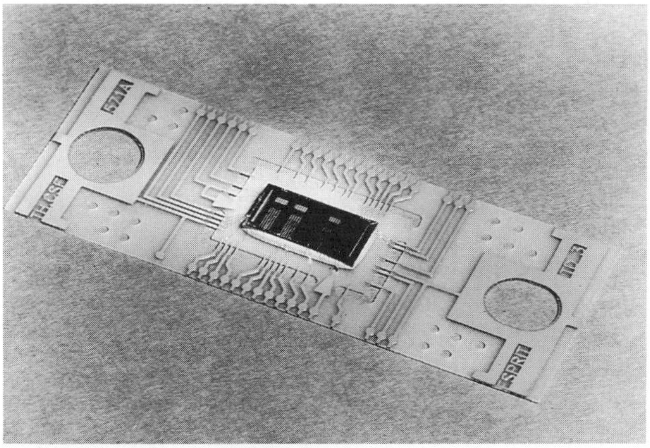

Figure 1. Wiring of chip on tape

## 2.2. ELECTRICAL TESTING AND BURN-IN

Electrical tests are performed on pads placed on the foil by means of probe cards. This process of testing prior to stacking is important because it determines the global yield of the block. It is common knowledge that the lack of a known good dice adversely contributed toward slowing down the development process of MCMs.

## 2.3. STACKING CHIPS ON FILM AND MOULDING

The process of stacking films is achieved with high precision (less than 10 $\mu$) by means of pre-existing locating holes on the tape. The locating pins are part of the mould in which flows the heavily loaded silicon coating resin (similar to epoxy resins used for the encapsulation of chips in plastic packages). After the polymerization and gelification cycle, a mould having dimensions slightly larger than those of the final block (a few millimeters more) is obtained.

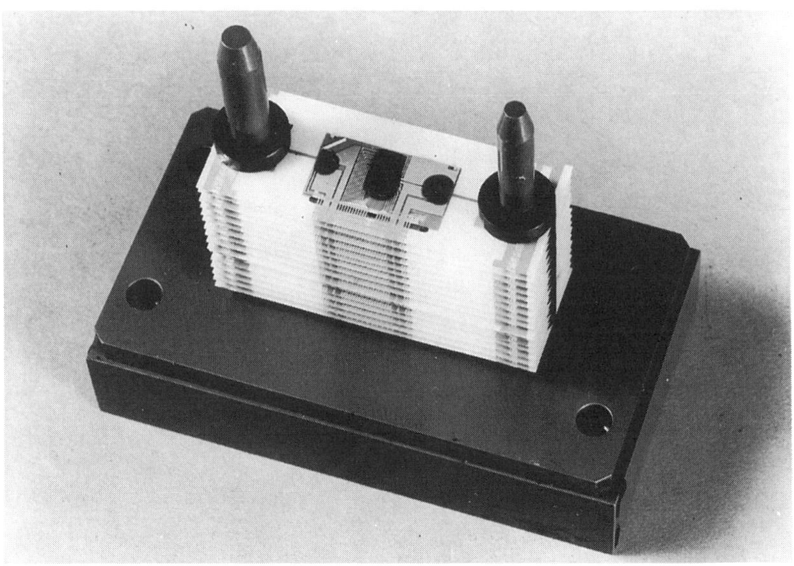

Figure 2. Stacking of 16 films with DRAM chips and centre pad - "Lead-on-chip" type wiring

## 2.4. SAWING THE BLOCK

The block is sawed with a diamond saw identical to that used to saw silicon wafers and is done at a distance of 0.2-0.4 mm from the largest chip. This depends on where the cutting line is positioned, either at the level of the gold wires (diameter of 25-32$\mu$), or at the level of the copper conductors etched on the film (35-75$\mu$). The copper conductor sectors are visible on two of the sides of the block presented in Figure 3. The cubic block is not the only possible form; a cylinder or any other form of revolution may be obtained by means of a slide lathe. Thanks to this possibility, cylindrical blocks may be stacked for use in medical and military applications.

Figure 3. Block featuring 8 SRAM chips after sawing - the copper conductor sections appear on 2 of the sides

## 2.5. PLATING GLOBALLY

### 2.5.1. *Collective Metallization of Block Sides*

"Depositing-under-vacuuum" techniques were considered too expensive to be used for the metallization of the six sides of the block. By referring to the photo in Figure 3, it is easy to see the underlying analogy between the metallization of the holes of printed circuits (copper conductor section embedded in epoxy resin) and that of the block. The same perfectly-mastered, metallization process was therefore used :

- a binder layer chemically deposited in nickel (approximately 2$\mu$).
- a conducting layer electrochemically deposited in copper (5-7$\mu$) and then a diffusion barrier in nickel and a protective layer in gold.

The interface between the gold wire section or the gold-plated copper conductor and the metallization of the cubic-shaped block's sides (Ni+Cu+Ni+Au) determined the quality of the interconnection. The metallurgy is the same as that used for the plated-thru holes of PC boards. Figure 4 presents a metallographic section of the interface.

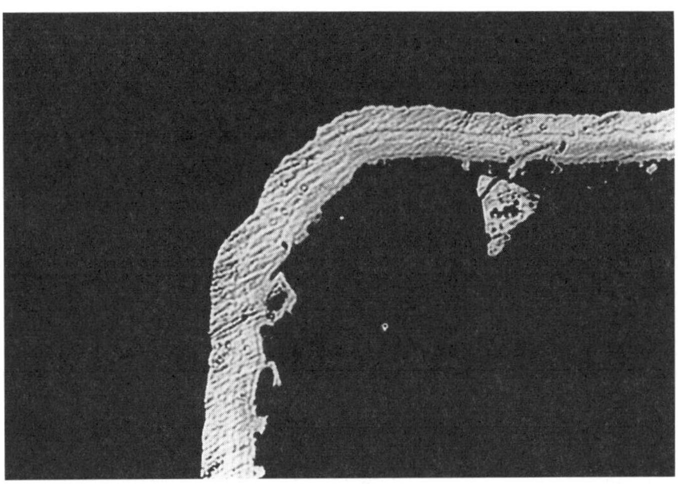

Figure 4. Metallographical section of the block's corner - The series of layers Ni+Cu+Ni+Au appear in clear colour on the photo

2.4.2. *Etching the Conductors on the Block's Sides*
The etching of the metallic layers is accomplished by means of a YAG laser beam. As one may observe on the photo in Figure 5, a ground plane may encircle all of the signal routings in order to limit crosstalk.

Figure 5. Functional blocks featuring 8-16 memory chips

The minimum width of the isolating kerf is 40-50$\mu$. A form recognition device working in conjunction with the laser allows one to know the position of the wire or conductor sections in order to automatically perform any possible positioning corrections. A current study [3] is being pursued to eventually allow the direct deposit of metal photo-assisted by laser for conductor widths of a few microns.

The flexibility of etching signal routings on one or more sides of the block allows one to functionally select the sides. That is,

- "signal routing" sides.

- "cooling" sides; the copper conductor sections on which are glued the chips appearing after sawing constituent so many mini heatsinks.
- "block-on-a-substrate" bonding sides (use of bonding pads "BGA", bonding grid, ...).

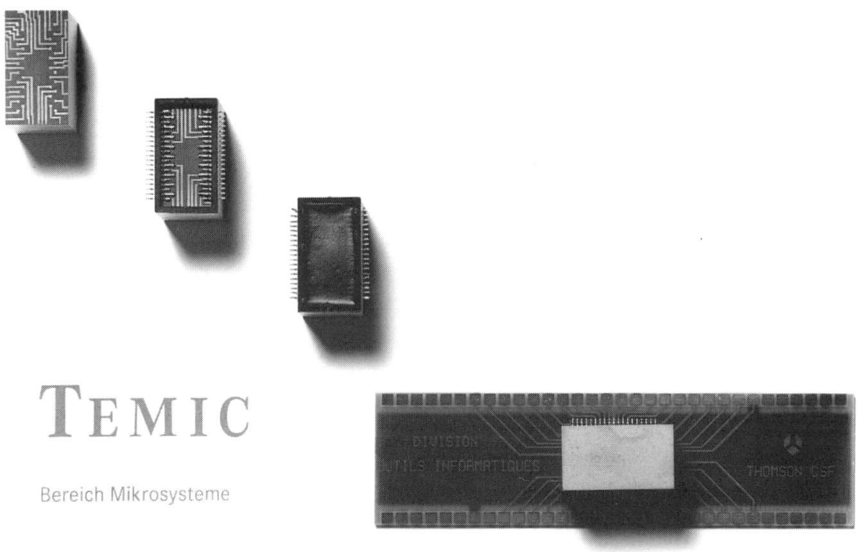


Figure 6. Block with leads soldered by "Temec"

## 3. Assessment of the 3-D Technique (MCM-V)

### 3.1. ESPRIT PROGRAMME

The 3-D stacking technique was realized, on the one hand, through Thomson-CSF inhouse programmes, and, on the other, through a major European ESPRIT programme "TRIMOD " [4] initiated in 1992, which is scheduled to terminate in July 1994. The team members of this ESPRIT consortium include the following firms and university : Alcatel-Espace (F), Implex (UK), Daimler-Benz/Temic (D), Thomson-

CSF (F), University of Sheffield (UK), and NMRC (IR).

Six types of demonstrations have been realized to date :

a) Three types of technological demonstrators, allowing to measure constraints, temperatures, corrosion, and contact resistance between wire or conductor sections and the lateral metallization on each of the eight levels of each of the 150 blocks engineered (1,200 test chips).

b) Three functional demonstrators, consisting of:

- Blocks of 16 16Mbit-DRAM chips stacked and defined by Alcatel-Espace.
- Blocks of 9 chips, 8 of which being 8Mbit-Flash memory chips associated with a command ASIC chip. These blocks are assembled four by four in a 104-pin PGA ceramic package, so that 36 chips can be placed in this package (defined by Implex).
- Mixed multichip block MCM-V, constituting a microcamera and featuring 9 chips, 40 discrete and passive components, and 1 microlens (defined by the University of Sheffield).

The simulations and measurements performed on the blocks featuring 16 x 16Mbit DRAM memory chips underscored very good thermal behaviour. As a result, with 50 mW of power per chip and a block side at 60°C, the temperature rise due to natural convection is not significant (Figure 7). This concurs with the very low thermal resistance of the "chip/copper mini heatsink under each chip/block conductor wiring/heatsink".

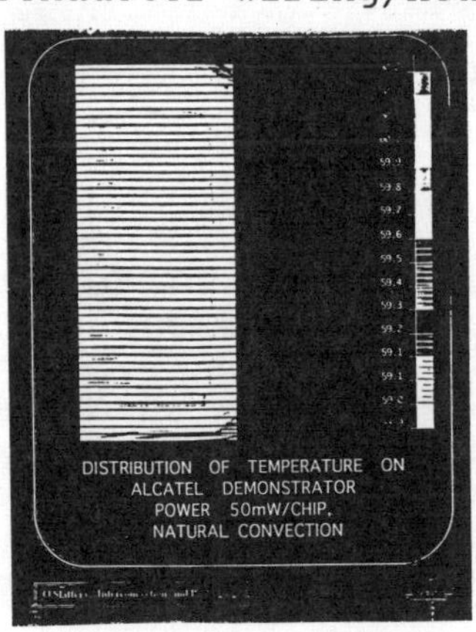


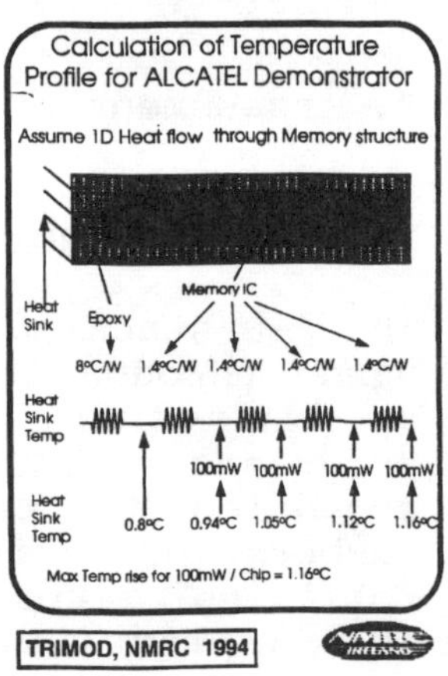


Figure 7.

The thermomechanical simulations performed by NMRC stressed the following points, which were taken into account in the design of the functional blocks:

- Dimensional symmetrization of the block.
- Choice of epoxy resins, increasing Young's modulus from 3,700 MPa to 9,400 MPa increases stress by 50% at corners and 100% at the centre.
- Reducing the moulding compound CTE from 35 ppm/°C to 18 ppm/°C reduces stresses by 100% at corners and 65% at the centre.
- Glass transition temperature has a significant effect on stress levels.
- Bonding resin/foil, resin/gold-plated copper, and resin/silicon is a key factor to prevent delaminations (phenomenon known in TSOP plastic packages).

Figure 8 presents the maximum plastic tensile stress vs. distance along plastic above centre of middle die to edge distance for various resins. The resin ref. EO1060 gives interesting results. The use of a new resin ref. EO4450 results in plastic tensile stress levels of an even lower magnitude.

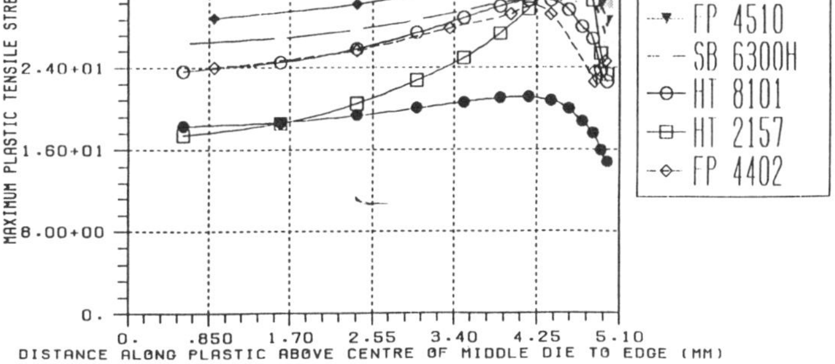


Figure 8. Stress vs. distance along plastic above centre of middle die to edge (mm).

The results of the assessment concerned 150 blocks, or 1,200 chips and 12,000 contact resistances, and will be totally completed by the end of July 1994. These results are excellent to date, especially for the resin EO4450 with a very high mineral content (70%).

TABLE 2.

|  |  |
|---|---|
| - Life testing | 3,000 hours |
| - Humidity testing | in progress |
| - HAST | in progress |
| - Temperature cycles | |
| -55°C to +105°C | 1,000 (to be continued) |
| -55°C to +125°C | 800 (to be continued) |
| -130°C to +20°C | 700 (tests performed to simulate a Martian year (ESA/CSEM)) |
| - Constant acceleration | 30,000 g's |

The stacking of the microprocessor chips (use of IBM and Motorola POWER PC 601 and 603 chips), DRAM memory chips, and buffer chips is being realized under a European programme for the development of synthetic diamonds [5]. The use of a diamond heatsink between each dissipating chip will allow evacuating the energy developed in the 2.4 $cm^3$ of the block, or 8-20W.

The team members of the consortium for this programme are Gec Marconi (UK), Daimler-Benz (D), and Thomson-CSF (F) and the programme is scheduled to terminate in 1995.

## 3.2. FRENCH PROGRAMME RELATED TO THE STACKING OF PLASTIC PACKAGES

A research programme sponsored by the French Ministry of Research and Space in 1993 will culminate by the development of a collective stacking technique for TSOP memory chips employing the same aforementioned technique.

Tests related to the stacking of film consisting of 64 TSOP packages are promising and will significantly reduce labour costs. This collective process will be subsequently applied to the coupling of bare chips.

## 4. Applications to Stacked Memory Chips

### 4.1. DRAM MEMORY CHIPS STACKED 16 BY 16

Memory blocks for space applications were developed for Alcatel-Espace (Figure 9). They consist of 16 x 16Mbit-DRAM memory chips manufactured by IBM Corbeil. The entire block therefore has a capacity of 256 Mbits for a volume of 3.4 $cm^3$, or a memory capacity/volume of 75 $Gbits/dm^3$. In addition, two generations of memory chips, specifically the advent of the 64 Mbit memory chip and then the 256 Mbit memory chip, could use the same type of stacking.

Figure 9. Block featuring 16 x 16Mbit-DRAM chips, or 256 Mbits in 18 x 19 x 10 mm

### 4.2. FLASH MEMORY BLOCKS

Blocks consisting of 8 Intel 8Mbit Flash memory chips associated with a command ASIC chip are mounted 4 by 4 in a hermetically sealed 104-pin PGA package. This product, designed by our license partner Implex, is currently undergoing qualification in the United States for application to airborne military avionics systems (Figure 10).

Figure 10. Blocks featuring 8 memory chips + 1 ASIC chip mounted four by four in a hermetically sealed package

## 5. Applications to Heterogeneous Multichips MCM-V

### 5.1. MICROCAMERA

The 3-D technology was first implemented in microsystems as a microcamera for application to portable video communication. The microcamera consists of a microlens, a sensor (a 312 by 287 photodiode array) in conjunction with image processing, storage, and compression electronics, or 9 chips and 40 discrete components distributed over 8 levels (Figure 11). It measures 14 x 11 x 10 mm, or 1.5 cm$^3$.

This application illustrates the capabilities of miniaturization offered by this technique.

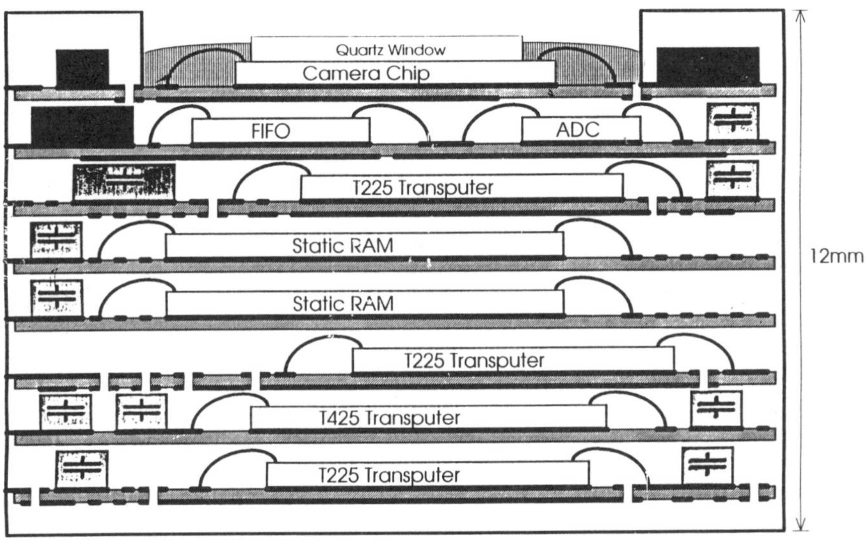


Figure 11. Schematic cut-out of the microcamera

## 5.2. AUDIO-DENTAL PROTHESES

### 5.2.1. *Elementary Microsystem Block for Medical Applications*

Concepts studied 20 years ago [6] on the conduction of audio-dental protheses have been reactivated employing 3-D techniques. The piezo-electric actuator works in conjunction with a command electronics. The supply of autonomous energy constitutes a real problem that we hope to solve by decoupling power supply and actuator module functions from the associated electronics for use in future microsystems envisioned under the MITI programme [7].

## 6. Special or Specific Applications of the MCM-V

### 6.1. HIGH FREQUENCY APPLICATIONS

The integration of chips operating in high frequency and low frequency bands is being investigated by Thomson-CSF in order to develop active antenna blocks as well as HF blocks.

### 6.2. APPLICATIONS TO SMART MUNITIONS

Those applications in which weight and volume are critical factors are encountered in smart munitions such as 155 mm howitzer shells, submunitions, small rockets, and smart bullets. In some cases the electronic device must operate after impacting in the cavity. Packaging specifications are extremely stringent :

a) high acceleration : 30 to 100,000 g's.
b) electrical contact maintained at all moments.
c) shape miniaturized and adapted to munition.
d) long term storage capability.
e) very low production cost.

Both positive and negative accelerations were tested up to 30,000 g's in our Munitions Division (TBA).

Long term storage capability is, in general, difficult to associate with low production costs due to the use of thermal accessways. So that plastic packaging or plastic encapsulation (MCM-V) may be used under stringent environmental conditions, the "deposited hermetic seal" concept was developed. It concerns depositing a mineral substance (SiOx or Si3N4) by means of the Plasma Enhanced Chemical Vapour deposition process at low temperature and in thicknesses on the order of 1 micron. This single or double layered diffusion barrier provides a hermetically sealed encapsulation around the block [8].

The miniaturization of the form is achievable thanks to the versatility of this 3-D technique, in particular, the possibility of obtaining cylindrically-shaped volumes, for example, by using a slide lathe. Figure 12 presents an outside view of the 3-D module. Three types of surfaces are observable:

- a cylindrical surface, supporting the interconnection conductors between the layers; the etching of these conductors (trimming) is readily realizable by laser, while photolithography would be practically impossible.
- a side of the cylinder, also supporting the interconnection between layers.
- a flat section, allowing the installation of connecting pads to the outside of the MCM-V, that is, soldered wires, contact by pressure, etc.

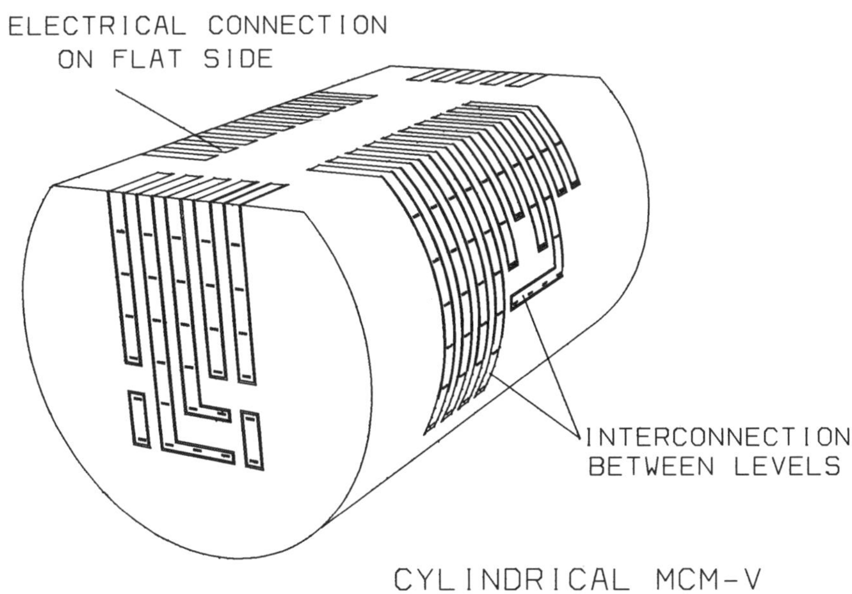


Figure 12. External view of a cylindrical 3-D module

The internal section presented in Figure 13 focuses on the components (chips, capacitors, ...) mounted on the various foils constituting the different levels. One can observe the diverse layers along with the conductors that are flush with the outside surface of the cylinder.

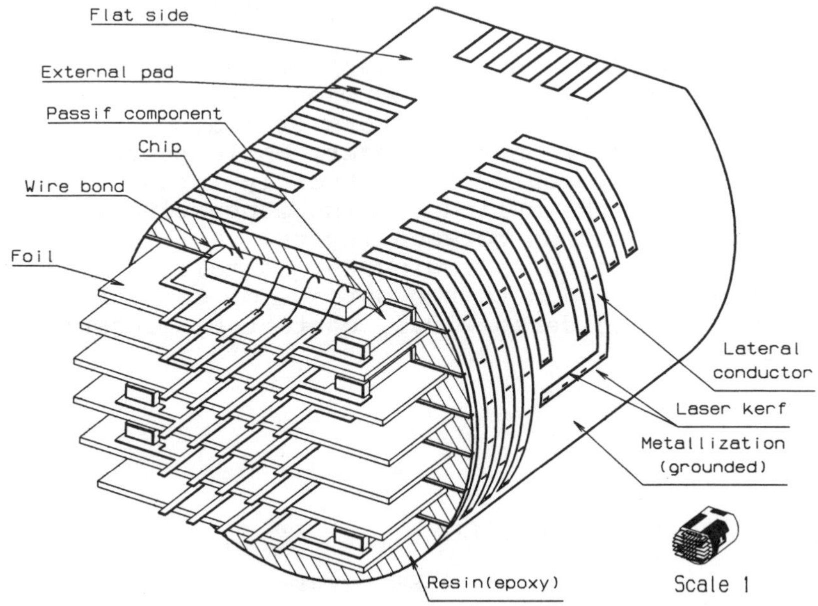


Figure 13. Cut-out view of an MCM-V module

Figure 14 illustrates the module MCM-V inside the munition's body (small rocket or small bullet, for example). Custom machining allows adapting the various diameters of the munitions with precision.

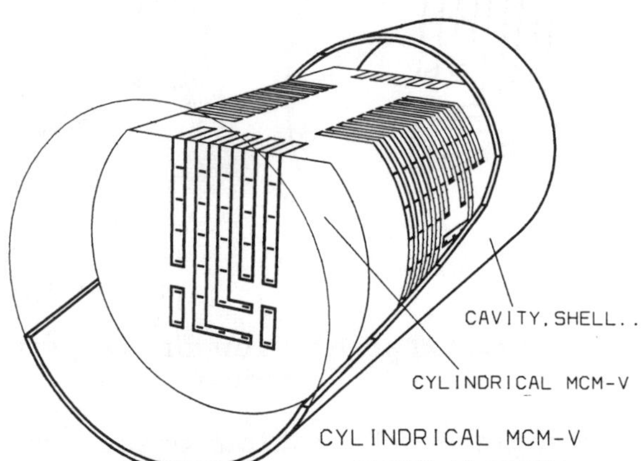


Figure 14. Cylindrical MCM-V placed in cavity

## 7. Extension to Microsystems - Perspectives

The simplified diagram in Figure 15 presents the various "building blocks" constituting a microsystems. The "electronic" building blocks represent an important part of the volume, but - as one can see - an "optical" part also appears which is used for the interconnection, the teletransmission of energy, and optical calculations. Finally, the heart of the microsystem is built around micromachines, which in general cannot be encapsulated in resin on account of the mobile parts [9].

In addition, the extension of this 3-D technique to microsystems, high temperature applications (higher than 150°C), or high-speed interconnections forced us to search for a technique which does not rely on plastic materials. To do this, it is sufficient to add a single step to the flowchart given in Table 1. All the other steps are conserved. In this case, all of the chips are interconnected by "beams" isolated by air (dialectric constant equal to 1), which increases transmission speeds and prevents the micromachines from being mechanically constrained. As a result, the device attains an almost nominal weight since it is only built with elementary components (including the "interconnection" component).

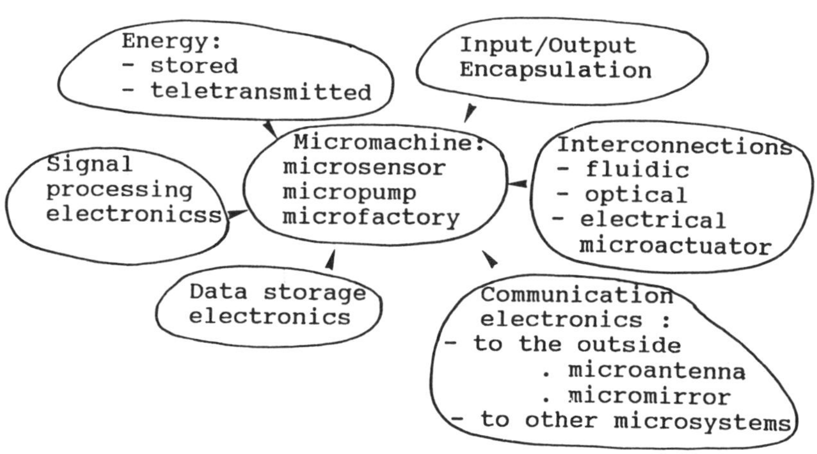


Figure 15. Simplified diagram of a microsystem

## 8. Conclusions

Vertical interconnection will surely allow moving into microsystems which are more autonomous (communication device and/or data storage device), smarter (associated signal processing), and denser. It is common knowledge since the development of integrated circuits that there exists a close relationship between the cost of a device and its weight or volume. The advent of microsystems and their associated electronics assembled in a 3-D configuration will considerably lower costs.

Starting from stacked memory applications, to memory chips associated with ASICs, then Multichip Modules Vertical MCM-V (microcamera, computing block on a Power PC 603 chip), and now microsystems for implanted medical protheses, it is evident that all major electronics fields are concerned because of the cost of these microsystems vs. volume.

Applications in high-speed 3-D electronics (with or without plastic encapsulation) using this 3-D technique for electronics installed in smart munitions will also open a very promising field.

## 9. References

1. Texas Instruments Semiconductor Technology Trends, Texas Instruments France Symposium 1990.
2. Suer, M.F. (1991) Pursuing 3D packaging, *Electronics Engineering Times*.
3. European Programme Brite-Euram 1994-97 Photo assisted catalysis of electroless deposition.
4. European Programme Esprit 1992-94 Tridimensional modules "TRIMOD" No. 6490.
5. European Programme Brite-Euram 1993-96 Diamond for applications in thermal management "DAITM".
6. Dahlin, G.C. and Allen F.G. (1973) Bone conduction thresholds of human teeth, *Journal of the Acoustical Society of America* 53, pp.1434-37
7. Challenge to the micro-universe micromachine; Introduction to micromachine centre MITI/NEDO 1992.
8. Val, C., Leroy, M. and Bourret, D. (1991) Hermetic plastic packages with application to ruggedized boards, Proc. of 11th IEEE/CHMT Conf.
9. Esteve, D. and Val, Alexandre CNRS/LAAS, personal communication.

# HIGH PERFORMANCE PACKAGING WITH MULTI LAYER CERAMIC MODULES

ECKARDT BIHLER
*IBM Germany*
*71044 Sindelfingen, Germany*

## 1. Introduction

With increasing performance and functional integration of semiconductor technology the packaging of the chips gains more importance for the design of complex microelectronic systems. To avoid a major loss of system performance a relevant discussion of the packaging technologies has to be started early in the design process.

Ceramic chip carriers are an important element of future packages, which have to fit to the continously improving semiconductor performance. Ceramic packages offer several advantages compared to conventional plastic packages. The superior thermal and mechanical properties ensure an excellent reliability. The high layer count and via density give a high wireability satisfying the demands of state of the art CMOS logic.

The further development of ceramic materials from alumina based ceramics to glass ceramic systems will bring better electrical performance needed to support the increased speed and interconnection density of future semiconductor technology.

In order to achieve the optimum performance at the lowest cost a comprehensive, concurrent design effort of the semiconductors, packages, interconnection and the system architecture is needed. The following paper will therefore start with a brief discussion of the requirements for electrical performance and then draw the necessary conclusions for the packaging technology.

## 2. Key Factors to Electrical Performance of Packages

### 2.1. SIGNAL ATTENUATION

Along the signal paths the signal will be attenuated due to the resistance of the wires. This is the dominant effect at low and medium frequencies. The total attenuation depend mainly on the wire length and the material. At frequencies above 1 GHz a.c. (200 MHz



*W. K. Jones et al. (eds.), MCM C/Mixed Technologies and Thick Film Sensors,* 19–22.

pulse equivalent) the skin effect in the conductor and the dielectric absorption in the insulator cannot be neglected anymore.

## 2.2. SIMULTANEOUS SWITCHING AND CROSSTALK

Especially for data processing applications with complex logic functions the number of simultaneous switching drivers will be a critical factor for the noise generated in the package. There are basically two sources of noise: 1. The noise on the power supply caused by the inductance of the supplying nets. It will increase with the cycle frequency and the number of simultaneous switching drivers. 2. The noise induced by the crosstalk between adjacent signal paths. The signals can be coupled by the magnetic or the electric field. The driver characteristics will determine, whether the lines are coupled predominantly by the current or the voltage. In both cases the layout of lines and vias, especially the line pitch determine the coupling coefficent. Decreasing the line pitch will increase the coupling. With increasing frequency the coupling becomes more important. In the case of capacitive coupling the dielectric permittivity of the insulator is an additional parameter.

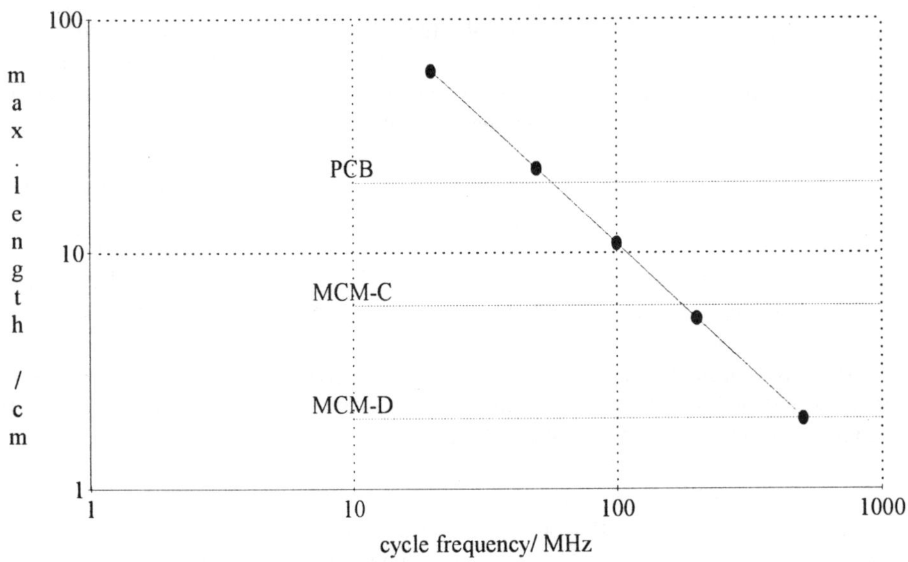


FIGURE 1: Decision matrix for high frequency design. Above the line the system is considered to be fast. A complete high frequency design has to be performed. Signal reflexions have to be taken into account. Below the line no special effort has to be considered concerning high frequency simulation and design.

### 2.3. REFLEXION AND SIGNAL DELAY

The transit time of a signal between driver and receiver is determined by the length of the connection and the speed of light, which is reduced in dielectrics by the square root of the dielectric permittivity. If the signal paths are long compared to the wavelength of the highest frequency (e.g. the rise time of the pulses), reflexions of the signal at interconnections and terminations have to be taken into account. In this case the design of the package has to include a high frequency simulation, which may be a hidden cost adder. For deciding on the neccesity of a high frequency simulation a rough estimate could be helpful: The maximum wire length, which could be estimated as half of the circumference of the module or board should be less than 10% of the wave length calculated from the cycle frequency of the system. Figure 1 shows this graph of decision line between "slow" and "fast" system design.

### 2.4. WIREABILITY

Beside the electrical characteristics of the signal and power nets the wiring capability of the package technology is an important aspect to decide on the best solution for a microelectronic application. The wireability is determined by the pitch of lines and via connections between wiring layers. Together with the maximum number of wiring layers this yields a limit of the maximum number of channels useable for interconneting the chips and the interfaces. The number of channels per unit length is calculated by the number of channels between two vias (can be derived from the line pitch) devided by the via pitch and multiplied by the signal layer count. (see Table 1)

| | unit | LTCC | Thinfilm | rigid PCB | Flex |
|---|---|---|---|---|---|
| line pitch | mm | 0,20 | 0,05 | 0,20 | 0,15 |
| via grid pitch | mm | 0,45 | 0,10 | 0,68 | 0,30 |
| wires/grid | | 1,00 | 1,00 | 1,00 | 1,00 |
| signal layers | | 30,00 | 3,00 | 8,00 | 4,00 |
| wiring capability | channels/mm | 70,00 | 30,00 | 12,00 | 14,00 |

TABLE 1: Typical data of different packaging technologies used to estimate the maximum wireability

## 3. Comparison of Packaging Technologies

### 3.1. DECISION CRITERIA

In order to decide for a given microelectronic application between the different possible packaging technologies it is necessary to consider a set of criteria. This is an important prerequisite to achieve an optimum balance between required performance and lowest cost.

In the second paragraph of this paper the key factors for electrical performance has been discussed in some detail. Within the frame of this paper it is not possible to cover the other criteria in similar detail. Additional criteria will be reliability, thermal performance, weight and size. To cover the economic aspects also the availability of components, processes and equipment have to be checked and evaluated. The costs of the various components and assembly processes have to be collected and summed up before a final decision is possible.

Table 2 summarizes most of the important data needed to compare alumina and glass based multi layer ceramics with printed circuit board (PCB) technology.

| | unit | HTCC | LTCC | FR4 |
|---|---|---|---|---|
| MATERIAL PROPERTIES | | | | |
| conductor | | moly/tungsten | silver | copper |
| dielectric | | Al2O3 | glass+Al2O3 | glass epoxy |
| dielectric constant | | 9.5 | 7.8 | 5 |
| thermal conductivity | W/mK | 20-30 | 2-4 | 0,2 |
| thermal coeff. exp. | ppm | 6.5 | 6.0 | 16 |
| MECHANICAL PROPERTY | | | | |
| layer thickness | mm | 0.2-0.5 | 0.2 | 0.1 |
| minimum via pitch | mm | 0.225 | 0.225 | 0.65 |
| bending strength *) | MPa | 500 | 280 | 450 |
| INTERNAL WIRING | | | | |
| line width | mm | 0.1 | 0.1 | 0.1 |
| wiring capability | channels/mm | 70 | 70 | 10 |
| ELECTRICAL CHARACT. | | | | |
| line capacitance | pF/cm | 2.2 | 1.8 | 1.3 |
| line resistance | Ohm/cm | 0.32 | 0.25 | 0.1 |
| line propagation | ps/cm | 103 | 95 | 80 |

*) 5 mm/s cross head speed

TABLE 2: Comparison of technical data for alumina based ceramic (HTCC), low temperature cofired ceramic (LTCC) and state of the art printed circuit board (PCB) technology.

## 3.2. LOW TEMPERATURE COFIRED CERAMICS

Low temperature cofired ceramic (LTCC) exhibits improved electrical characteristics. The reduced sintering temperature of approx. 850 centigrade at ambient atmosphere allows the usage of noble metals with better conductivity. Additional postfiring of thickfilm resistors or sensors expands the usage of this ceramic for complex circuits and microsystems. The decreased dielectric permittivity of the glass/alumina mixture gives reduced signal capacitance and coupling. This is of importance for advanced, high density CMOS logic and fast applications.

The planar surface of multi-layered, cofired ceramics together with the high density of via interconnects allows direct flip chip attach. This type of interconnection is suitable for high chip I/O count and HF applications demanding for less inductance. The surface can be polished in order to deposite additional thin film redistribution layers, if necessary.

# HIGH FREQUENCY LTCC MODULES

P. JOHNSSON, C.VAKTNÄS, N. BILLSTRÖM
*Ericsson Radar Electronics AB*
*P.O. Box 1001*
*S-431 84 Mölndal, SWEDEN*

## 1. Introduction

The Low Temperature Cofired Ceramic (LTCC) technology has been given increased attention over the last years [1]. This is because LTCC has proven to be a viable substrate technology for many applications where low cost and high reliability are required. The major part of these applications are digital circuits operating at moderate speed.

Ericsson Radar Electronics has used LTCC technology during several years in military computing applications and for telecom applications at 2.5 GHz. Due to the good results obtained with these circuits, LTCC is now studied for applications at even higher frequencies.

## 2. Experimental design

In order to investigate the high frequency properties of the low temperature cofired tape as a substrate material, a test circuit was designed. An amplifier at 4.5-6.5 GHz was designed with different substrate materials. The amplifier circuit consists of 2 GaAs FETs, 2 resistors and 7 capacitors. The test circuit does also contain a 50 ohm microstrip of 40 mm length.

The test circuit comes in two alternatives (Fig 1). The first one, called "normal", has all the circuitry in one single layer on top of the substrate with the backside as a ground layer. The other one, called "via", has the DC-circuitry in the interior and the high frequency circuitry on the top side with an internal ground plane.

The test circuit has been fabricated with two LTCC types and with a cyanate ester PCB. The tapes come from DuPont and Ferro EMD. DuPont tape has only been used in one type of circuit, the one with two ground planes. The signal ground plane is gridded for the two LTCC substrates and solid for the cyanate ester.



*W. K. Jones et al. (eds.), MCM C/Mixed Technologies and Thick Film Sensors*, 23–28.

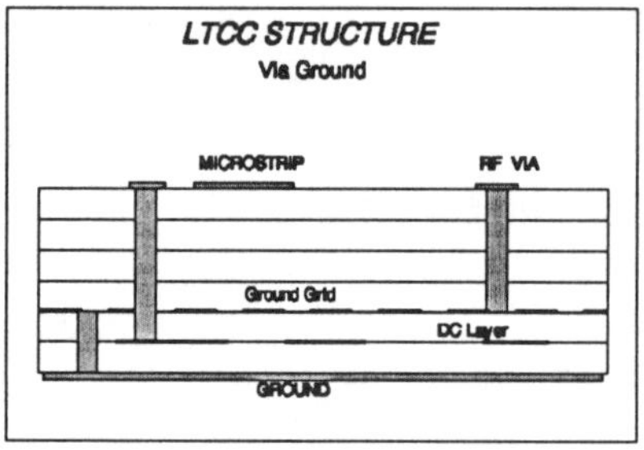


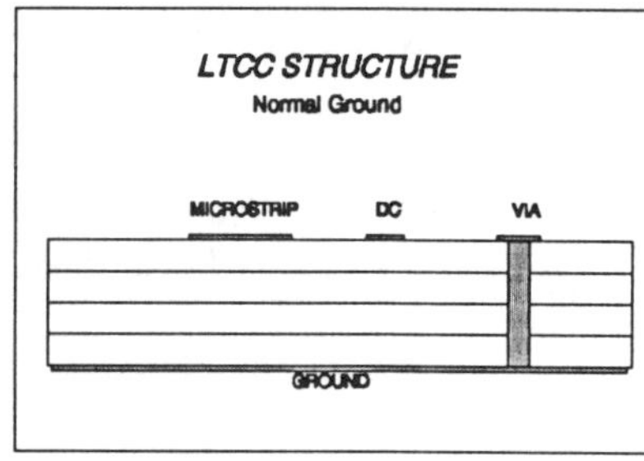


*Figure 1.* Cross section of substrates. To the left the multilayer design with an interior signal ground plane (via ground). To the right, the backside is the signal ground plane (normal ground.)

The electrical properties of the three materials are found in table 1. The Ferrotape A6 is marketed as a tape suitable for high frequency applications. This tape contains more glass than the DuPont. Cyanate ester is a polymer material. The cyanate ester laminate is reinforced with woven glass fibre and manufactured with normal PCB techniques.

TABLE 1. Electrical properties of the substrate materials

| | Dielectric const. (5 GHz) | Eff. dielectric const (50 Ohm) | Dissipation factor | Distance to ground (mm) | Width of 50 Ohm line (mm) |
|---|---|---|---|---|---|
| Ferrotape A6 | 5,9 | 4,3 | 0,005 | 0,51 | 0,77 |
| DuPont 851 | 7,9 | 5,6 | 0,008 | 0,38 | 0,44 |
| Cyanate ester | 3,8 | 2,9 | 0,010 | 0,30 | 0,64 |

## 3. Manufacturing

### 3.1. DUPONT 851

The DuPont substrates were manufactured by a French company called SOREP. Silver was used as conducting material in inner layers and in vias. Top conductors and backside were printed with gold paste. The resistors were incorporated in the substrate with screenprinting.

On the conductor edges it is possible to see the screen pattern. The thickness is about 8 microns and the width of the 50 Ohm line is measured to 400 microns, which is 10 % narrower than it was designed for.

### 3.2. FERROTAPE A6

The Ferro substrates were manufactured by a German company named Siegert Electronic. Silver was used as conducting material in inner layers and in vias. Top conductors and backside were printed with gold paste. Resistors were screenprinted.

There was problem in finding a European substrate manufacturer that had any experience with the Ferro tapesystem. Part of the production of the Ferro substrates was done at Ferro EMD in California (sintering of resistor paste).

The top gold print was found to be very thin, only 3 to 4 microns and there were rather large pores in the conductors. The Ferrotape is also very brittle. It is very easy to snap off corners if care isn't exercised when handling the substrates.

### 3.3. CYANATE ESTER

The cyanate ester substrates were manufactured by Photo-Chemie, a Swiss PCB manufacturer. Copper is the conducting material. The laminates were Isola Duraver Ce-Cu

## 4. Electrical measurements

### 4.1. MICROSTRIP MEASUREMENTS

The S-parameters of a 50 ohm microstrip has been measured. In figure 2 and 3 the insertion loss for the three materials are given together with the simulated result. As can be seen from the graphs, the measurements on the circuits with two ground planes exhibit ripple for the LTCC materials starting at frequencies just below 3 GHz. This is probably due to the fact that the signal ground plane is gridded. The grid openings are 600 microns square, and the plane is approximately filled to 75 %. In the cyanate ester circuit, which doesn't show any ripple, the signal ground plane is solid. It is possible that a grid with smaller openings would decrease the ripple but that can be difficult from a manufacturing point of view. A solid groundplane would cause problems with warpage and mechanical stress in the substrate.

The insertion loss per wavelength is given in table 2.

TABLE 2. The insertion loss per wavelength (dB) for a 50 ohm microstrip with normal ground plane.

| | 1 GHz | 10 GHz | 17 GHz |
|---|---|---|---|
| Ferro measured | 0,43 | 0,22 | 0,18 |
| Ferro simulated | 0,43 | 0,25 | 0,24 |
| DuPont simulated | 0,51 | 0,39 | 0,36 |
| CE measured | 0,88 | 0,58 | 0,58 |
| CE simulated | 0,53 | 0,39 | 0,34 |

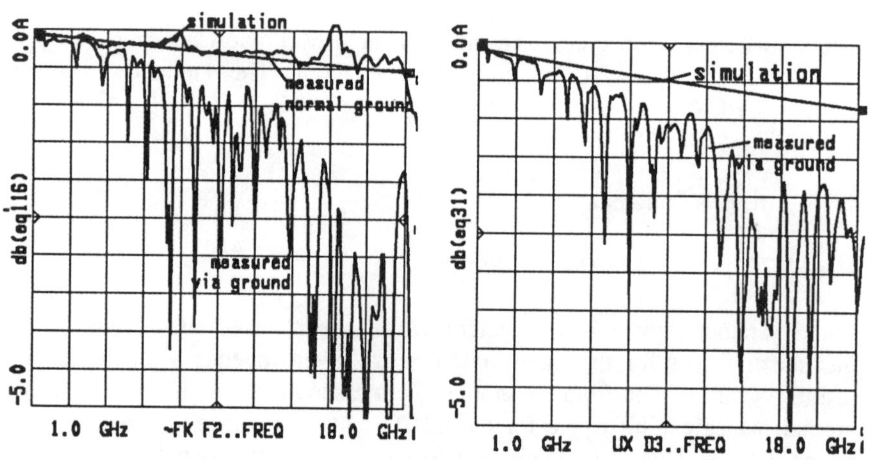


*Figure 2*. The insertion loss in dB for a 18 mm microstrip on Ferrotape A6 (left) and the insertion loss for a 18 mm microstrip on DuPont 851 (right). Scale on the vertical axis is from 0 dB to -5 dB. Frequency range is 1 to 18 GHz.

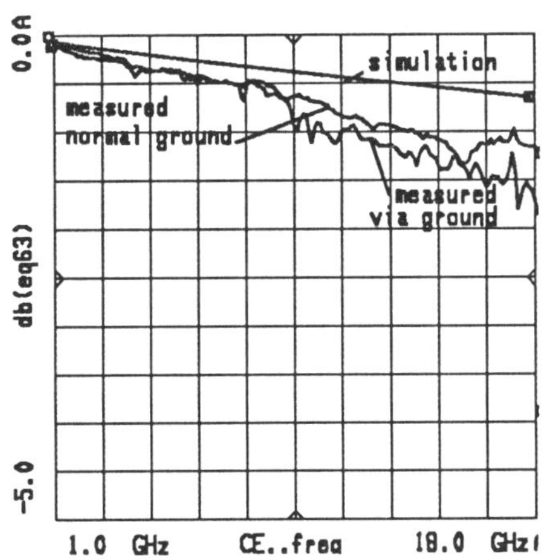


*Figure 3.* The insertion loss in dB for a 20 mm microstrip on cyanate ester.

## 4.2 AMPLIFIER MEASUREMENTS

The two-stage amplifier described above has been measured and the result has been compared to software simulations (figs 4 and 5). The similarity between measured and simulated data is good for all substrate types except for the LTCCs with via ground, where the ripple that was found in the microstrip measurement still can be seen. The ripple over the band is to large to be OK in these two cases. Cyanate ester is the only substrate that gives good data over the band in the via ground design.

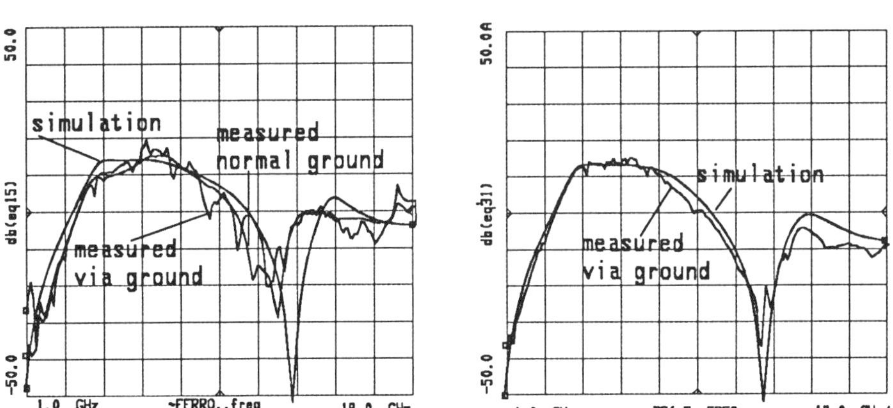


*Figure 4.* The gain in dB for the amplifier on Ferrotape A6 substrate (left) and the gain for the amplifier on DuPont 851 (right). Scale on the vertical axis is from -50 dB to +50 dB. Frequency range is 1 to 18 GHz.

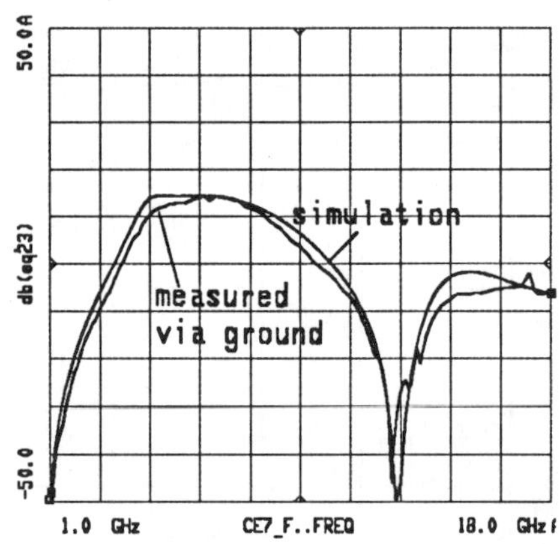


*Figure 5.* The gain in dB for the amplifier on cyanate ester. Scale on the vertical axis is from -50 dB to +50 dB. Frequency range is 1 to 18 GHz.

## 5. Conclusions

We have seen that the use of a gridded ground plane can cause problems at frequencies above 2-3 GHz. The gridded ground plane gives ripple in the measurements on the microstrips. The gridded ground plane does also increase the characteristic impedance, a fact that has been observed by others [2]. If a gridded ground plane is going to be used it is advisable to try to make the grid openings as small as possible.

We found a lower loss for the Ferrotape A6 compared to both DuPont 851 and cyanate ester.

We also found that cyanate ester seems to work well in multilayer structure at high frequencies. In designs were you can tolerate a higher loss, cyanate ester would be a good substrate choice. Another favorable property of cyanate ester is the short turnaround compared to the longer processing times of the LTCCs.

## 6. References

1. Cicognani, J. et al. "LTCC-An Economical and Industrial Interconnect and Packaging solution for MCM´s", Proceedings of the 9th European Hybrid Microelectronic Conference, Nice 1993, 150-157.

2. Vickes, H. "Characterization of Low Temperature Cofired Ceramic Structures for Microwave Applications", Proceedings of the 30th ISHM-Nordic Conference, Oslo 1992, 229-240.

# Analysis and optimization of circuit interconnect performance

Joris Peeters, Eric Beyne

*IMEC, Kapeldreef 75 - 3001 Leuven (Belgium)*

## 1 Introduction : a system approach for interconnection design

If the interconnection between two integrated circuits is designed for high speed data transmission, not only the physical line parameters matter, but also the on-chip driver and receiver characteristics. The behaviour of the interconnection line is primarely determined by the interconnect technology which is used. Low loss and generally relatively long printed circuit board interconnections behave differently than lossy MCM-type interconnections and require a different approach. For digital signal transmission on uniform interconnections, lines can be described using their equivalent distributed network parameters and the interconnection length. Driver characteristics depend on the IC technology which is used. Further on, the driver behaviour is determined by its circuit design (e.g. : the number of driver stages) and by the dimensions of the driver transistors.

Designing interconnection schemes for optimum performance therefore requires a combined analysis of interconnection and terminal circuitry. For lossy lines and non-linear terminal networks, interactions between lines and driver / receiver circuits becomes very complicated and requires a global numerical analysis using an accurate numerical simulation tool. The network simulation program Transplus has been developed for this purpose. Numerical simulations are however time consuming. For that reason an analytical study is performed based on a simplified interconnection network. In this way, by means of an approximated analysis of line to driver / receiver interactions, some general design guidelines can be formulated to support a detailed numerical analysis.

## 2 Analytical study of single line performance parameters

*2.1 Ideal situation : fundamental switching modes*

An analytical study of interconnection network characteristics is always based on a simplified representation of the network components. Several examples are met in general packaging handbooks. Interconnections are generally ideal and lossless. Figure 1 shows a basic representation of such an ideal interconnection circuit. Interconnections are represented by their characteristic impedance $Z_o$ and phase velocity $v_p$, expressed by :

$$Z_o = \sqrt{\frac{L}{C}} \qquad \text{and} \qquad v_p = \frac{c_o}{\sqrt{\varepsilon_r}} = \frac{1}{\sqrt{LC}} \qquad (1)$$

with L and C respectively the inductance and capacitance p.u.l., $c_o$ the speed of light in vacuum and $\varepsilon_r$ the (effective) relative dielectric constant (it is commonly assumed that no magnetic properties of the dielectric material are involved). Driver and receiver circuits are approximated by an equivalent linear impedance, respectively $Z_S$ and $Z_L$. In most cases, only a real value is considered for source and load impedance ; a capacitive load is represented as an open circuit. Interaction between the line and termination at the sending or receiving end is expressed by the reflection coefficient $\rho$, which is the ratio of the reflected voltage wave amplitude $V_r$ to the incident amplitude $V_i$ :

$$\rho = \frac{V_r}{V_i} = \frac{Z - Z_o}{Z + Z_o} \qquad (2)$$

where Zo is the characteristic amplitude defined by (1) and Z is the (real) source or load impedance. If losses are involved, their effect is approximated by a constant voltage



*W. K. Jones et al. (eds.), MCM C/Mixed Technologies and Thick Film Sensors*, 29–34.

amplitude attenuation during propagation, neglecting the dielectric conductance and the frequency dependence of the resistance :

$$\frac{V(l)}{V(l=0)} = \exp(-\alpha l) \text{ with } \alpha = \frac{R_{DC}}{2Z_o} \tag{3}$$

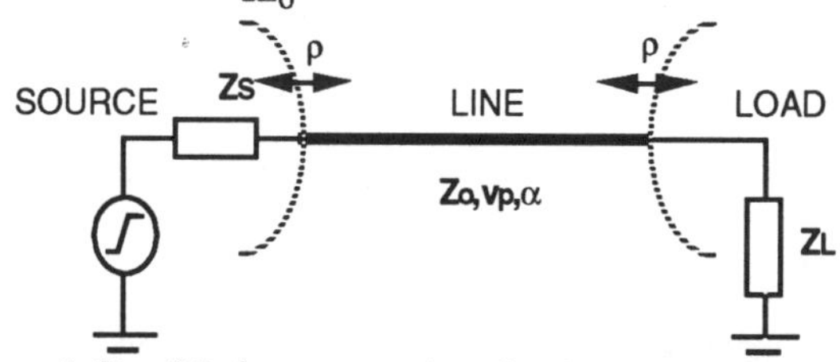


*Figure 1. Simplified representation of an interconnection network.*

Equations (1) to (3) are well suited to study interconnect configurations addressing low loss PCB type lines. They often lead to an optimum solution of lines with a matched termination. In this way, interconnection and driver / receiver characteristics are decoupled and can be optimised on a separate basis. This will generally lead to over-dimensioning of the driver size, yielding short delay times at the cost of high power consumption. To apply a simplified analysis to the MCM type interconnection schemes, the potential advantage in delay time and power consumption can be demonstrated only if the components of the simplified interconnection system are refined, for better correspondence with the real world. This refinement involves in the first place a more complete description of the non-linear driver behaviour. Secondly, the effect of line losses on the driver behaviour should be investigated.

*2.2 CMOS invertor drivers*

For the transmission of digital signals, two different situations can be considered. The first case is known as first incidence switching, corresponding to the situation where the transmitted signal at the receiving end has sufficient amplitude to drive the receiver network. First incidence switching is commonly considered as the only possibility for high speed digital signal transmission. It requires over-driving of the line, which implies driver impedance that is several times smaller than the interconnection characteristic impedance. To avoid delay and possibly erroneous transitions due to reflections on long lines, first incidence switching requires matched terminations at the receiving end. This situation is opposed to the situation of under-driven interconnections - a driver impedance which is several times larger than the characteristic impedance - a situation which can be referred to as multi-reflection switching. Under-driven interconnections require an open receiving line end to built up the signal by means of multiple reflections at both line ends, so no termination resistors are applied in this case. For a large driver and receiver impedance, this situation approximates an RC type loading behaviour.

The concept of under- and over-driven lines refers to the source impedance of the line driver. Since we are dealing with non-linear devices, this impedance is a function of the output signal, and will not be constant during switching. Further on, the non-linear driving behaviour depends on the driver technology and the transistor configuration. The multi-stage CMOS buffer invertor is still the most commonly used driver type. Lets consider the interaction of the final CMOS output buffer stage with an interconnection. Figure 2 shows the sink (load current) of an NMOS (PMOS) transistor at maximum gate amplitude as a function on the load voltage, for different values of the channel width. The output amplitude at the initial transmission can be determined with the help of a load curve. The slope of the load line corresponds with the characteristic impedance of the interconnection line.

For large output buffer dimensions the point of intersection is located in the triode region. A high amplitude initial step is generated, corresponding to the over-driven situation. In this case, the driver equivalent internal impedance relates to the slope of the line that connects the the point of intersection of driver and load curves with the origin. If the point of intersection is located at halve of the supply voltage, the equivalent driver impedance is equal to the line impedance, the driver is "matched". At maximum gate amplitude ($V_{gs} = V_{dd}$), intersection is normally still in the triode region. In case of small transistor dimensions, the point of intersection is located in the saturation region. The initial amplitude of the wave launched onto the interconnection is no longer determined by the marginal output impedance of the driver - the slope of the tangent approximates infinity - but by its saturation current :

$$V_{s,initial} = Z_o \, I_{sat} \qquad (4)$$

For fast digital signal transmission, the first reflections often are deterministic for the actual propagation delay, since they establish a major part of the transmitted signal. Further reflections will be attenuated in amplitude because of line loss, dispersion and incomplete reflection levels at both the sending and the receiving line end. For a simplified analysis, the use of a fixed current source, which equals the saturation current, is a better model for the driver than the one of a large impedance value.

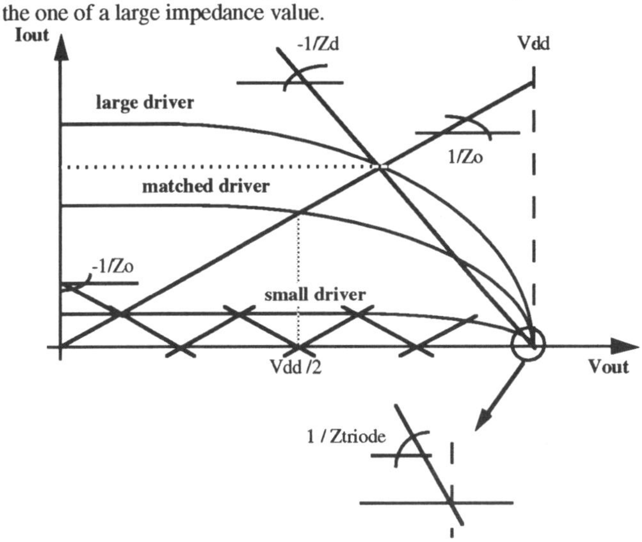


*Figure 2. Load diagram addressing first incidence switching, multiple reflection switching and matched driver impedance.*

*2.3 Delay time and power consumption*

Using the constant current source equivalent circuit for "under-driven" lines, some useful expressions can be derived for delay time and power consumption that hold for multiple reflection switching, and a comparison can be made with the first incident switching approach. Suppose that interconnection delay is defined as the time needed to attain a threshold voltage $V_t$ on the line. Assuming a constant current driver source with saturation current $I_{sat}$ and unit reflection coefficients at both line ends, the voltage level increase at each reflection is given by expression (2) and the required number of reflections to attain $V_t$ is given by :

$$n = \frac{V_t}{Z_o I_{sat}} \tag{5}$$

The delay associated with a single transmission along the interconnection can be derived using the expression for the propagation speed (1). The total interconnect delay of n transitions is then given by:

$$T_d = n\frac{d}{v_p} = \frac{V_t C d}{I_{sat}} \tag{6}$$

with C the line capacitance p.u.l. and d the line length. Equation (6) corresponds with the well known delay expression associated with the delay of loading a capacitance by means of a constant current. This shows clearly that the multi-reflection case can be treated using a capacitive approach. The power consumption associated with driving the line with a signal amplitude V at frequency F becomes :

$$P = \frac{F\, Cd\, V^2}{2} \tag{7}$$

The expression of the delay time (6) limits the signal bit rate that can be transmitted on a line using a driver with certain pre-defined dimensions. Supposing that a symmetrical signal is transmitted, consisting of alternating positive and negative signal edges, and assuming equal saturation currents for NMOS and PMOS transistors, the maximum bitrate becomes the inverse of twice the interconnection delay $T_d$. Combining this with the expressions of propagation delay and characteristic impedance (1) and assuming a threshold voltage $V_t$ which is equal to half of the power supply voltage level $V_{dd}$, the following expression for the maximum bit rate of a particular driver-line combination can be determined :

$$F_{max,Isat} = \frac{I_{sat}}{2\, Cd\, V_t} = \frac{I_{sat}\, Z_o\, c_o}{\sqrt{\varepsilon_r}\; d\; V_{dd}} \tag{8}$$

with $c_o$ the speed of light in vacuum. At maximum frequency, the power consumption becomes :

$$P_{Fmax} = \frac{I_{sat}\, V_{dd}}{8} \tag{9}$$

There is a second limit for the bit rate involving multiple reflection transmission. Reflections are needed to build up the signal on the line, but may not give rise to ringing signals. Erroneous switching might be the result, and anyhow, the time required to transmit a signal will increase again because of the higher settling time. Therefore, the output driver current is limited to the case of adapted source impedance and the highest possible bit rate corresponds to a minimum global interconnect delay which equals four times the line delay :

$$F_{max} = \frac{c_o}{4d\sqrt{\varepsilon_r}} \tag{10}$$

For first incidence switching, the interconnect delay equals the delay of one propagation along the interconnection. The important difference with multiple reflection transmission is that this interconnect delay does not oppose any limit to the signal bit rate that can be transmitted along the interconnection. Once a bit has been launched on the line, the next bit can be transmitted, independent of the fact whether the first bit already arrived at the line output or not. In this way, maximum bitrate is theoretically decoupled from the line length. In reality however, the global effect of line loss and dispersion, which is a function of the line length, will still require a minimum bit hold time and signal period, limiting the maximum bit rate.

In the case of long, low loss interconnections, first incidence switching provides a way to brake through the limit established by (10). The power consumption associated with driving terminated lines can be expressed in the following way :

$$P_t = t_{on} V_{dd} I_{tri} \tag{11}$$

with $t_{on}$ the average (relative) "on" time of the driver. The driver current can be determined graphically using a load diagram such as in figure 2. Assuming an average on-time of 50 % (e.g. a symmetrical clock signal), comparison of equation (9) with equation (11) shows the potential advantage of multiple reflection switching over first incidence switching for power consumption.

*2.4 Typical application areas of the fundamental switching modes*

Figure 3 summarises some results concerning the required driver output current versus the required bit rate, in case of first incidence and multiple reflection switching. The first approach (FIS) requires an output current determined by the necessary step amplitude at the output $V_t$ and the characteristic impedance of the line $Z_o$. In a first approximation, maximum bit rate is independent of the driver current. This will not be the case for multiple reflection switching (MRS) : expression (8) shows a linear relationship. The practical frequency limit is indicated by (10), corresponding to the adapted source impedance case. The vertical line at frequency $F_{max}$ crosses the multiple reflection curve at current $V_t/2Z_o$ ($V_t$ is the receiver threshold voltage), the requirement of no reflection after incidence on the driver side corresponds with current $V_{dd}/2Z_o$. For larger driver currents, the highest bitrate decreases again because of an increased settling time (overshoot).

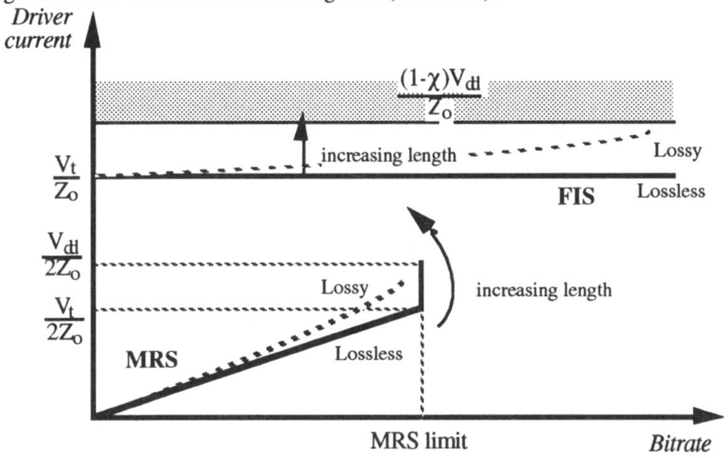


*Figure 3. Typical application areas of FIS and MRS as a function of driver current and bti rate.*

The effect of line length increase is indicated by means of the arrows on the figure. The FIS curve is theoretically independent, the MRS curve rotates counter clockwise and the MRS limit moves to the left. In this way, it is made clear that multiple reflection becomes less interesting for increasing line lengths.

The effect of losses can be introduced using equation (3). The required current for first incidence switching increases with a factor equal to $1/\exp(-\alpha d)$. Remark that, for long lossy lines, the FIS curve reaches the current limit and enters the forbidden zone (determined by the relative voltage loss $\chi$ in the driver). This means that for lossy lines, like the general MCM case, first incidence switching can not be applied any more above a particular line length. For short lines however, we enter in the zone where multiple reflection switching can be applied, with an important power consumption advantage. Unlike first incidence switching, multiple reflection switching turns out to be less sensitive to line loss. The smaller the driver current, the more transmission resembles a capacitive load behaviour, with an RC-constant mainly determined by the line capacitance and the driver current (internal impedance). The dotted curves on figure 3 show schematically what happens if losses are involved.

It can be concluded that, in the case of MCM type interconnections, the important advantage of first incidence switching without bit rate limit is fairly reduced. Generally, multiple reflection has to be applied with absence of parallel termination. Driver current can therefore be reduced. Bitrate limitations have to be countered by increasing the characteristic impedance, reducing the relative dielectric constant, and especially, reducing the line length. The latter suites well one of the characteristic features of MCM interconnections, which is reducing the average interconnection length by increasing the interconnect density and the liquidation of the single chip package level in the packaging hierarchy.

## 3. Numerical analysis of circuit interconnect performance

The theoretical analysis of interconnect performance gave a clear view on the typical application areas of the fundamental switching modes. Addressing in the first place lossy MCM-D type interconnections, MRS is the interesting - and for long lines also the only possible - approach to consider. For simplicity, several effects, like the non-linear driver behaviour, influence of frequency dependent transmission line propagation and attenuation characteristics, etc. have not - or not completely - been taken into account in the theoretical analysis. An accurate analysis including all these effects is only possible using a numerical transient analysis simulation program. The program should be able to handle non-linear devices as well as interconnections with frequency dependent characteristics. The interconnection simulation program Transplus [4] offers these features. An extensive numerical analysis of interconnection circuits revealed some interesting results.

Numerical simulations showed that lossy MCM interconnections allow high bit rate signal transmission using multiple reflection switching (without parallel line termination). The important advantage of high density MCM interconnections is the reduction of the line length. The increased line losses do not condition the interconnection delay time as long as the total line resistance is comparable to the characteristic impedance of the line. Among different MCM technologies, MCM-D offers the highest performance level. Due to its high via density and reduced via dimensions, no additional redistribution area is required at the top interconnection level to connect the devices. In this way, compared to MCM-L interconnections, the average interconnection length can be reduced 2.5 times. This is especially beneficial for power consumption : simulations showed a reduction of almost a factor two.

Performance can be improved by the use of high density MCM substrates, but the real advantage is met if also circuits are redesigned for co-optimisation of the total interconnection system, which consists of the interconnection lines and the on-chip driver/receiver circuits. MCM interconnections allow in the first place a reduction of the driver dimensions, yielding a reduction of the power consumption, crosstalk and of switching noise.

The influence of a reduction of the power supply voltage level was investigated especially addressing interconnection performance. Important reductions of the power consumption (-55%), crosstalk (-33%) and switching noise (-67%) were obtained. The increase of the signal delay time is acceptable if the dimensions of the driver transistors are adapted (+38%).

## 4. Conclusion

Two transmission modes can be defined for interconnections of digital systems : first incidence switching and multiple reflection switching. By means of a simplified analysis, typical application areas were defined and performance parameters were compared. Accurate performance analysis of (lossy) interconnections however requires a simulation program that can handle non-linear devices and frequency dependent transmission line characteristics.

## High Performance Interconnect on Cofired Ceramic

D.LAMBERT , K.KURZWEIL
*Bull SA*
*Rue Jean Jaurès, B.P.68*
*78340 Les Clayes-sous-Bois*
*France*

### Abstract

The European Economic Community is sponsoring technical projects carried out in cooperation by European Companies and academic institutions. APACHIP -Advanced Packaging for High Performance - has significantly contributed to the advancement of packaging technology in Europe, covering various aspects of high performance packaging. This paper illustrates one of the specific project tasks aimed at building a complex MCM-D (Polyimide-Copper on multilayer ceramic substrate) module populated with TAB bonded CMOS VLSI chips .

Key words : APACHIP, Esprit Projects, MCM-D, TAB, multilayer ceramic substrate .

### European Community Esprit Programme

The European Economic Community - EEC - has launched in 1984 dedicated R&D Programmes to strengthen the scientific and technological base of the European information and telecommunication industries. Three main objectives have been established for the series of three consecutive ESPRIT programmes with the first one initiated ten years ago :

. to contribute towards providing the European IT industry with the basic technology it needs to meet the competitive requirements of the 1990s .
. to promote European industrial cooperation in IT .
. to contribute to the development of internationally accepted standards.



*W. K. Jones et al. (eds.), MCM C/Mixed Technologies and Thick Film Sensors,* 35–43.

The actual Esprit Programme is implemented by projects selected from public calls for proposals. The Programme comprises collaborative pre-competitive research and development projects, carried out across national borders by European Community companies, universities and research institutes, freely grouped into projects consortia for an optimized synergy. Typically, fifty per cent of the consortia working budget is covered by the European Economic Community.

ESPRIT Programmes embrace through specific projects various aspects of IT industry ranging from design to manufacturing activities. Semiconductor and packaging developments fit into the Microelectronic Activities of the Esprit Programme concentrating on the development of process technologies, equipment and design tools required by European electronics system manufaturers. Most of the Microelectronics semiconductor process projects incorporate application-specific packaging developments, reliability methodologies and quality assurance procedures.

The APACHIP - Advanced Packaging for High Performance - project carried out from 1988 into early 1993 was one of 47 Microelectronics projects launched as a part of Esprit II Programme .

**APACHIP - Advanced Packaging for High Performance : Esprit II Project**

Eight Partners from four European countries - BULL (F), GEC-Marconi (UK), Hoechst-Ceramtec (D), MCTS (F), NMRC (IRL), Siemens Nixdorf (D), Framatome Connectors (F), Technical University Berlin (D) - have formed the APACHIP Consortium to develop high performance packaging technology for advanced complex VLSI circuits and to demonstrate its availability through functional demonstrators (1).

APACHIP project results have a major impact on the IT, industrial control and telecommunications switching sectors. High-performance CPUs benefit from high density multi-chip assemblies developed and characterized as a part of the project.

In addition to contributing to the development of more efficient electronic assemblies and sub-assemblies, the project has provided advanced tape automated bonding process TAB, tapes, new generations of single-chip and multi-chip ceramic packages, high-density substrates and connectors, large thermal test-chips for R&D and process control purposes, improved electrical and thermal modelling and design, and a non-destructive analytical service based on scanning acoustic microscopy.

**Air cooled MultiChip Module**

The purpose of this paper is to highlight the development of an electrical air cooled MultiChip Module based on polyimide-copper multilayer (MCM-D) as interconnection circuit on cofired ceramic substrate, this substrate provides power and ground distribution. The module has to interconnect up to 16 CMOS VLSI chips bonded to the module using Tape Automated Bonding (TAB), each having 316 I/O's at a pitch of 124,3 um and dissipating less than 4 Watts. The module is connected to the board by a peripheral connector of 888 contacts. The main electrical characteristics of the module are also specified to ensure the functional module to meet the required performance (2).

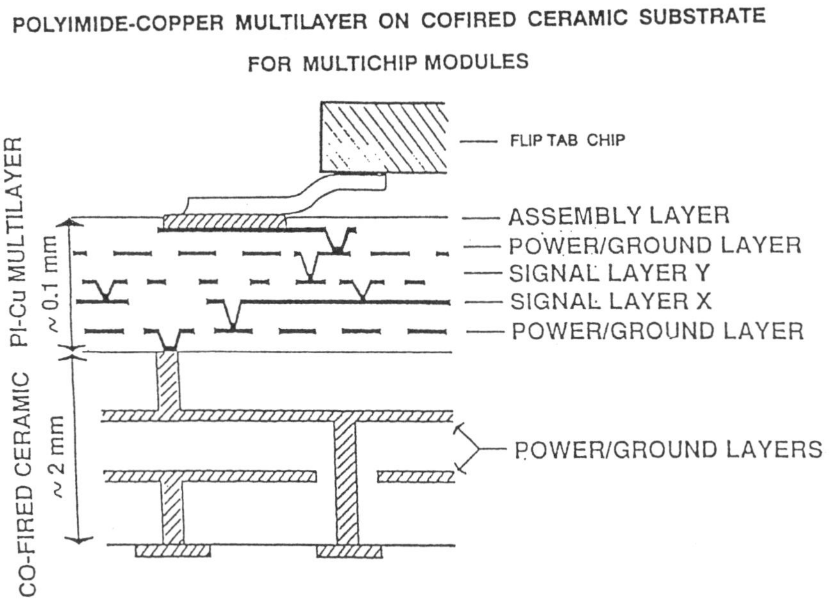


Figure 1 . MCM ceramic & deposited cross sections .

From these requirements the main technological objectives have been defined as shown in table 1

| | Project Objectives | |
|---|---|---|
| | Initial | Final |
| Conductor pitch (um) | 125 | 80 |
| Conductor width (um) | 30 | 20 |
| Via hole diameter (um) | 40 | 30 |
| Via hole pitch (um) | 125 | 80 |
| Total number of layers | 4 | 5 |
| MultiChip Module size (mm) up to | 80 * 80 | 100 * 100 |

Table 1 . Main module parameters .

Before the project start, only the feasability of a multilayer with 125 um track pitch has been demonstrated .
The module construction involved close teamwork and synergy between the participating companies in the project : Hoechst CeramTec for the cofired ceramic substrates, Framatome Connectors for the module connector, MCTS for the TAB tape, the LEMO-INPG and TUB for the electrical modelling and NMRC for Scanning Acoustic Microscopy evaluation and thermal test chips .

**Cofired ceramic substrate**

The main characteristics of the ceramic substrates were :

- Size : 101 mm * 101 mm * 2,1 mm .
- Internal metal planes providing Power and ground and a part of decoupling capacitance .
- Connector side : 4 * 222 I/O's = 888 peripheral pads ( 4 * 148 = 592 signal pads ) .
- Polyimide Copper multilayer side : 592 signal filled vias and 2 times 84 vias providing power and ground .
- Pads and vias are electroless plated with Nickel and Gold .
- Three notches allow to precisely insert the substrate in the connector.

**Polyimide Copper multilayer technology**

OUTLINES OF THE PROCESS

The technology is based on thin film of metal and dielectric deposited in sequential operations on 100 mm square ceramic substrates .The thickness of the films are in the range 5 to 20 microns; the metal lines and the via holes are defined by photoresist patterning; the copper tracks are obtained by a semi-additive galvanic process; the dielectric layers are provided by polyimide spin-on coating and curing and the via holes are obtained by plasma etching of the polyimide or developing photosensitive polyimide; the top assembly layer consists of electroplated tin-lead on copper with a polyimide "solder mask" .

LITHOGRAPHY

Metal tracks and via holes are defined by standard photolithographic processes using liquid positive photoresist, chromium mask and contact printing . Unlike the silicon substrates, the ceramic substrates have a surface roughness (Ra = 0,4 um, Rt = 4 um) and a flatness ( 40 um ) which are involved in the photolithographic process .

## METAL LAYERS

The copper is electroplated into spaces in the photoresist patterns . It is a semi-additive process since the seed layer is deposited by magnetron sputtering . This seed layer, covering the whole dielectric area with its pattern of via holes, consists of a thin adhesion layer (Chromium 90 nm) and 250 nm thick copper layer . The copper thickness is 5-6 um; the resistivity is 1,9 microOhms.cm . After the electroplating, the photoresist is stripped and the seed layer has to be removed . It has been found that the safer mean to keep the integrity of the copper connections is the use of an Ion milling of the sub layers by an Argon beam . To obtain a good insulation between the copper areas, the sub layers to be etched back must be on a smooth surface, like a poyimide layer, and not directly on ceramic substrate . Ion milling technique is also used as a cleaning method for the bottom of via holes before the deposit of the seed layer . The high connectivity of the polyimide copper technology enabled us to route the interconnections of the functional demonstrators with a single logic pair : total length of interconnections : 22 m .

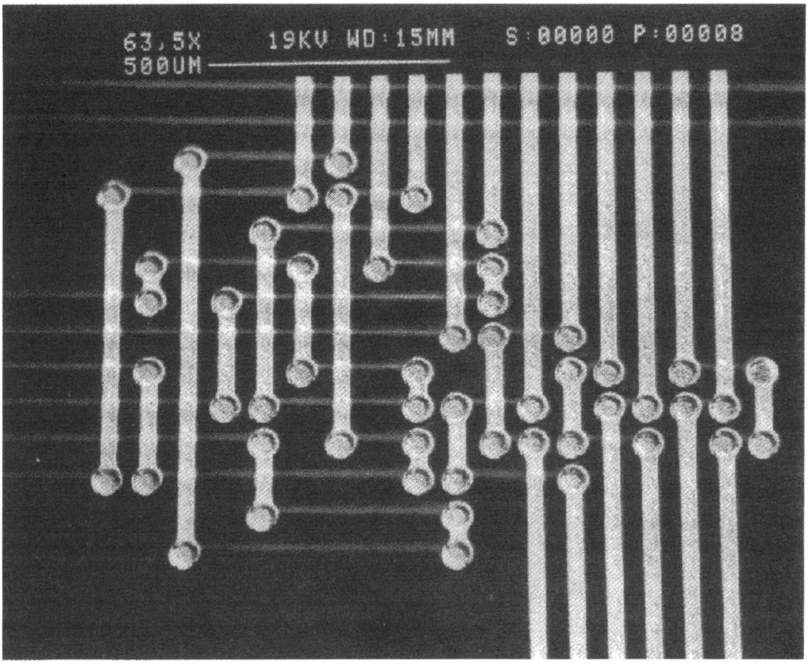


Figure 2 . Copper signal levels (80 um pitch)

## DIELECTRIC LAYERS

The dielectric layers are provided by polyimide spin-on coating and curing The PI 2525 (Du Pont) which was the the first available on the market, has been used for all the demonstrators . The photoneece 3840 (Toray) which is photosensitive, has been also extensively used . Depending on the levels and taking into account the line impedance requirements, the final thickness of the polyimide layers are 10, 15 and 20 um at +/- 0,5 um .

## VIA HOLES AND BAKING OF POLYIMIDE

The via holes are obtained by Reactive Ion Etching of the polyimide , conditions have been found out to obtain via holes with a 65-70° slope, or developing Photosensitive polyimide . The resistance of contact (via) is lower than 3 milliOhms .

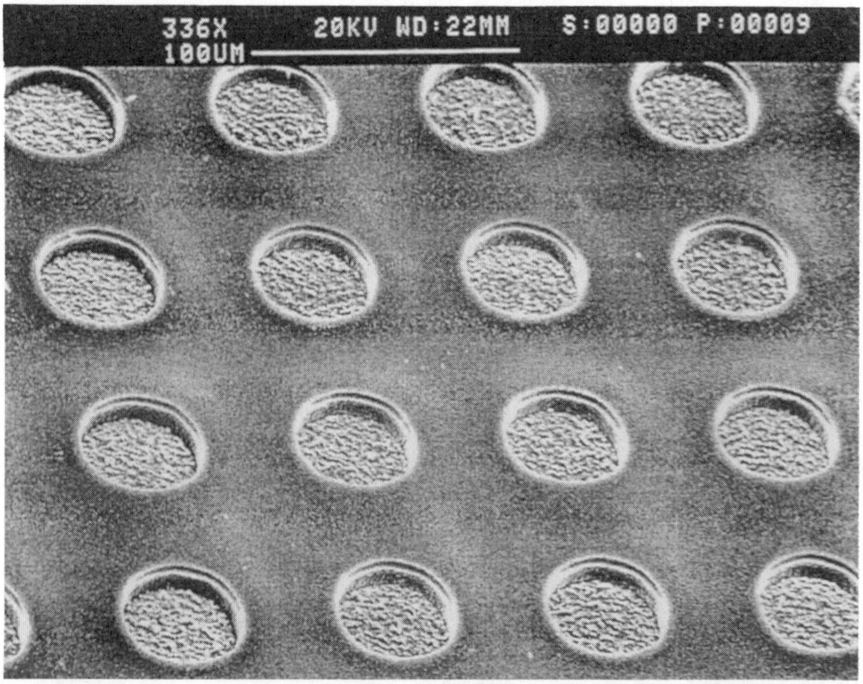


Figure 3 . Via holes (40 um), polyimide tappered Reactive Ion Etched .

The polyimides used within the project have a maximum moisture absorption between 2 and 2,5% (in weight), the different heat treatments have been studied and optimized . It is really a key point to successfully achieve a good polyimide copper multilayer substrate .

## TAB

TAB is a well established assembly technology . It has numerous advantages for constructing complex MCM due to its inherent capability to simplify testing and burn in . VLSI evolution towards large devices and narrow pitch allows TAB to compete with more conventional chip assembly techniques as wire bonding .

| | Objectives of APACHIP |
|---|---|
| Chip size | 12 to 15 mm |
| Chip I/O pitch | 125, 100 um |
| Number of I/O's | 400 |
| Tape format | 70 (105) mm |
| Si/Sub. ratio | 0,5 |
| Numb. of chip/Sub. | up to 16 |

Table 2 . TAB parameters

The TAB development has been undertaken on configurations where OLB pitch was the same as the ILB pitch . An innovative evolution of TAB development has been named "bumpless ILB" consisting in performing ILB on devices without bumps, chip size : 12 mm square, 316 I/O's, ILB pitch = OLB pitch : 125 um .

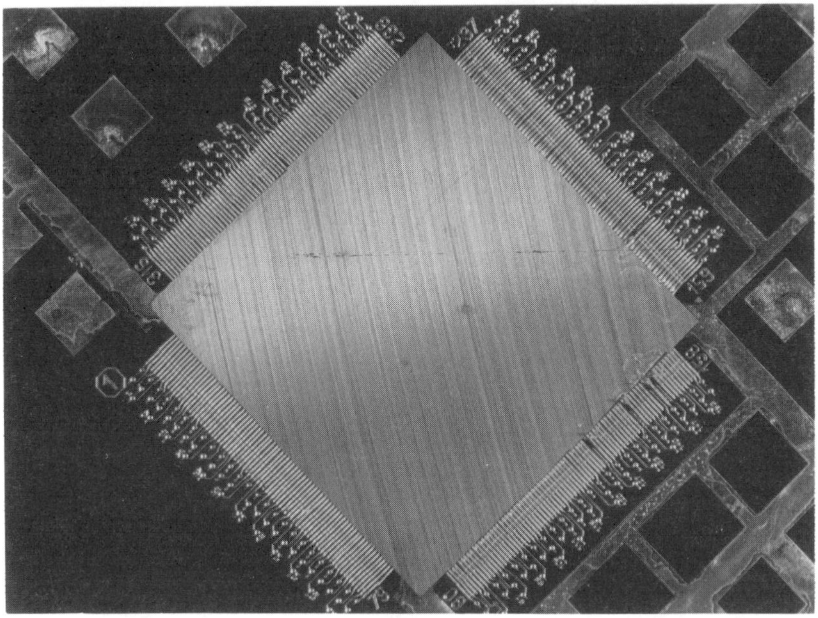

Figure 4 . VLSI (316 I/O's) Flip TAB bonded on MCM .

The TAB technology has been brought to a very high state of the art level, compared to equivalent developments in US and Japan .

**Electrical measurements**

The purpose of the built MCM-D module is to explore the possibility to transfer information between two processors in a serial way at a througput of at least 1 Gbit/s, thus providing a high performance and highly integrated solution for data transfer problems from chip to chip or board to board .
The measurements show that the propagation time by unit length is higher than expected with 7ns/m, due to the design of the mesh ground, this could be improved by a redesign of the power and ground planes .
The signal measurements show as expected, that due to the short length of the lines and the edge rate of 1,5 ns, the edges are clean with no overshoot or undershoot .
The usable frequency reaches 1,5 Gbit/s, which is 50 % above the objectives.

Figure 5 . Air cooled MultiChip Module .

## Conclusion

The MCM-D module with functional devices has been built and characterized for its thermal and electrical performances. Its properties have been measured in accordance with results of simulation and modelling. The electrical performance of the MCM-D module with a large number of active devices exceeds the necessary requirements imposed by the majority of present applications. Due to the comparatively high cost of MCM-D's, they remain presently limited in applications to sectors where the prime performance is mandatory. The progress in multilayer ceramic technology obtained during the MCM-D substrate development, have been sucessfully applied into the realization and use of a high performance ceramic single chip packages .

## Acknowledgement

The technical results were obtained by Bull SA in a close and mutually beneficial cooperation with other partner Companies in the APACHIP Project Consortium. The work has been carried out thanks to the financial support from the DG XIII of the European Economic Community.

## References

1. P. Chantraine et al. "Achievements in High Density Packaging", Esprit 91 Conference, Proceedings, p. 101 .
2. "APACHIP Project Final Summary" - February 1993, Copy can be requested from the authors .

# NEW ASPECTS IN THE RELIABILITY DESIGN OF HIGH DENSITY INTERCONNECTS (IN MCMs)*

*Avoiding Migrated Short Failures: A Review*

G. HARSÁNYI, Ph.D.

*Technical University of Budapest, Dept. of Electronics Technology*
*1521 Budapest, Hungary, Phone: 361-1811520; Fax: 361-1666808*

**Abstract.** *Metals can exhibit dendritic short-circuit growth caused by electrochemical migration in conductor-insulator structures. A theoretical and practical review is given about the most important problems in connection with MCMs. Various material systems based on different technologies (MCM-C,D,L) are discussed. Both material and layout design ideas are given for the solution of the reliability problems.*

**Keywords**: electrochemical migration, dendrites, reliability of MCMs, metallization failures

## 1. Introduction

Recently, in connection with the production of high density interconnection systems in MCMs, the claim to conductor-systems with high resolution and high reliability has emerged. The possibilities of integration are determined not only by the technological bases but also by those physical and chemical processes that can cause resistive shorts between neighbouring metallization strips. One of these phenomenons is the electrochemical migration. This can be defined as a transport of ions between two metallization strips under bias through an aqueous electrolyte. Electrodeposition also occurs forming dendrites or dendrite-like deposits. Ultimately, such a deposit can lead to a short circuit in the device and can cause catastrophic failure. The conditions are: a film of polar liquid (usually water) to form an electrolyte, bias, and operating time [1].

Migrated resistive shorts occur rarely in the practice and only under extreme conditions. A device can operate for many hundreds of hours under normal operating conditions, and then, after a short exposure to special environmental conditions, fail [2-3].

Various material systems and packaging solutions have been developed in the last decades, mainly in the area of multichip thin and thick-film hybrids, to minimize or eliminate the failure possibilities. On the other hand, several material systems were banished from the high reliability applications.

In MCMs, however, very fine line and high density interconnection systems have been developed, thus new material structures have been applied which are not

* This research is supported by the Hungarian National Scientific Research Fund (OTKA, project No. F007365)



*W. K. Jones et al. (eds.), MCM C/Mixed Technologies and Thick Film Sensors,* 45–51.

new in the electronics technology but new in high reliability applications. Therefore, the knowledge of the failure possibilities became more critical.

Another important point of view is that in some special application fields, it is simply not enough to reduce the failure possibility, these are absolutely not allowed. Two main reasons can be mentioned here:

- the consequences of an electrical failure are in real meaning catastrophic
- the repair of the failed parts is not possible, the whole module must be replaced for relatively high costs (this is exactly the situation with MCMs).

## 2. Theoretical Models

The classical model of electrochemical migration has been well known for several decades, however, a few anomalous or newly discovered phenomenons need to make some revisions and add supplementary models.

The *conventional model* describes the process as a transport of metal ions from an anodic site, through an aqueous electrolyte, toward a cathodic site where the electrodeposition of the ions occurs forming dendrites [1]. Figure 1.a and 1.b show the possible processes: the water decomposition and metal migration, respectively. Silver is famous for its inclination to form migratory shorts. However, it is less known that several other metals also show this phenomenon, i.e.: copper, lead, and tin [4-5].

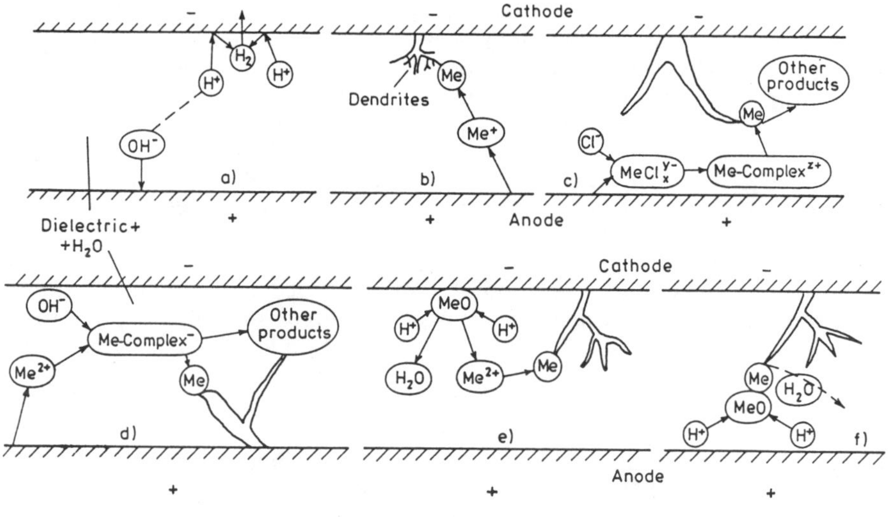


*Figure 1. Electrochemical processes between metallization strips: a) water decomposition, b) classical model of electrochemical migration, c) halogen induced migration through complex formation processes, d) anodic dendritic growth through negative complex-ion formation, e) migration of reduced dielectric constituents, f) "virtual migration"*

*Halogen ion-* (e.g. $Cl^-$) *induced migration* can occur in the case of strongly precious metals (gold, platinum and palladium) by the formation of primary halogen and secondary hydroxid-halogen complex ions (see Figure 1.c) [2,5].

A few metallization types may show an *anodic dendritic growth*, the cause of which is still not well understood. An intermediate complex-ion formation process might be supposed resulting negative ions, as shown in Figure 1.d. This anodic type dendrites have been found at nickel and copper metallization systems [3,6].

Recently, it has been demonstrated that not only metallic components, but also *oxides* can take part in the formation of migrated shorts, after a chemical reduction process - independently that they are constituents of conductors or dielectrics (see Figure 1.e) [7]. Thus the classical model should be completed: metal ions forming dendrites can originate not only from the anode by ionic dissolution but from insulating constituents as well. The materials having been found to take part in dendrite formation corresponding to this model are the oxides of copper, bismuth, lead and iron. Other materials may also show this ability.

If the critical oxide component is a constituent of the dielectrics between the metallization strips, a "*virtual migration*" can also occur (see Figure 1.f) which means the movement of local chemical reduction sites through the dielectrics - resulting in the formation of thin dendrite-like metallic paths, without any real metal-ion migration [8]. This latter process has a much lower activation energy level than the former ones because the ionization and ion migration steps are avoided.

## 3. Practical Aspects

There are three design stages of MCMs in which the possibility of migrated shorts can be minimized:

- in the design of packages and related processing steps
- in the material design of the joints and interconnecting network
- in the layout design.

### 3.1. PACKAGING DESIGN

Moisture protection and removal are old but always persistent problems in the packaging of microcircuits [9]. Hermetic metallic packages have given the best solution. They were developed for individual chips and for high reliability multichip hybrids [10]. However, they are still not well suited for MCM applications. Hermetic ceramic packages, which are the ancestors of MCM-Cs, showed many reliability problems due to the presence of moisture and ionic contaminates inside the package. These originate mainly from the materials and the processing used for the lid sealing [11].

Plastic packaging and any combination of plastics are not enough for moisture protection because these materials are permeable for gas molecules. Although good results can be achieved with silicone RTV materials, they give only a reliability improvement in comparison with carbon based polymers and can not give a safe solution [5,12].

The seriousness of the moisture removal can be demonstrated by the following data. A safe (moisture condensation free) operation of a circuit inside a ceramic package at -50 $^{o}$C allows a humidity level of 10ppm, which means an atmosphere with 0,05% RH at room temperature. The same for -75 $^{o}$C is possible only with a moisture concentration level of 0,3ppm.

Washing in CFCs and subsequent vacuum bake-out, hot air-knife and convection oven drying were appropriate processes for moisture removal. The replacement of CFCs by water soluble cleaning agents have made the situation more complicated. The problems are summarized in the following items [9]:

- the cooling through the vaporization may lead to ice formation hindering the further removal,
- contaminate replacement is possible from water droplets,
- water forms a meniscus between adjacent surfaces (e. g. between flip chips and substrates).

Recently, a patented thermal/vacuum drying process and apparatus have been developed to remove all traces of moisture, which means also the monitoring of the water content inside the packaging chamber [9].

Contaminates, porous surfaces, holes and adjacent surfaces act as moisture condensation sites and may result continuous water films even when being far from the saturation conditions. In the presence of NaCl contamination, for example, condensation is possible above 75% RH [13].

## 3.2. MATERIAL DESIGN

Material design considerations are of key importance. Studying MCM structures, there are four levels within the structures where migrated resistive shorts may occur.

### *3.2.1. Component (Chip) Level*

Silicon chips generally apply aluminium metallization system which can show open circuit failures due to electromigration and corrosion but does not exhibit electrochemical dendritic short circuit formation [14]. The great variety of other type chip components should be discussed separately.

### *3.2.2. Electrical Joints between Chips and Interconnection Systems*

Wire bonding gives the most reliable solution, however, this is the less convenient one. Soldering of high density pin systems or flip chips is becoming more and more dangerous with the increase of the pin resolution due to the possibility of lead and tin dendrite formation between the soldered pins [3,15]. Solder replacement by conducting epoxies and polymers means a return to the big problems of silver migration because these materials are mainly filled with silver [6]. That means, soldering and gluing has no high reliable solutions for MCMs at present.

### *3.2.3. Interconnection*

The interconnection systems themselves have quite different behaviour in the MCM-C, L and D technologies, respectively.

*MCM-Cs.* Various metallization systems have been studied already in thick film hybrids that are also mentioned previously and the results can be adopted to LTCC systems. The best results can be obtained using precious metal compositions except silver, or applying ternary Pd-Pt-Ag alloys [5].

Porous ceramic parts are especially dangerous [7]. Stable non-reducable oxides should be used as binding insulating materials. Most dangerous are those oxides that are reducible in a hydrogen evolution and remain in the state of oxides after the firing of MCMs. Thus, at reactive type oxides which take part in the physical-chemical reactions during firing, the migration lifetime is determined by the excess amount of metal-oxide. Lifetime data can be improved by composition and technology changes (e.g. $Bi_2O_3$, PbO, CuO). At non-reactive type oxides, which do not take part in any reaction during firing, lifetime data can not be improved by changing the composition and the firing parameters. The use of this oxide-types should be avoided (e.g. $Fe_2O_3$).

HTCC structures have not been studied in connection with their migration behaviour yet. There is no information about the behaviour of Mo and W. Thus, a detailed analysis is needed in the near future [13].

*MCM-Ls and MCM-Ds.* They apply mainly copper metallization with plastic insulation: mostly epoxy and polyimide, respectively.

Both dielectrics are inclined to adsorb and absorb water molecules and even an aging process through an irreversible chemical reaction has been detected [15]. Copper is also known to grow dendrites. The best solution is the application of aluminium instead of copper which is a usual method in MCM-Ds, but less applicable in MCM-Ls. Therefore, the latters still seem not to be suitable for high reliability interconnection systems.

*3.2.4. Output Leads*

They mean the fourth level of interest in MCMs. Migrated shorts can often be observed between the pins of brazed ceramic structures because brazing alloys contain Ag and Cu and the plating of the pins may contain Ni and Sn. Leadframes are often made of Cu-alloys with a plating that contains Ni, Ni-Sn or Ag. Gold plated 'covar' leads in the metal packages gave the best results, however, they are still not widely applied in MCMs [16].

## 3.3. LAYUOT DESIGN

An appropriate layout design can compensate the weak points of the moisture removal, packaging methods and material systems.

As it is well known, dendritic growth can only occur between metallization strips under DC bias [5]. In signal processing circuit units, the ratio of such wiring pairs is quite moderate. The DC voltage difference is generally in the order of a few Volts, maximum 10-20 Volts. The possibility of migrated short circuit formation becomes significant when the distance between the stripes is under 1 mm (excepting a few strongly migrating metallization systems such as silver) [5]. Thus, it is possible to increase the reliability by the application of a few simple layout design rules, such as:

- Components having neighbouring leads under DC level difference should be omitted during the component selection.
- A special care should be taken during the layout design to avoid the application of neighbouring metallization strips under DC bias. A few mm distance must be kept between them. This distance should be increased by the increasing level of DC bias voltage.
- Similar rules are effective for the own output leads of the module.

The mentioned rules are very simple but very important and should be built in into the layout design CAD softwares.

Layout design methods have several advantages in comparison with those of the material and processing design:

- They may result in a reliability improvement by several orders of magnitude, whilst the application of high cost materials and complicated processing steps can not always achieve this
- Their application does not need special additional costs
- They may also be effective when material design has no solutions (see soldering, etc.).

## 4. Summary

A review has been given about the reliability design aspects of MCMs in order to decrease the possibility of migrated short circuit formation during operation. A theoretical survey has been given about the physico-chemical models of electrochemical migration. The design methods have been discussed as practical solutions of the problem. The conclusion can be drawn that moisture removal and protection needs complicated pre-packaging processes and high cost packaging solutions, material design has only high cost solutions and for some problems nothing at all, whilst simple layout design methods are of low cost and much more effective. The best solution that can be suggested is to build in the rules into the MCM-CADs.

## 5. References


1. Kohman, G.T., Hermance, H.W., and Downes, G.H. (1955) Silver Migration in Electrical Insulation, *Bell System Technical Journal* **34**, 1115-1147.
2. Shumka, A. and Piety, R.R. (1975) Migrated-gold resistive shorts in microcircuits, *Proc. 13th Int. Reliability Physics Symp.*, 93-98.
3. Benson, R.C., Romenesko, B.M., Weiner, J.A., Nall, B.H., and Charles, H.K.,Jr. (1988) Metal Electromigration Induced by Solder Flux Residue in Hybrid Microcircuits, *IEEE Trans. Comp., Hybrids, Manuf. Technol.*, CHMT-**10**, 363-370.
4. DerMarderosian, A. (1978) The Electrochemical Migration of Metals, *Proc. Int. Society of Hybrid Microelectronics*, 134-136.
5. Ripka, G. and Harsányi, G. (1985) Electrochemical Migration in Thick-Film ICs, *Electrocomp. Sci. Technol.*, **11**, 281-290.
6. Coleman, M.V. and Winster, A.E. (1981) Silver Migration in Thick Film Conductors and Chip Attachment Resins, *Microelectronics Journal*, **4**, 23-29.
7. Harsányi, G. (1993) Dendritic Growth from Dielectric Constituents: a Newly Discovered Failure Mechanism in Thick Film ICs? *Int. J. Microcircuits & Electronic Packaging* **16**/3., 207-216.
8. Harsányi, G. (1992) New Type Short Circuits at Fritless Thick Film Conductors - Formed from Reduced Oxides, *Intern. Symp. on Microelectronics, ISHM, San Francisco*, 140-143.
9. Leech, C.S.Jr. (1993) Moisture Removal From Hybrid Electronic Components and Assemblies, *Intern. Symp. on Microelectronics, ISHM, Dallas*, 104-107.
10. Neff, G.R. (1986) Hybrid Hermeticity and Failure Analysis, *Hybrid Circuit Technology*, April 1986, 19-24.
11. Thomas, R.W. (1976) Moisture, Myths and Microcircuits, *IEEE Trans. Parts, Hybrids, Packaging*, PHP-**12**, 167-171.
12. Kookootsedes, G., Antonen, M. and R. (1986) Selecting Protective Materials For Coating Hybrid Circuits, *Hybrid Circuit Technology*, May 1986, 12-15.
13. Warren, G.W., Wynblatt, P., and Zamanzadeh, M. (1989) The Role of Electrochemical Migration and Moisture Adsorption on the Reliability of Metallized Ceramic Substrates, *J. Electr. Mater.*, **18**/2, 339-353.
14. Koelmans, H. and van de Ven, E.D.G.T. (1976) The Cathodic Corrosion of Aluminium, *J. Electrochem. Soc.*, 1976/1 143-144.

15. Krumbein, S.J. (1988) Metallic Electromigration Phenomena, *IEEE Trans. Comp., Hybrids, Manuf. Technol.*, CHMT-**11**, 5-15.
16. Dumoulin, Ph., Seurin, J.P., and Marce P. (1982) Metal Migrations Outside the Package During Accelerated Life Tests, *Proc. Electr. Comp. Conf.*, 229-236.

# MCM-D technology with active and passive substrates

by

Hans Hentzell
IMC
Teknikringen 3
S-583 30 Linköping Sweden

## Summary

For Multi Chip Module (MCM) technology several types of substrates like laminate board, ceramic, metal, glas and silicon may be used. Traditionally these have only been a substrate for interconnecting metal wires. All components, active and passive are either mounted or screen printed on the surface. During resent years however there has been a development towards integration of "embedded" passive components for both LTCC and MCM-D technology and mount the ICs on top. With the use of silicon as a substrate material it is now possible to increase the integration by adding active components in the substrate. Micro Circuit Engineering (MCE) in UK developed a MIL-1553/1760 Data Bus with an active substrate (all CMOS logic). No extra metallisation is needed for this system. Lucas Electronics together with IMC are developing a CPU MCM with the memory in the substrate and IMC together with Saab Combitech developed an Intelligent Power Module for automotive applications with protection circuits in the substrate. The Data Bus MCM is produced in large volumes today

## Introduction

Multi Chip Module technology is becomming more and more important in electronic applications. For the MCM-D technology we forsee that the development will go in two directions. For volume applications the flat panel technology using Silicon or other substrate materials will become very important. For high complecity or high speed applications active silicon substrates may be a competitive solution. Using active silicon as a substrate it is possible to integrate functions like logic, memory, test and protection circuits and/or high speed devices in the substrate and mount other components on top. In order to use an active silicon substrate the adventage of the extra functions and the smaller size must balance the extra cost for the substrate. In this paper we will describe three applications where active substrates are beeing used:

* CMOS logic: MIL-1553/1760 Data Bus from Micro Circuit Engineering in UK
* Memory: CPU module with memory in the substrate from IMC/Lucas Electronics in UK
* Test and protection circuits: Intelligent power module from IMC/Saab Combitec



*W. K. Jones et al. (eds.), MCM C/Mixed Technologies and Thick Film Sensors*, 53–56.

**MIL-1553/1760 Data Bus**

Micro Circuit Engineering Limited in UK has designed and built a MIL-1553/1760 Data Bus Remote Terminal, Bus Controller & Sub-system Interface. The device comprises an active substrate that contaions digital processing functions and memory, onto which two analogue power transcievers are mounted.

The substrate is designed and built by the same methodes that wuold be used for a conventional custom ASIC and is fabricated using a standard IC foundry service. In order to achieve high yield in the substrate design the layout of the small the active areas are separeted on the substrate. The additional tracks needed for inteconection between different areas on the substrate and interconnection to the transcievers do not use extra metal and dielectric layers. The active substrate MCM replaces a previous passive substrate MCM-D

| | | MT19004.Active Substrate | MTA51001 Passive substrate |
|---|---|---|---|
| Substrate size | | - 14*14 mm | 25,4*25,4 mm |
| Technology | - | 1,2 micron CMOS | Al-metal/Silicon dioxide |
| Integration | - | 379000 transistors | 2 layer, 25 micron track width |
| Memory | - | 3k*16 SRAM | - |
| Wire bonds | - | 198 | 494 |
| Surface IC´s | - | 2 of 4*5 mm | 12 |
| Die Attach | - | Thermo-plastic | Polyimide |
| Peak Power | - | 0,75 A and 5V | 0,75A and 5V |
| Environment | - | -55 to + 125 C | same |
| Qualification | - | MIL-std 883 | MIL-std 883 |
| Status | - | Volume production | |

A true comparison between the two versions cannot be drawn since the active substrate design has increased digital functionality as well as 3k*16 of SRAM. After reliability testing it was concluded that the MCM satisfactorily completed the MIL-STD-883, Methode 5008 tests.

It may be noted that MCM-D with active substrate is the fifth generation of the same system

**CPU module with integrated memory**

Many MCM CPU applications contain a processor, glue logic and memory. The CPU and the glue logic need the high interconnection capability of the MCM-D technology whereas the memory chips cover a lot of area and have few I/O´s.

In the JESSI project Silicon Hybrids Lucas Electronics evaluate the possibilities to use SRAM memory with 1M*8 usable memory in the substrate (40*40 mm) with a 68020 processor, glue logic and a global memory manager mounted on the substrate.

A similar CPU MCM without active substrate will have substrate size of 42*84 mm, i.e. the CPU MCM with active substrate is half the size.

Figure 1 shows a schematic cross section of the CPU MCM.

Looking from the bottom the substrate will have SRAM memory and two metal layers for interconnection between memory cells, specifict to the type of memory used. The next two metal layers will be specific for the interconnection between the logic circuits. The fifth layer will be used for assembly.

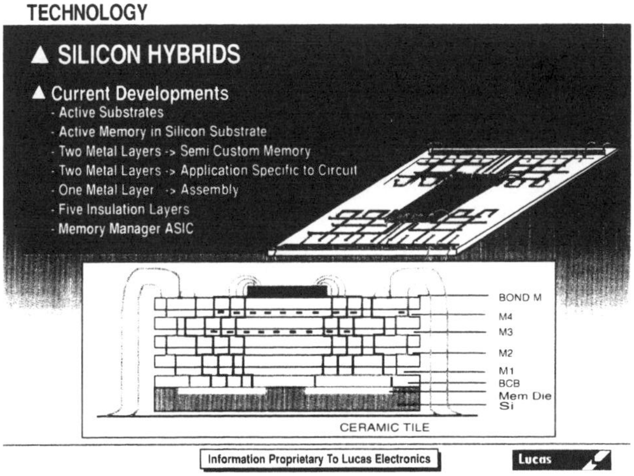


Figure 1. Crossection of MCM-D with memory in the substrate

**Intelligent power module with active substrate**

A power module with a processor was implemented in MCM-D technology using an active substrate. As a comparison the same system was built using MCM-C technology. The MCM-D uses a process with three metal layers, thin film resistors and thin film capacitors integrated in the metallisation structure. Protection cirquits like power resistors and diodes are integrated in the Silicon substrate.

Included in the module are a microcontroller HC11, two 50 V DMOS FETs with 10 A rms capability, a regulator, six digital and analogue ICs along with several discrete capacitors, all mounted on the the silicon substrate. The resistors, capacitors and diodes integrated in the substrate fulfill a number of functions: very high ESD protection; high voltage transient protection; current sensing; analogue feedback; decoupling; analogue filtering and temperature sensing.

For resistors and capacitors IMCs A/D process with $CrSi_2$-resistors and SiOxNy capasitors

The two versions of the Intelligent power module was tested for 15 kV ESD protection and recieved 100 V/m EMC protection without any failures.

## Discussion

The examples given above describe three different cases where active silicon substrates may be used. In the case of the Data Bus application it was shown the the solution with active substrate is more reliable and cost competitive. This shows that although the substrate has been processed with a full CMOS process the the solution may be competitive with the traditional MCM-D solution.

For the CPU MCM with memory in the substrate it is still too early to draw any conclusions concerning the adventages and disadventages compared to the CPU with passive substrate.

The Intelligent power module was half the size of the MCM-D solution but after packaging the difference in size is not very large. The cost of the MCM-D with active substrate is about 3 times higher than the MCM-C version. For the Intelligent power module the complexity is so small that the system can be built in traditional MCM-C technology and there is no adventage in using the more advanced technology.

## Aknowledgment

Micro Circuit Engineering Ltd, Saab Combitech AB and Lucas Electronics Ltd are thanked for allowing me to use information about their MCM products in this presentation.

# HIGH PERFORMANCE CERAMIC MODULES AND PACKAGES

PAUL DANNER

PACIFIC MICROELECTRONICS CORP.
10575 S.W. CASCADE BLVD.
PORTLAND, OR. 97223

## ABSTRACT

WITH THE INCREASED REQUIREMENTS FOR HIGH DENSITY MULTICHIP PACKAGING A LOT OF WORK HAS BEEN DONE WITH VARIOUS TECHNOLOGIES. MILITARY APPLICATIONS ARE LOOKING FOR LOWER COST ALTERNATIVES THAN WHAT HAS BEEN AVAILABLE PREVIOUSLY.

WITH THE ADVENT OF LOW TEMPERATURE DIELECTRIC TAPE, FINER GRIND THICK FILM CONDUCTOR INKS AND HIGHER RESOLUTION EMULSIONS A MUCH HIGHER DENSITY CERAMIC MULTICHIP MODULE IS NOW POSSIBLE. THIS MODULE IS APPROACHING WHAT HAS BEEN ACHIEVED WITH DEPOSITED COPPER ON POLYMER MULTILAYER CIRCUITS.

BALL GRID ARRAY PACKAGES HAVE BEEN AROUND FOR SEVERAL YEARS, BOTH IN CERAMIC AND LAMINATE TECHNOLOGIES AND ARE NOW BEING ACCEPTED AS A VIABLE ALTERNATIVE. A LOW COST METHOD TO COMBINE CERAMIC MULTICHIP MODULE TECHNOLOGY WITH BALL GRID ARRAY ASSEMBLY IS NOW AVAILABLE. THE METHODS ARE UNIQUE IN THE PLACEMENT OF THE SOLDER SPHERES, THE ALMOST SPHERICAL DIMENSIONS OF THE CONNECTIONS AND THE EXTREME RUGGEDNESS OF THE RESULTING PACKAGE. THE IMPROVED THERMAL CHARACTERISTICS, EASE OF ASSEMBLY AND LOW COST WILL MAKE THIS PACKAGING TECHNIQUE A PRACTICAL SOLUTION FOR TODAYS DEMANDS.

THIS PRESENTATION WILL EXPLORE HOW THESE DENSITIES



*W. K. Jones et al. (eds.), MCM C/Mixed Technologies and Thick Film Sensors, 57–67.*

ARE ACHIEVED, THEIR ELECTRICAL, THERMAL AND MECHANICAL PERFORMANCE AND THE COST SAVINGS THAT CAN BE REALIZED.

## LOW TEMPERATURE CERAMIC TRANSFER TAPE

THE DECISION TO USE THE LOW TEMPERATURE TRANSFER TAPE OVER THE LOW TEMPERATURE CO-FIRE TAPE IS AS FOLLOWS:

A) STRENGTH - USING A STANDARD 96% ALUMINA SUBSTRATE AS A BASE AND LAMINATING THE TRANSFER TAPE TO IT PROVIDES OVER TWICE THE FLEXURAL STRENGTH OF LOW TEMPERATURE CO-FIRE TAPE.

B) SHRINKAGE - THE TRANSFER TAPE IS LAMINATED TO A PREFIRED SUBSTRATE AND WHEN FIRED THE SHRINKAGE IS IN THE "Z" AXIS ONLY PROVIDING TIGHT TOLERANCE DIMENSIONS IN THE "X" AND "Y" AXIS. THIS IS A GREAT ADVANTAGE FOR AUTOMATIC DIE ATTACH AND WIRE BONDING.

C) THERMAL DISSIPATION - BOTH THE LOW TEMPERATURE CO-FIRE AND THE TRANSFER TAPE ARE POOR THERMAL CONDUCTORS. HOWEVER CAVITIES CAN BE FORMED IN THE TRANSFER TAPE ALLOWING THE SEMICONDUCTOR DEVICES TO BE MOUNTED DIRECTLY TO THE 96% ALUMINA SUBSTRATE WHICH HAS OVER 10 TIMES THE THERMAL CONDUCTIVITY OF THE TAPE ITSELF AND WITH THE USE OF BEO AS THE BASE SUBSTRATE THE THERMAL CONDUCTIVITY IS 100 TIMES THAT OF THE TAPE.

THE TRANSFER TAPE PRODUCTION PROCESS DEVELOPED AT PACIFIC MICRO USES COMMERCIALLY AVAILABLE DIELECTRIC TAPE WHICH IS SUPPLIED IN ROLLS OF VARIOUS WIDTHS AND LENGTHS. SUPPLIERS OFFER VARIOUS DIELECTRIC CONSTANTS AND WILL CAST TO A SPECIFIC THICKNESS.

AFTER A DESIGN IS COMPLETED ON THE CAD SYSTEM A DISC WITH ASCII DATA FOR THE VIA LOCATIONS IS USED TO PROGRAM THE LASER. AS MANY AS 18,000 VIAS (40 PER SQUARE MILLIMETER PER LAYER) PER SUBSTRATE HAVE BEEN PRODUCED. THIS SAVES A LOT OF LABOR FOR MANUAL DATA ENTRY AND ELIMINATES ANY HUMAN ERRORS. THE LASER WILL DRILL THE ALIGNMENT TOOLING HOLES , THE

VIA HOLES AND EXCISE THE TAPE LAYER FROM THE ROLL. UP TO TWELVE METAL LAYERS HAVE BEEN LAMINATED AND FIRED, ON A SUBSTRATE 150 MM SQUARE, WITH BETTER THAN ± 25 MICRON REGISTRATION ACCURACY. WITH A CO2 LASER VIAS HAVE BEEN DRILLED DOWN TO 100 MICRONS IN DIAMETER AND WITH A EXIMER LASER VIAS HAVE BEEN SUCCESSFULLY DRILLED DOWN TO 50 MICRONS IN DIAMETER. USING A PROPRIETARY PROCESS 50 MICRON DIAMETER VIAS 100 MICRONS DEEP HAVE BEEN SUCCESSFULLY FILLED WITH METAL.

WHEN THE SUBSTRATE AND THE TRANSFER TAPE ARE DRILLED AND CUT TO SIZE THE ASSEMBLY PROCESS IS READY TO BEGIN.

THE FIRST STEP IS TO PRINT THE METAL ON THE PREFIRED SUBSTRATE, USUALLY A GROUND PLANE, AND FIRE IT. THEN LAMINATE AND FIRE THE FIRST TAPE LAYER, PRINT AND DRY THE VIA FILL, PRINT, DRY AND FIRE THE METAL LAYER AND THEN LAMINATE THE SECOND TAPE LAYER AND REPEAT THE STEPS UNTIL IT IS COMPLETE.

IN THE LAMINATING PROCESS A COMPLIANT MATERIAL IS REQUIRED TO KEEP FROM FRACTURING THE PREFIRED SUBSTRATE. ALL RUBBER TYPE MATERIALS WHEN COMPRESSED WOULD SQUEEZE OUT IN THE "X" AND "Y" DIRECTIONS AND DISTORT THE TAPE CAUSING CRACKING AND DELAMINATION IN THE OUTER EDGES. AFTER QUITE A BIT OF RESEARCH A MATERIAL WAS FOUND THAT WAS COMPLIANT IN THE "Z" DIRECTION ONLY ALLOWING COMPLETE LAMINATION FROM EDGE TO EDGE AND FORMING INTERNAL CAVITIES WITH NO DISTORTION.

A TACKY ROLLER IS USED TO REMOVE ANY DUST ON THE LAYERS PRIOR TO STACKING FOR LAMINATION. ANY DUST LEFT INSIDE THE LAYER WILL CAUSE AN EXPLOSION WHEN FIRED FORMING A MINIATURE VOLCANO.

THE LAMINATING PROCESS IS DONE AT 70 DEGREES C AND AT A PRESSURE OF 56KG/SQ.CM. FOR TWO TO THREE MINUTES. FIRING OF THE TRANSFER TAPE IS DONE WITH THE STANDARD 850 DEGREES C RESISTOR PROFILE EXCEPT IT IS HELD AT 600 DEGREES C FOR 20 MINUTES TO ALLOW FOR BINDER BURNOUT.

REQUIREMENTS, BETTER MECHANICAL STRENGTH, IT MEETS THE RADIATION REQUIREMENTS OF THE MILITARY AND AEROSPACE INDUSTRIES AND IS A PROVEN, WELL ESTABLISHED TECHNOLOGY.

ADDITIONAL TAPE MATERIALS ARE AVAILABLE SUCH AS HIGHER "K" CAPACITOR DIELECTRICS FOR BURIED CAPACITORS, AN OPAQUE BLACK DIELECTRIC FOR OPTICAL APPLICATIONS, AND FERROMAGNETIC MATERIALS.

## FINE LINE PRINT

DURING THIS DEVELOPMENT A FINER LINE AND SPACE SCREEN PRINTED CONDUCTOR WAS DESIRED TO COMPLEMENT THE SMALLER VIA SIZE NOW POSSIBLE WITH THE TRANSFER TAPE. IN TODAYS PRODUCTION ENVIRONMENT 125 MICRON LINE AND SPACE IS QUITE COMMON.

WITH THE TRADITIONAL THICK FILM EMULSIONS AND THE PARTICLE SIZE OF THE CONDUCTOR PASTES THAT HAVE BEEN AVAILABLE 75 MICRON LINE AND 100 MICRON SPACE SEEMED TO BE THE LIMIT THAT COULD BE ACHIEVED. HOWEVER A FINER GRIND GOLD CONDUCTOR PASTE IS NOW AVAILABLE AND BY SEARCHING OUTSIDE THE THICK FILM ARENA A HIGHER RESOLUTION EMULSION WAS DISCOVERED THAT COULD EASILY RESOLVE 50 MICRON DIAMETER DOTS.

SOME ADVANCED SCREEN PRINTING TECHNIQUES WERE ALSO DEVELOPED SUCH AS:

A) ELIMINATE THE RIGID SQUEEGEE AND LET IT TILT AND FIND ITS OWN PARALLELISM WITH THE SUBSTRATE.
B) ELIMINATE THE POSITIVE DOWN STOP AND USE AIR PRESSURE TO CONTROL THE SQUEEGEE PRESSURE.
C) USE DIFFERENT SQUEEGEE ANGLES OTHER THAN 45 DEGREES.
D) USE DIFFERENT ANGLES FOR THE SCREEN MESH, PRINT DIRECTION AND SUBSTRATE REGISTRATION.

USING THE FINE GRIND GOLD, THE HIGH RESOLUTION EMULSION AND THE ADVANCED PRINTING TECHNIQUES SUCCESSFUL PRINTING WAS ACHIEVED WITH 50 MICRON LINES AND 75 MICRON SPACES.

RESISTORS CAN BE PRINTED AND LASER TRIMMED ON TOP OF THE TRANSFER TAPE WITH THE SAME STABILITIES AS ON ALUMINA. THE OFFSET OF THE RESISTOR VALUES IS LESS THAN 10% AND THE TCR'S SHIFT 20 TO 30 PPM MORE NEGATIVE. 10% TOLERANCE RESISTORS CAN BE BURIED IN THE INNER LAYERS WITH PREDICTABLE AND REPEATABLE VALUES.

## Electrical Parameters

| | Low K Transfer tape | Standard Transfer Tape |
|---|---|---|
| Dielectric Constant (K) | 4.5 | 8.0 |
| Dissapation Factor | <0.4% | <0.09% |
| Insulation Resistance | $>10^{12}\ \Omega$ | $>10^{12}\ \Omega$ |
| Breakdown Voltage | >1200 volts/25µ | >1200 volts/25µ |
| Fired Thickness (specify) | 50-100µ | 50-100µ |

THE ELECTRICAL PARAMETERS CHART SHOWS SOME OF THE CHARACTERISTICS OF A STANDARD AND A LOW "K" DIELECTRIC TRANSFER TAPE. THE LOW "K" TRANSFER TAPE IS APPROACHING WHAT CAN BE ACHIEVED WITH ORGANIC MATERIALS FOR HIGH FREQUENCY PERFORMANCE.

**ADVANTAGES**

SOME OF THE INHERENT ADVANTAGES OF THE GLASS FILLED CERAMIC TRANSFER TAPE OVER THE POLYMER/COPPER MULTICHIP MODULE PROCESS IS AS FOLLOWS: IT IS NOT HYGROSCOPIC AND THE DIELECTRIC CONSTANT DOES NOT CHANGE WITH HUMIDITY, IT IS HERMETIC AND NEEDS NO SPECIAL PROTECTION FROM THE ENVIRONMENTS, IT CAN WITHSTAND A MUCH BROADER TEMPERATURE RANGE, IT HAS MUCH GREATER THERMAL CHARACTERISTICS FOR POWER

## Conductor Performance

| Gold Conductor | |
|---|---|
| Sheet Resistance | 3mΩ/sq |
| Thermosonic 25µ Gold Wire Bond Strength | 5 to 6 grams |
| **Platinum Gold Conductor** | |
| Sheet Resistance | 50mΩ/sq |
| Solderability | Excellent |
| Adhesion On 2mm Sq Pad | 1.5 to 2.0 Kg (ceramic plug pulls out) |

THE CONDUCTOR PERFORMANCE CHART SHOWS THAT BOTH THE BONDABLE GOLD AND THE SOLDERABLE PLATINUM GOLD IS EQUIVALENT TO THE PERFORMANCE ON A 96% ALUMINA SUBSTRATE. THIS PERFORMANCE IS GENERALLY BETTER THAN TYPICALLY SEEN ON SCREEN PRINTED DIELECTRICS.

## COSTS

WITH THE VOLUME OF TRANSFER TAPE PRODUCTION GOING UP, THE COST IS COMING DOWN AND APPROACHING THAT OF SCREEN PRINTED DIELECTRICS.

WITH THE ELIMINATION OF PINHOLES AND A MUCH SMOOTHER SURFACE THE YIELDS ARE VASTLY IMPROVED MAKING IT VERY COMPETITIVE WITH THE SCREEN PRINTED PROCESS.

USING A 112.5 MM SQUARE SUBSTRATE THAT WILL YIELD A 100 MM SQUARE USABLE AREA AND 4 PALLADIUM SILVER METAL LAYERS THE COST IS BELOW $1.60 PER SQUARE CENTIMETER AND SOON WILL BE LESS THAN $1.00. THE POLYMER/COPPER PROCESS TODAY IS REPORTED TO COST $5.00 TO $20.00 PER SQUARE CENTIMETER MAKING THE CERAMIC TECHNOLOGY VERY COMPETITIVE NOT ONLY IN PRICE BUT ALSO IN PERFORMANCE.

## HIGH PERFORMANCE BGA PACKAGES

IN MICROELECTRONIC APPLICATIONS SOLDER BONDS ARE COMMONLY USED TO ATTACH THE PACKAGE TO A SUBSTRATE SUCH AS A PRINTED CIRCUIT BOARD. A BALL GRID ARRAY PACKAGE IS CONNECTED TO A PRINTED CIRCUIT BOARD, BOTH ELECTRICALLY AND THERMALLY, BY THE USE OF MULTIPLE SOLDER BALLS IN AN ARRAY. THE PACKAGE IS PLACED IN REGISTRATION WITH THE PRINTED CIRCUIT BOARD AND HEATED UNTIL THE SOLDER BALLS FLOW AND COLLAPSE TO A LIMITED DEGREE TO EFFECT CONNECTION TO TERMINALS ON THE PRINTED CIRCUIT BOARD OR OTHER SUBSTRATE.

THIS PATENT PENDING PROCESS PROVIDES A METHOD OF FABRICATING A HIGH DENSITY ARRAY OF SOLDER BALLS ON A SUBSTRATE WHILE ALSO PROVIDING VERY PRECISE CONTROL OF THE LOCATION AND ULTIMATE SIZE OF THE ARRAYED BALLS. THIS METHOD IS EASIER TO PERFORM THAN PREVIOUSLY KNOWN METHODS, INVOLVING FEWER STEPS AND DRAWING FROM CONVENTIONAL THICK FILM FABRICATION TECHNIQUES. THE PROCESS INCLUDES THE STEP OF PRECISELY PERFORATING SHEETS OF DIELECTRIC TRANSFER TAPE TO CREATE A GRID OF CAVITIES TO MATCH A GRID OF TERMINAL PADS ON THE SUBSTRATE.(FIGURE 1) THE PERFORATED TRANSFER TAPE IS THEN LAMINATED AND FIRED TO THE SUBSTRATE, EXPOSING A RESPECTIVE TERMINAL PAD IN EACH OF THE CAVITIES.(FIGURE 2)

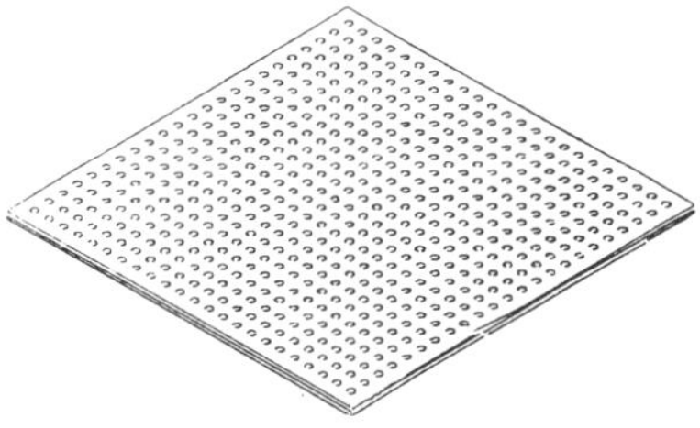

FIG. 1

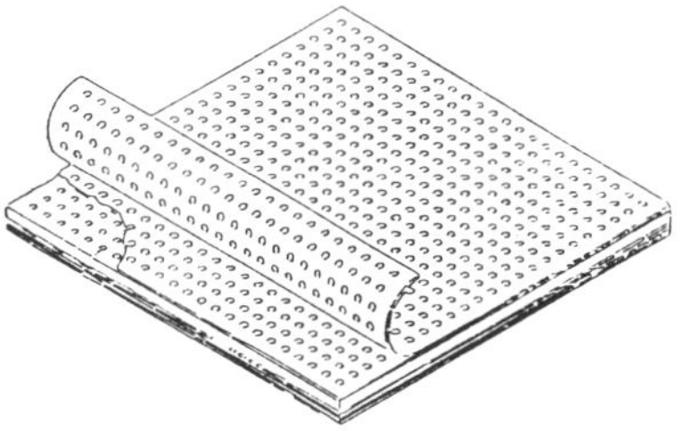

FIG. 2

SPHERICAL SOLDER BALLS ARE THEN PLACED IN THE CAVITIES OF THE TRANSFER TAPE (FIGURE 3) AND HEATED UNTIL THE BALLS FLOW TO FILL THE HOLES AND BOND TO THE TERMINAL PADS OF THE SUBSTRATE.(FIGURE 4) SURFACE TENSION MAINTAINS THE SPHERICAL SHAPE OF THE SOLDER BALLS WHILE THEY ARE FLOWING.

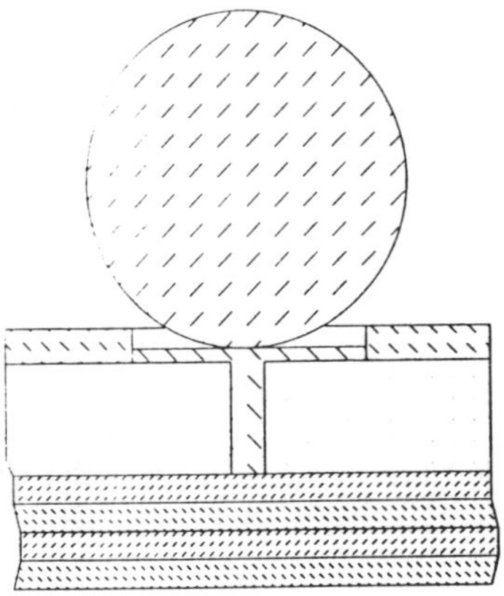

FIG. 3

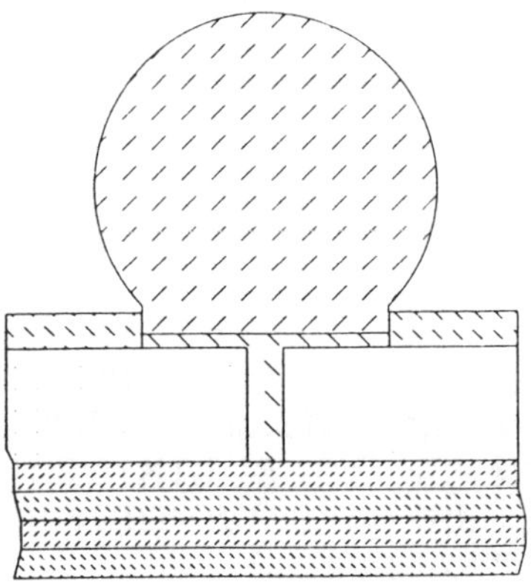

FIG.4

THE SUBSTRATE WITH THE BALL GRID ARRAY IS READY FOR MOUNTING ONTO A CIRCUIT BOARD BY POSITIONING THE SUBSTRATE WITH ITS BALL GRID ARRAY FACING AN ARRAY OF ELECTRICAL CONTACTS OR TERMINAL PADS ON THE CIRCUIT BOARD WITH THE SAME GRID ARRAY. THE ASSEMBLY IS HEATED TO REFLOW THE SOLDER BALLS AGAIN TO MAKE CONNECTION WITH THE ELECTRICAL CONTACTS ON THE CIRCUIT BOARD.

AFTER THE LAMINATING AND FIRING OF THE DIELECTRIC TAPE, WITH THE UNIFORM CAVITIES, TO THE SUBSTRATE THE SOLDER BALLS CAN THEN BE PLACED. THE SOLDER BALLS ARE PLACED ON THE ARRAY OF CAVITIES MECHANICALLY WITH A VACUUM CARRIER WHICH HAS A SIMILAR ARRAY OF PICKUP NOZZLES, AND AFTERWARDS THE DIELECTRIC HOLES WILL HOLD THE SOLDER BALLS IN PLACE WITH A TACKY FLUX. THE SUBSTRATE IS THEN HEATED TO FLOW THE SOLDER BALLS TO FILL THE CAVITIES IN THE DIELECTRIC TAPE AND BONDS WITH THE TERMINAL PAD TO CREATE A MECHANICAL AND ELECTRICAL CONNECTION. SURFACE TENSION ON THE SOLDER MAINTAINS THE SPHERICAL SHAPE OF THE SOLDER BALLS DURING FLOW AND WHILE THEY COOL. IT IS IMPORTANT THAT THE HOLE DIAMETERS IN THE DIELECTRIC TAPE BE PRECISE RELATIVE TO ONE ANOTHER SO THAT AFTER FLOW THE HEIGHT OF ALL THE SOLDER BALLS IS VERY UNIFORM. THIS UNIFORMITY IS IMPORTANT FOR ACCURATE AND RELIABLE CONNECTION OF THE BALL GRID ARRAY TO TERMINALS ON A PRINTED CIRCUIT BOARD OR OTHER SUBSTRATE.

SEVERAL SOLDER ALLOYS ARE SUITABLE FOR USE AS SOLDER BALLS SUCH AS:

| | |
|---|---|
| 95Sn/5Ag | 42Sn/58Bi |
| 98.75In/1.25Al | 95Sn/2.5Ag/1.5In |
| 25Sn/75Pb | 35In/65Pb |
| 62Sn/36Pb/2Ag | 10.5Sn/89.5Pb |
| LEAD(PURE) | 60Sn/40Pb |
| 5Sn/95Pb | 89Pb/11Sb |
| 50Sn/50In | 63Sn/37Pb |
| 10Sn/2Ag/88Pb | |

IT IS RECOMMENDED THAT NO ADDITIONAL SOLDER PASTE BE USED TO ATTACH THE BALL GRID ARRAY PACKAGE TO A PRINTED CIRCUIT BOARD. PRINTING SOLDER PASTE IS A DIFFICULT PROCESS TO CONTROL AND WILL ONLY ENCOURAGE SOLDER SHORTS. A COMPATIBLE FLUX SHOULD BE APPLIED BY BRUSHING OR SPRAYING TECHNIQUES AND THE AMOUNT OF SOLDER ALREADY ON THE BALL GRID ARRAY WILL BE SUFFICIENT TO MAKE THE CONNECTION.

THE DIELECTRIC TRANSFER TAPE WAS USED TO CONTAIN AND SHAPE THE SOLDER BALL AND IT ALSO HELPED TO GREATLY INCREASE ADHESION. SHEAR STRENGTHS WERE MUCH HIGHER THAN ANTICIPATED. (FIGURE 5) THE TEST NOT ONLY PULLED THE SOLDER OFF THE TERMINAL PAD BUT ALSO SHEARED THE SOLDER ACROSS THE CAVITY IN THE DIELECTRIC TAPE.

| TEST # | FORCE |
|---|---|
| 1 | 1.27Kg |
| 2 | 1.09Kg |
| 3 | 1.15Kg |
| 4 | 1.16Kg |
| 5 | 1.33Kg |
| 6 | 1.19Kg |
| 7 | 1.26Kg |

AVERAGE = 1.2Kg
MAXIMUM = 1.33Kg
MINIMUM = 1.09Kg
RANGE = .24Kg
STD.DEV.= 0.08709

FIG.5

ANOTHER ATTRIBUTE OF THE BALL GRID ARRAY PACKAGE IS THE ABILITY TO TRANSFER HEAT THROUGH THE SOLDER BALLS. A TEST WAS CONDUCTED TO MEASURE THE TEMPERATURE RISE, FOR A "J" LEADED SUBSTRATE AND COMPARE IT TO A BALL GRID ARRAY SUBSTRATE. THE TWO SUBSTRATES WERE EACH ATTACHED, BY SOLDER REFLOW TECHNIQUES, TO A 2 INCH SQUARE FR-4 PRINTED CIRCUIT BOARD WHICH WAS THEN ATTACHED TO ALUMINUM PLATE THAT WAS 1/2 INCH THICK. EACH ONE INCH SQUARE SUBSTRATE HAD A 1/2 INCH SQUARE SCREENED ON 10 OHM RESISTOR TO BE USED AS A HEAT SOURCE. AS SHOWN IN THE GRAPH OF FIGURE 6 THE BGA SUBSTRATE WAS 45% LOWER IN TEMPERATURE RISE THAN THE "J" LEADED SUBSTRATE AND THE BGA WITH THERMAL PIPES WAS AN ADDITIONAL 15% LOWER.

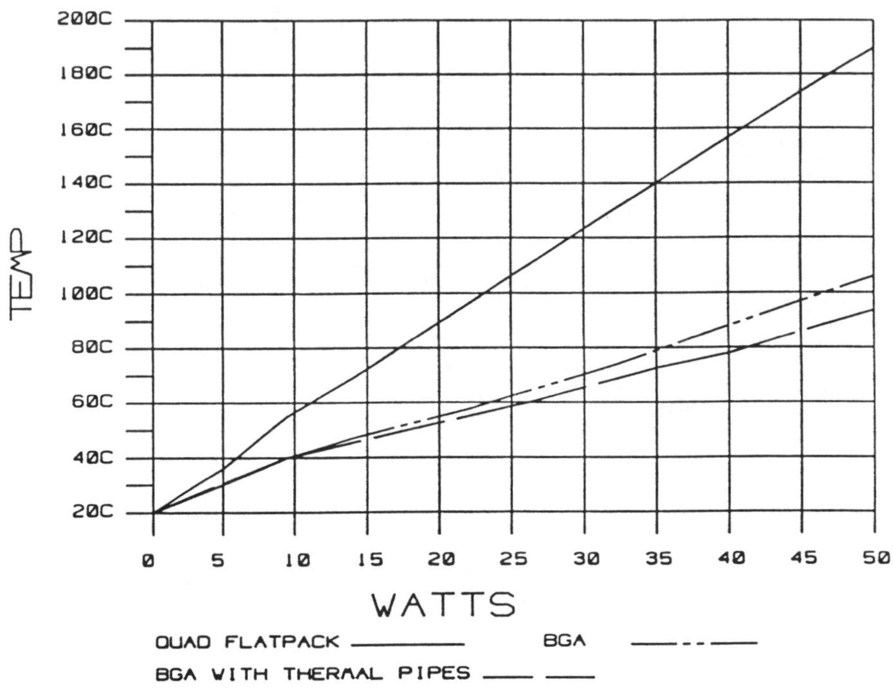


FIG. 6

## CONCLUSIONS

THE POLYMER/COPPER MULTICHIP MODULE CAN PRODUCE DENSITIES TWO TO THREE TIMES THAT OF THE CERAMIC TRANSFER TAPE PROCESS BUT WITH THE LOWER COST AND BETTER MECHANICAL AND ENVIRONMENTAL PERFORMANCE OF THE CERAMIC MULTICHIP MODULE IT SHOULD FILL THE NEED FOR MOST APPLICATIONS.

THE TRANSFER TAPE PROCESS OF LAMINATING ONTO A PREFIRED SUBSTRATE SOLVES THE PROBLEMS INHERIT IN THE COFIRE PROCESS. THESE PROBLEMS ARE FLEXURAL STRENGTH, SHRINKAGE VARIATION IN THE "X" AND "Y" DIRECTIONS AND THERMAL DISSIPATION.

THE CERAMIC BGA PACKAGE PROVIDES HIGH PERFORMANCE FOR HIGH SPEED/FREQUENCY APPLICATIONS, EXTENDED TEMPERATURE RANGE, RUGGED STRENGTH, HIGH POWER DISSIPATION AND EASE OF ASSEMBLY.

# LASER PROCESSING IN MCM-C TECHNOLOGIES

Z. ILLYEFALVI-VITÉZ, PH.D.
*Department of Electronic Technology,*
*Technical University of Budapest,*
*Goldman t. 3, H-1521 Budapest, Hungary*

## 1. Introduction

Different laser technologies can be used in the processing sequence of ceramic substrates and packages applied for multi-chip modules. An interesting new possibility is the application of aluminium nitride which exhibit specific properties (e.g. good thermal conductivity and chemical resistivity) that are important for substrate and packaging materials. Compatible technologies developed for aluminium nitride ceramics improve the applicability of this material. The possibilities and limitations of laser processing technologies for AlN ceramics and some experimental results are discussed.

## 2. Ceramics for Multi-Chip Modules (MCMs)

As electronic devices and modules become more complex, dissipate more power and operate at higher frequencies, the effect of the substrate and the package on the operation of the circuit becomes more and more important [1]. The properties of the materials used in packaging determine how well the package can meet its requirements.

In MCMs the chips that produce large amounts of heat require high thermal conductivity paths for removing the heat in order to maintain the chips at safe operating temperatures. Thermal expansion differences between the chips and the package result in stresses during assembly as well as during operation that can cause the failure of the chips or the package. In high frequency applications the substrate becomes part of the circuit, thus, its dielectric constant is important. Furthermore the packaging materials should have high insulating resistance, should be physically and chemically inert, thermally stable etc. All these requirements can be met by the application of advanced ceramics as substrate and packaging materials [2].

Oxides, such as alumina and beryllia, are ideal insulating materials and have been used for years [3]. The substrate should also provide a thermal conductive path from devices or components that generate heat. Beryllia has been the material of choice for many years in providing the necessary thermal conductivity that helps control the temperature of these devices and components. Aluminum nitride is also a good insulating substrate material and is considered as an alternative to beryllia. Other



*W. K. Jones et al. (eds.), MCM C/Mixed Technologies and Thick Film Sensors*, 69–76.

materials that may be used for thermal management include silicon carbide and aluminum nitride. In Table 1. the typical property values for the most popular packaging materials are listed.

TABLE 1. Properties of advanced ceramic packaging materials

| Property | Units | AlN | SiC | BN (HPGrade) | BeO | $Al_2O_3$ (99%) |
|---|---|---|---|---|---|---|
| Thermal conductivity | W/mK | 180 | 70 | 25 | 260 | 25 |
| Electrical resistivity | ohm-cm | $\geq 10^{14}$ | $\geq 10^{11}$ | $\geq 10^{11}$ | $\geq 10^{14}$ | $\geq 10^{14}$ |
| Dielectric constant | RT-1MHz | 8.6 | 40 | 4.1 | 6.7 | 9.9 |
| Dielectric loss | RT-1MHz | 0.0005 | .05 | 0.0045 | 0.0004 | 0.0004 |
| Thermal expansion coefficient | $10^{-6}$/°C (RT-400°C) | 4.2 | 3.8 | 0-3 | 7.2 | 7.1 |
| Density | g/cm$^3$ | 3.30 | 3.21 | 1.90 | 2.85 | 3.89 |
| Bending strength | MPa | 310 | 441 | 53 | 221 | 341 |
| Hardness, Knoop | GPa | 11.8 | 27.5 | 2.7 | 9.8 | 19.6 |
| Youngs Modulus | GPa | 331 | 407 | 43 | 262 | 345 |

Aluminium nitride offers excellent thermal management when used as the base for multi-chip modules. The high thermal conductivity of AlN and close thermal expansion match to silicon result in minimum stress and excellent heat removal. In addition, co-fired multilayer aluminium nitride has excellent electrical properties for power and ground. Signal layers can be located above the aluminium nitride in multilayer organic systems or, in some cases, direct thin film signal interconnections onto the aluminium nitride. The technology also lends itself well to built-in decoupling capacitors, as well as copper-tungsten via interconnection.

## 3. Technologies compatible with aluminium nitride

### 3.1. THICK FILM MATERIALS

Thick film materials were developed and reported which are compatible with AlN substrates [4]. A family of glass compositions chemically compatible with AlN was the key to the successful development of these pastes. The glass compositions also provide excellent wetting and mechanical bonding characteristics on AlN substrates. A proprietary chemical bonding mechanism was identified. This mechanism is different to the most commonly observed spinel-formation mechanism utilized on alumina substrates. Using the developed glass compositions and reactive bonding compounds, the following were developed: a series of gold (Au), platinum/gold (Pt/Au), and silver/palladium (Ag/Pd) conductors with good adhesion characteristics; a crossover and multilayer hermetic dielectric (compatible with Au-containing conductors) which displays excellent electrical properties (dielectric constant, k, of 5.5); and low-value sheet resistance resistors ranging from one to ten ohms with equivalent performance to commercial resistors on alumina substrates.

$RuO_2$ thick-film resistors for aluminum nitride ceramics with controlled temperature coefficient of the resistor (TCR) and reliabilities were developed [5]. The resistors, in the range of approximately 10 ohms to 10 kohms sheet resistance were obtained using a crystallized glass powder and a $RuO_2$ powder, binder materials, and a conducting material. The glass had a low PbO content of 6 wt %. The resistors, to which $MnO_2$ was added to the above combination, had excellent TCRs of ±250 ppm/°C at 30 ohms to 30 kohms sheet resistance. These resistors showed small resistance changes, which were less than +1.0% after a thermal cycle test (-55 to 150 °C, 1000 times) and high-temperature storage test (150 °C, 1000 h).

An Ag-Pd thick-film conductor for aluminum nitride ceramics was developed [6]. The conductor showed adhesion strength and reliability almost comparable to those of commercial conductors for alumina ceramics. The influence of physical properties of the lead oxide glass frit on conductor performance was studied, and the most suitable glass was selected for the frit material. The softening point of frit glass has the largest effect on conductor performance; this temperature should be approximately 500 °C. If the softening point was either too high or too low, compared to the sintering temperature of the conducting metals, poor adhesion strength resulted. Use of glass with too high a softening point caused poor wettability but resistance to erosion by Pb-Sn solder, while it resulted in good wettability but poor erosion resistance when the softening point was too low. Additionally, Cr-, Ti-, and Zr-oxides were identified as wettability-promoting agents. The molten glass did not penetrate into the AlN ceramics; rather, it remained on the surface. This behavior differed from that in the case of alumina ceramics, for which the molten glass penetrated along the grain boundaries.

## 3.2. BRAZING FOR PACKAGING

Aluminum nitride is a potential candidate for replacing aluminum oxide as a substrate material for electronic circuit packaging. One of the requirements for such a material is that it can be metallized and joined to produce hermetic enclosures for semiconductor devices [7]. A technique for brazing AlN using a nonactive metal braze has been investigated. The process involves the in situ decomposition of an active metal hydride. This process improves the wetting of the AlN and led to the development of strong bonding between braze and ceramic. In addition to joining ceramic to ceramic, braze joints of AlN to copper and to a low expansion iron-nickel lead frame alloy were made. Residual stress resulting from a mismatch of thermal expansion coefficients between AlN and copper caused cracking in the ceramic upon cooldown from the brazing temperature. No cracking occurred in the ceramic when joined to the iron- nickel alloy. The results obtained are important for the realization of AlN as a ceramic packaging material for semiconductor devices.

### 3.3 LASER TRIMMING

Problems and their countermeasures in laser trimming of thick-film metal resistors on aluminum nitride substrates were studied [8]. Trimming was made possible by selection and combination of suitable Q-switching frequency, laser power, and number of irradiations. For example, under the conditions of 1 kHz, 2.0 W, and double irradiation, resistance values could be controlled to within ±0.3% of target values. Resistance changes of trimmed metal resistors were less than +5.5% and comparable to those on alumina after thermal cycle tests, high-temperature storage tests, and high-temperature and humidity storage tests. Since the insulation characteristics of an irradiated portion strongly depended on the irradiation conditions, suitable conditions were selected for practical applications. The degree of insulation degradation after the laser irradiation was different for the two types of AlN substrates studied. This is attributed to the different amounts of free Al formed on the AlN surfaces. Further, AlN has a higher absorption factor and smaller reflectivity than alumina for a YAG laser beam. Thermal and electrical damage of AlN are affected by these properties.

When an AlN ceramic is irradiated by a laser beam during the scribing and resistor-trimming processes, the resistivity of the ceramic decreases or its values become nonuniform [9]. After analyzing the laser-irradiated area of the ceramic by X-ray photospectroscopy and thin-film X-ray diffraction, a very small amount of metallic aluminum was detected in the irradiated area. To avoid this problem the laser irradiation should be carried out under oxygen atmosphere so that the precipitated aluminum becomes gamma aluminium oxide; or the ceramic should be treated thermally after irradiation so that the precipitated aluminum becomes alpha aluminium oxide.

### 3.4. VIA FORMATION BY LASER

A high-speed laser drilling process was developed for producing fine-pitch vias of dimensions equivalent to fine-line screen printed conductors [10-12]. Drilling speeds as high as 150-220 vias per second have been demonstrated on thick film and Green Tape dielectrics using an Nd:YAG laser. Throughputs greater than 300 vias/s could be achieved when multiple layers of Green Tape were drilled at once. There are a number of laser parameters which are critical to achieving high via formation rates while maintaining the accuracy of the via positioning. In addition, the electric composition and its thickness determine the beam power and number of pulses needed to drill each via. The quality of the drilled vias was improved by including a laser absorbing pigment in the dielectric formulation. Overall speed was also dependent upon design parameters such as via pitch, the symmetry of the circuit design, and the drilling sequence.

## 4. Investigations on laser processing of aluminium nitride ceramics

Experimental investigations were carried out with the aim of laser processing of aluminium nitride. Different laser technologies were used to process the AlN substrate of a Leaded Ceramic Chip Carrier (LCCC). The scheme of the LCCC is shown in Figure 1.

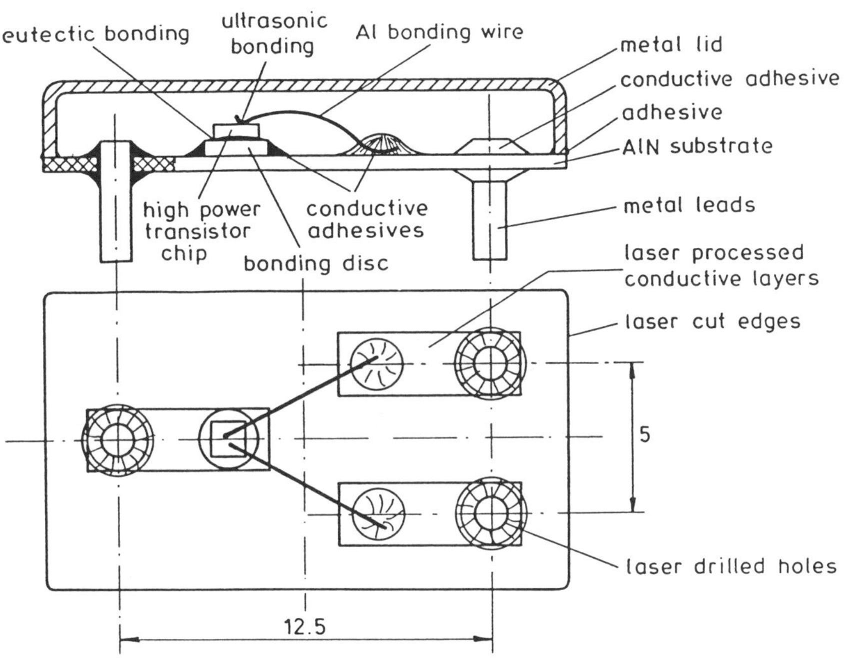


Fig. 1. The scheme of the AlN chip carrier

Cutting and drilling of the AlN substrates were performed by $CO_2$ laser. The processing parameters were optimized on the basis of light and scanning electron microscopic analysis of the cuts and holes. As an example, the cross-section of a cut, which was prepared by too high laser intensity, is shown in the photo of Figure 2. Near optimum laser parameters provided smoother cuts, as it is presented by the side view of a cut after separation (Figure 3.). The lighter part of the cut was not affected by the laser beam, it is a broken, original AlN surface. From the darker area of the surface, however, the heating effect of the laser beam removed away some parts of material, and changed the surface layer from AlN to aluminium. This part of the surface is conductive with a sheet resistance of about 0.5 ohm.

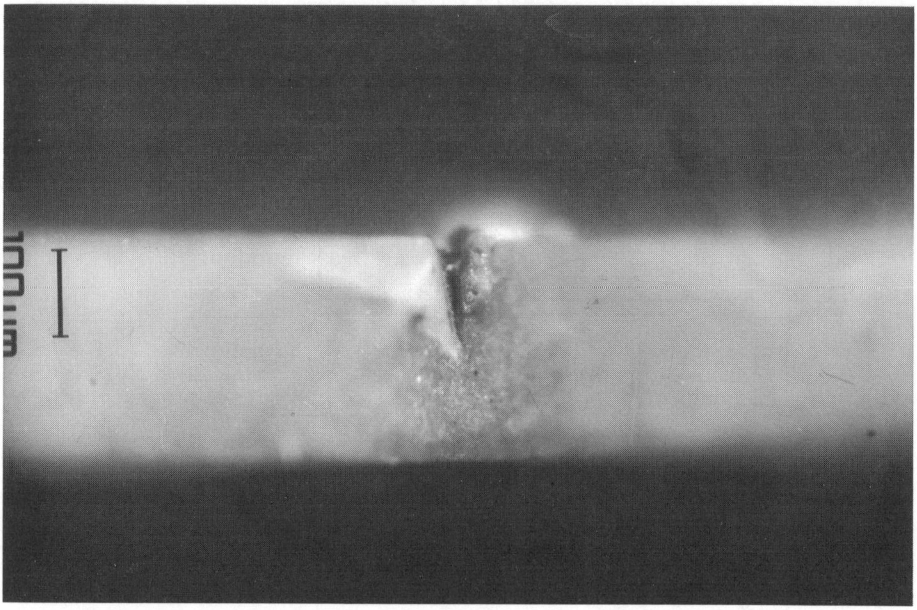

Fig. 2. The cross section of a cut in AlN prepared by $CO_2$ laser processing

As an innovative application of laser technologies to chip carriers, XeCl excimer laser radiation was used to write the conductive pattern onto the surface of the substrate. The material of the conductive layer was aluminium resulting from the decomposition of the aluminium nitride on the laser affected areas. Thorough investigation was carried out to find the proper processing parameters, the physical/chemical basis of the process, the geometrical and material characteristics of the resultant layers etc. Surface profilers, light microscopes and SEM microscopic material analysis were used as analytic methods. As a result it was found that 40-60 laser pulses with an energy density of about 3 $J/cm^2$ and with a repetition rate lower than 10 Hz provide near optimum condition for the formation of an aluminium layer with a sheet resistance of 0.2-0.4 ohm.

A high power transistor chip was mounted onto the patterned substrate applying the conventional wire bonding and conductive adhesive technologies. The application of laser soldering or welding for wire bondings, however, will be considered at a later time.

The produced devices demonstrate the applicability of different laser technologies for the flexible manufacturing of AlN ceramic chip carriers.

Fig. 3. Side view of a cut of an AlN substrate with an Al layer on the heat affected zone

Some initial research was carried out to determine whether excimer laser processing could be applied for drilling holes into green tapes and metals.

In green tapes, that are about 100 μm thick prefired alumina foils, the ablation rate caused by excimer laser radiation was very small. From theoretical point of view, it was interesting that rough edged through holes were created by the acoustic pressure of the plasma generated on the surface.

## 5. References


1. Palmer, R.H.; Reed, D.J.; Adlam, E.J.; Maney, J.P.: A High Power RF Package Using Aluminium Nitride Ceramic - Prospectus of the Carborundum Company, 1994
2. Holmes, R.E.: Choices widen for ceramic substrates - Electronic Packaging and Production, Vol: 31, Iss: 7, p. 56-8, Date: July 1991
3. Advanced Semiconductor Packaging - Prospectus of the Carborundum Company, 1992
4. Allison, K.; Shaikh, A.; Leandri, D.; Raymond, M.; Ackerman, G.; Hankey, D.L.: Thick film materials for applications on aluminum nitride substrates - Proceedings of the 5th International Microelectronics Conference, p. 153-60, 25-27 May 1988, Tokyo, Japan
5. Kurihara, Y.; Takahashi, S.; Yamada, K.; Endoh, T.; Kanai, K.: Thick film resistors for AlN ceramics - IEEE Transactions on Components, Hybrids, and Manufacturing Technology, Vol: 14, Iss: 1, p. 199-203, Date: March 1991
6. Kurihara, Y.; Takahashi, S.; Yamada, K.; Endoh, T.: Ag-Pd thick film conductor for AlN ceramics, IEEE Transactions on Components, Hybrids, and Manufacturing Technology, Vol: 13, Iss: 2, p. 306-12, Date: June 1990
7. Harris, R.D.; Enck, R.C.; Fields, J.L.: Dielectric properties of sintered AlN - Advanced Electronic Packaging Materials. Symposium, p. 229-34, 27-29 Nov. 1989, Boston, MA, USA
8. Norton, M.G.; Kajda, J.M.; Steele, B.C.H.: Brazing of aluminum nitride substrates - Journal of Materials Research, Vol: 5, Iss: 10, p. 2172-6, Date: Oct. 1990

9. Kurihara, Y.; Takahashi, S.; Yamada, K.; Kobayashi, R.; Kanai, K.; Endoh, T.: Laser trimming of thick film metal resistors on aluminum nitride substrates - IEEE Transactions on Components, Hybrids, and Manufacturing Technology, Vol: 14, Iss: 1, p. 204-10, Date: March 1991
10. Takahashi, M.; Kurihara, Y.; Yamada, K.; Kanai, K.; Kurihara, K.: Changes of characteristics in AlN ceramics surface by laser irradiation - Electronics and Communications in Japan, Part 2 [Electronics], Vol: 73, Iss: 8, p. 105-14, Date: Aug. 1990
11. D'Ambra, D.M.; Needes, M.C.A.; Needes, C.R.S.; Wang, C.B.: Via formation in green ceramic dielectrics using a YAG laser - 1992 Proceedings. 42nd Electronic Components and Technology Conference (Cat. No.92CH3056-9), p. 1072-81, 18-20 May 1992, San Diego, CA, USA
12. M.; Gonzalez, R.; Ortega, N. Application of laser technology in drilling and scribing low temperature cofired ceramic - Hybrid Circuits, Iss: no.21, p. 10-14, Date: Jan. 1990
13. Rabil, M.E.; Zablotny, G.; Johnson, W.: A case study: laser-drilled ceramics - Hybrid Circuit Technology, Vol: 6, Iss: 7, p. 10-12, Date: July 1989

# LTCC TECHNOLOGY: WHERE WE ARE AND WHERE WE'RE GOING

Charles Q. Scrantom
Scrantom Engineering, Inc.
3550 Cadillac Avenue
Costa Mesa, California 92626

## 1. Abstract

New electronic systems for military/aerospace will continue to challenge both packaging engineers and technology due to the increased performance requirements, higher power densities, and limited space available. This challenge mandates the use of unique packaging techniques such as multi-chip modules (MCM) that must not only provide the increased circuit density but also the reliability, electrical and mechanical performance, thermal management, and hermeticity.

Recent interest in MCM technology has focused on different packaging technologies such as laminate, ceramic, and thin film technologies. One such ceramic technology, low temperature co-fired ceramic (LTCC) tape technology, offers significant benefits over other established packaging technologies for high RF and fast digital applications that require hermeticity and good thermal management.

This paper will address the current status of LTCC technology as well as new and promising developments that could make this the leading MCM package technology.

## 2. Expected MCM Growth

All market indicators are predicting tremendous growth in the MCM market. Dataquest has projected that in North America alone the MCM market will grow from \$350 million in 1993 to about \$10 billion by the year 2000. The military's share of the market will grow from about \$16 million to \$1.1 billion by the year 2000. So important is the MCM market, ARPA has awarded \$40 million to one MCM consortium comprised of the key military electronics companies. It is interesting to note that much of the funds will be used to improve the technical merits of laminate technology (MCM-L) and reduce the cost of thin film (MCM-D) technology closer to that of a technology that already exists -- LTCC tape technology (MCM-C)!!



*W. K. Jones et al. (eds.), MCM C/Mixed Technologies and Thick Film Sensors, 77–87.*

## 3. Introduction To LTCC

LTCC tape systems are composed of a glass/ceramic dielectric tape provided in rolls with shrinkage-matched metallization pastes. Unlike thick film technology where sequential printing, drying, and firing of each layer is required, LTCC technology, like PWB technology, processes all the different layers in parallel (i.e. all layers are punched, printed, and dried in parallel). As a result optimum yields and cost effectiveness are achieved since each process is not dependent on the previous process.

Once all the layers are processed, they are stacked, laminated, and co-fired at temperatures between 800°C-900°C to form a high density, fully integrated substrate or package (package + electrical interconnections). An integrated package is ideal for MCM applications since the components mount directly down on the package surface rather than on a separate ceramic or laminate insert that is then mounted in a standard HTCC package resulting in additional wire bonding and lower reliability.

During the firing process, the tape will shrink between 12%-16%±0.2% in the X- and Y-axis and 15%-25%±0.5% in the Z-axis (depending on manufacturer). Assuming balanced metal loading and good process control, the shrinkage is consistent; however, with certain systems, there is a narrow processing window which the fabrication must take place.

## 4. MCM Packaging Using LTCC Technology

Besides cost effectiveness and integrated packaging, LTCC technology offers other features that are ideal for MCM packages.

**Fine Lines And Spaces**
**Low Resistance Metallization**
**Excellent High Frequency Characteristics**
**Allows For Unique Package Designs**

### 4.1. Fine Lines and Spaces

While the definition of fine lines and spaces has changed with each improvement of technology, the thick film printing process in LTCC technology is easily capable of 0.006" lines and spaces, and 0.004" lines and spaces are do-able. If finer lines and spaces are required, two methods are available to provide line widths and spaces on the order of 0.0015"-0.002" on the outer layers only. A thin film deposition process can be used on the outer layers to produce finer lines and spaces; however, it is very expensive, and the surface of the fired tape must be extremely smooth for good adhesion. Polishing will help, but very small pits will remain and cause some adhesion problems.

Another method to achieve finer lines and spaces on the top layer is a photoimagable thick film gold post-fire process. Dupont has introduced a material system called FODEL® which enables one to resolve 0.0016"/0.002" lines and spaces, respectively. The FODEL® material system requires printing a gold area where the fine resolution is required. After drying, the fine line pattern is exposed onto the gold area by a high intensity ultraviolet light source. The pattern is developed using a sodium carbonate spray process followed by a water rinse and drying. The resulting patterin is then fired through a belt furnace at 850°C, like a standard thick film print. Although other thick film etchable gold processes are available, the sodium carbonate developer used with FODEL® does not attack the LTCC, and so no additional masking and stripping is required.

While fine lines and spaces are available with LTCC technology, it comes at a significant cost. One must weigh the option of adding layers and using wider lines/spaces or using the more expensive thin film or photoprintable thick film processes. In many cases where fast speeds or high frequencies are used, there is justification for fine lines and spaces; however, in the majority of applications, additional layers and wider lines and spaces can be used and significant cost savings can be realized.

## 4.2. Low Resistance Metallization

What makes LTCC technology both desirable and undesirable for many applications is the use of expensive high conductivity metals such as gold and silver. MCM applications requiring high speed digital and/or low level, high frequency analog processing cannot tolerate the IR losses associated with the metallizations used in high temperature co-fired ceramic (HTCC) technology. But the metallizations in HTCC do offer many cost advantages, and in many high impedance applications they have more than adequate conductivity.

Without sacrificing the electrical performance, LTCC tape manufacturers have addressed the cost issue by offering mixed metallization systems and a copper system. Mixed metallization systems use lower cost silver on the inner layers and gold on the exposed layers. Intermetallic migration between silver and gold is prevented by a barrier via fill connecting the outer and inner layers.

Lower cost copper systems are now becoming available with LTCC. While the metallization is lower cost, the fabrication process requires a wet nitrogen process which can be costly to install and maintain if not supported by significant production.

The manufacturers of LTCC tape systems will continue to work with their metallizations to reduce costs; however, they will always be subject to the cost fluctuations of the world commodity market. As a result, emphasis will be placed on mixed metallization systems of gold/silver, copper systems, and mix of copper/silver/gold that is solderable and wire bondable.

## 4.3. Excellent High Frequency Characteristics

LTCC tape systems exhibit excellent high frequency characteristics making it ideal for applications such as packaging T/R modules for phased array radar systems. Certain tape systems such as Ferro A6 and Dupont 901 have low dielectric constants (K = 5.2-5.9), low insertion loss, and good loss tangent characteristics. In addition, the tape is cast in consistent 0.005"/0.008" thicknesses, so dielectric thicknesses are predictable to within ±.5%.

While the low dielectric tape is desired for impedance control, many tape manufacturers are experimenting with combining the low dielectric with a very high K tape. This will enable designers to print high value capacitors on the inner layers of the substrate, thereby allowing many GaAs MMIC devices to be smaller. Efforts continue on these tape systems for both military and commercial applications.

Since LTCC technology is expected to be used in many GaAs MMIC packaging applications, it is noteworthy that LTCC's Cte closely matches that of GaAs (7.0 vs. 6.5).

## 4.4. Unique MCM Package Designs

Being able to process each layer allows the designer to create unique package designs without high tooling costs. Custom pin grid arrays (PGAs), ball grid arrays (BGAs), and quad flat packs are easily manufactured using LTCC technology. These packages can be made with cavities, steps, thick film printed resistors and/or capacitors, surface mount pads, seal rings, lead frames, and pins.

Since each LTCC tape layer is individually fabricated, cavities and steps are formed by simply punching the cavity's dimensions on the applicable tape layers and stacking the layers to the required cavity depth. In addition, steps for close pitch wire bonds can be incorporated just by changing the cavity dimensions.

LTCC lends itself to many value-added post fire processes which adds to the uniqueness of the MCM package. Thick film resistors can be printed on the top and bottom layers and trimmed to specific values within ±1%. While it is desirable in many applications to bury precision resistors on the inner layers, the technology is only to a point where resistors can be printed (no trimming) and co-fired to an accuracy of ±30%.

Where the values of resistors or capacitors exceed the limits of thick film printing processes, surface mount components can be added to the top or bottom surfaces. Surface mount pads that interconnect to the inner layers can be printed as a post fire process and the surface mount chip components can be added during a final process.

Tape manufacturers have also developed low temperature braze systems for LTCC which enable seal rings and lead frames/pins to be attached to the MCM package. Hermetic packages (PGA, QFP, BGA) are manufactured by routing the I/Os to surface pads. In the case of PGAs, pins are attached to the pads/vias on the bottom layer, and lead frames are attached to pads along the top or bottom perimeter of the package on QFPs. Integrated LTCC packages allow for routing the I/Os through the inner layers of the package to pads outside the hermetic seal ring, therefore no glass-to-metal seals are required. As a result, all leaks caused by stressing the glass-to-metal seal during the lead form and assembly process are eliminated.

Typical braze systems for LTCC utilize a gold-tin, gold-germanium, or gold-indium eutectic that reflows between temperatures from 380°C-580°C. The braze systems also include adhesion and barrier materials to facilitate the brazing process.

While the braze systems provide for hermetic packages with lead frames or pins, the gold content can drive the price of the package quite high. Manufacturers are looking at alternative materials to attach seal rings and lead frames/pins that would be more cost effective.

## 5. Perceived LTCC Limitations And New Developments

### 5.1. Shrinkage

This paper has presented LTCC as a versatile, high density packaging technology that is ideal for many MCM applications. While there is no doubt that LTCC technology offers many benefits over other packaging technologies, there are two characteristics that are perceived as limitations of this technology -- shrinkage control and thermal conductivity. While these are valid concerns, it should be noted that industry has addressed these issues by improved fabrication processes and technology.

Fabricating large multi-layer structures (>5" x 5") that incorporate fine lines and spaces (<0.006") can result in alignment problems in the X- and Y-axis due to the shrinkage tolerance of ±0.2%. This poses a particular problem with fine pitch connectors that traverse across the width of the substrate. The cumulative tolerance of 0.2% could result in the connector leads not aligning with the pads on the substrate.

Tape manufacturers have addressed the shrinkage issue by improving their tape casting processes and refining the processing windows. Manufacturers of LTCC substrates and packages are fabricating 5" x 5.5", 20+ layer substrates with consistent and repeatable results.

To eliminate shrinkage altogether, some manufacturers have promoted a "tape-on-substrate" technology. Shrinkage is virtually eliminated by laminating and firing each layer of tape on a substrate made of alumina, BeO, or AlN. While this eliminates any component assembly alignment problems associated with shrinkage, "tape-on-substrate" is a serial process resulting in higher costs, and the ability of the LTCC layer stack-up is limited.

Other manufacturers such as Ragan Technologies have developed a zero shrink (ZST ™) LTCC tape system. This system requires no base substrate and has shrinkage of less than 1% with a tolerance of ±.02%. Preliminary test results on the tape's physical and electrical properties appear promising, and the metallization system is close to being released.

### 5.2. Thermal Conductivity

As the power densities of MCMs continue to increase, the thermal conductivity of the package technology must be able to conduct and spread the heat to maintain the product reliability. Although offering a thermal conductivity significantly higher than that of organic laminate material, LTCC's thermal conductivity of 2.0-2.5 W/m-K is a limitation to MCM designs dissipating many watts of power.

Many thermal management techniques are being utilized to improve the thermal conductivity through the substrate (Z-axis) as well as spreading the heat across the area of the substrate (X- and Y-axis). Probably the most common method of conducting the heat through the Z-axis is through thermal vias. Thermal vias are through holes that are filled with gold or silver metallization and are strategically placed beneath the die pads of hot components. The high thermal conductivity of the metals in the vias reduces the thermal impedance in the Z-axis and can improve the thermal conductivity to better than 70 W/m-K. The effectiveness of thermal vias is summarized below.

**Table 2. The Effect Of Thermal Vias On Thermal Conductivity (Compliments Of The Dupont Design Guidelines)**

| Via % | Via Dia. | Pitch | W/m-K Ag Vias | W/m-K Au Vias |
|---|---|---|---|---|
| 0.75 | 0.008" | 0.082" | 4.291 | 3.744 |
| 1.23 | 0.008" | 0.064" | 6.035 | 5.134 |
| 2.18 | 0.008" | 0.048" | 9.504 | 7.905 |
| 3.0 | 0.008" | 0.042" | 12.441 | 10.252 |
| 5.0 | 0.008" | 0.032" | 19.409 | 15.807 |
| 10.0 | 0.008" | 0.022" | 38.968 | 31.429 |
| 12.5 | 0.008" | 0.020" | 47.240 | 38.035 |
| 18.0 | 0.008" | 0.017" | 67.066 | 53.858 |
| 20.0 | 0.008" | 0.016" | 72.898 | 58.547 |

Once the heat is transferred through the thermal vias it is necessary to spread the heat uniformly across the X- and Y-axis. The most common method of spreading the heat is by applying a thick film layer of gold on the back side. For applications requiring better heat spreading, some manufacturers are looking at applying a higher conductivity thin film layer or mounting the substrate on higher conductive materials such as copper tungsten (CuW) and copper-molybdenum-copper (Cu-Mo-Cu). What makes CuW and Cu-Mo-Cu ideal for many high power applications is not only their thermal conductivity (190 W/m-K and 160 W/m-K, respectively) but also their Cte that closely matches that of GaAs (6.5 x $10^{-6}$/K). The one disadvantage of using these materials as heat spreaders is the additional weight they add to the modules.

**Table 3: Thermal and Expansion Properties Of Different Materials (Extracted From Polese Company Literature)**

| Material | Thermal Expansion (x $10^{-6}$/K) | Thermal Conductivity (W/m-K @ 25°C) |
|---|---|---|
| Copper | 17.8 | 398 |
| W | 4.6 | 178 |
| Mo | 5.5 | 138 |
| CuW: | | |
| 85/15 | 7.2 | 190 |
| 80/20 | 8.3 | 200 |
| 90/10 | 6.5 | 180 |

| | | |
|---|---|---|
| CuMo: | | |
| 85/15 | 7.0 | 160 |
| Alumina | 6.7 | 17 |
| AlN | 4.5 | 200 |
| BeO | 7.6 | 250 |
| SiC | 3.7 | 272 |
| Si | 4.2 | 151 |
| GaAs | 6.5 | 54 |

In some applications where weight and thermal conductivity are critical, different packaging options are available. One option is using some of the new metal matrix composite materials such as silicon carbide particulate reinforced aluminum (SiCp/Al) by Lanxide, or other light weight metal alloys from Brush-Wellman and Loral. These materials have excellent thermal conductivities and Cte characteristics. A second option that was mentioned earlier to achieve shrinkage control, "tape-on-substrate", will also provide improved thermal conductivity since the base substrate can be alumina (17 W/m-K), AlN (200 W/m-K) or BeO (250 W/m-K).

Packages fabricated totally from AlN are also an alternative. Besides its excellent thermal conductivity, AlN's Cte matches closely with silicon ($4.2 \times 10^{-6}$/K); however, a possible limitation to high frequency applications is its 8.5 dielectric constant and high resistance metallization.

Like LTCC, AlN can be cast as a green tape; however, instead of co-firing the multi-layer packages in air at 850°C, AlN packages are formed using a hot press at 2000°C in a dry reducing atmosphere to prevent oxidation. Because of the high temperatures, higher resistance refractory metals such as W and Mo/Mn must be used.

While certainly demonstrating excellent thermal conductivity characteristics, AlN is still a very expensive alternative. Efforts to reduce the cost, and development of a high thermal conductive low temperature co-fire ceramic tape that can utilize thick film precious metallization continues.

In typical MCM applications, there are only a handful of components that dissipate large amounts of power. In most cases the thermal management can be handled through thermal vias and a heat spreader, but in instances where thermal vias cannot meet the thermal management requirements Ragan Technologies has been able to embed slugs of BeO into zero shrink LTCC tape and form a hermetic oxide bond between the tape and the BeO slug. This technique not only will solve the most severe thermal management issues, but it is also very cost effective.

| | | | | LTCC TAPE SYSTEMS AND VENDORS | | | | | | | |
|---|---|---|---|---|---|---|---|---|---|---|---|
| | DUPONT | DUPONT | FERRO | ESL | ESL | ESL | ESL | ESL | RAGAN | RAGAN | SHOEI |
| | 951 | 901 | A6 | D-101C-70 | D-101C-70B(A) | D-121C-70 | D-111C-70 | D-101F-70 | ZST HI K | ZST LO K | |
| | | | | | | | | | | | |
| CO-FIRE | YES | YES | YES | YES | YES | YES | YES | YES | YES | YES | YES |
| COLOR | BLUE | WHITE | WHITE | WHITE | BLACK | GREEN | BLUE | WHITE | WHITE | WHITE | WHITE |
| DIELECTRIC | | | | | | | | | | | |
| CONSTANT (K) | 7.8 | 5.3 | 5.9 | 8.0-9.0 | 8.0-9.0 | 100 | 4.3-4.7 | 8.0-9.0 | 7 | 5 | 7.4 |
| THERMAL | | | | | | | | | | | |
| CONDUCTIVITY | 3 W/m-K | 3 W/m-K | 2 W/m-K | N/A | N/A | N/A | N/A | N/A | 5 W/m-K | N/A | 2.6 W/m-K |
| THERMAL | | | | | | | | | | | |
| EXPANSION | 5.8 PPM/C | 3.2 PPM/C | 7 PPM/C | 7 PPM/C | 7 PPM/C | 7 PPM/C | 7 PPM/C | 7 PPM/C | 7 PPM/C | 7 PPM/C | 7.7 PPM/C |
| LOSS TANGENT | 0.15% | <.3% | <.2% | <.3% | <1% | <2.0% | <.3% | <.3% | N/A | N/A | N/A |
| | | AT 1MHZ | AT 10 MHZ | AT 1 MHZ | AT 1 MHZ | AT 1 MHZ | AT 1 MHZ | AT 1 MHZ | | | |
| INSULATION | | | | | | | | | | | |
| RESISTANCE | >10E12 OHMS | >10E12 OHMS | >10 TERAOHM/LAY. | >10E12 OHMS | >10E12 OHMS | 10E12 OHMS | 10E12 OHMS | 10E12 OHMS | 10E12 OHMS | 10E12 OHMS | 10E14 OHMS |
| BREAKDOWN | | | | | | | | | | | |
| VOLTAGE | >1000V/mil | >500V/25u | >3000V/Layer | >1800V/25u | >1500V/25u | 800V/25u | >1500V/25u | >1700V/25u | >1200V/25u | >1200V/25u | >4000V/100u |
| FIRED DENSITY | | | | | | | | | | | |
| (%) | | | | | | | | | | | |
| FIRED DENSITY | | | | | | | | | | | |
| (THEORETICAL) | 3.1 gm/cc | | 2.5 gm/cc | 3.16 gm/cc | 3.2 gm/cc | 6.2 gm/cc | 2.3 gm/cc | 3.16 gm/cc | 3.1 gm/cc | 3.1 gm/cc | 3.14 gm/cc |
| FLEXURAL | | | | | | | | | | | |
| STRENGTH | 320 MPa | 230 MPa | 130 MPa | N/A | N/A | N/A | N/A | N/A | N/A | N/A | N/A |
| SURFACE | | | | | | | | | | | |
| ROUGHNESS | <10 uinches | <10 uinches | <15 u"/"CLA | N/A | N/A | N/A | N/A | N/A | N/A | N/A | N/A |
| THICKNESS | | | | | | | | | | | |
| GREEN | .0038"/.005".0085" | 127 microns | 127 microns | 100-130 microns | 100-130 microns | 100-130 microns | 100-130 microns | 300-350 microns | 127 microns | 127 microns | 120 microns |
| FIRED | | | | | | | | | | | |
| SHRINKAGE: | | | | | | | | | | | |
| Z-AXIS | 15%+/-.5% | 18%+/-.5% | 25%+/-.5% | 16%+/-1% | 16%+/-1% | 17%+/-1% | 16%+/-1% | 16%+/-unknown | 25%+/-.5% | 25%+/-.5% | 19+/-.3% |
| X-, Y-AXIS | 13%+/-.2% | 16%+/-.2% | 15%+/-.2% | 12%+/-.5% | 12%+/-.5% | 16%+/-.5% | 13%+/-.5% | 12%+/-.5% | .1%+/-.02% | .1%+/-.02% | 16+/-.3% |
| | | | | | | | | | | | |
| METALLIZATION | AU/AG/AU-AG | AU/CU | AU/AG/AU-AG | AU | AU | AU | AU | AU | AU | AU | CU |

| | TAPE-ON-SUBSTRATE VENDORS AND CHARACTERISTICS | | |
|---|---|---|---|
| | | | |
| | **ESL** | **ESL** | **EMCA** |
| | **D-111C-25** | **D-101C-25** | **T8800** |
| COLOR | **BLUE** | **WHITE** | **BLUE** |
| DIELECTRIC | | | |
| CONSTANT (K) | 4.1-4.6 | 8.0-9.0 | 7.0-10.0 |
| THERMAL | | | |
| CONDUCTIVITY | N/A | N/A | 3.41 W/m-K |
| THERMAL | | | |
| EXPANSION | 7 PPM | 7 PPM | 7 PPM |
| LOSS TANGENT | <.4% | <.5% | <.1% |
| | AT 1MHZ | AT 1 MHZ | AT 1 KHZ |
| INSULATION | | | |
| RESISTANCE | >10E12 OHMS | .10E12 OHMS | >10E11 OHMS |
| BREAKDOWN | | | |
| VOLTAGE | >1200V/25 microns | >1400V/25microns | >1000V/mil |
| FIRED DENSITY | | | |
| (%) | | | |
| FIRED DENSITY | | | |
| (THEORETICAL) | 2.26 gm/cc | 3.06 gm/cc | 3.06 gm/cc |
| FLEXURAL | | | |
| STRENGTH | N/A | N/A | N/A |
| SURFACE | | | |
| ROUGHNESS | N/A | N/A | N/A |
| THICKNESS | | | |
| GREEN | 100-130 microns | 100-130 microns | 105 microns |
| FIRED | | | |
| SHRINKAGE: | | | |
| Z-AXIS | 40-45% | 40-45% | N/A |
| X-, Y-AXIS | 0% | 0% | 0% |
| | | | |
| METALLIZATION | AU | AU | AU |
| | | | |

## 6. Summary

The market for MCMs will grow tremendously in the coming years and the type of packaging technology will play an important part in the maximum density, and electrical and mechanical performance of the MCM. LTCC technology offers many features that are beneficial for high reliability MCM applications at a cost effective prices.

Development of new tape systems will bring the cost down while improving performance. Zero shrink tape; better thermal management techniques; and low loss, high/low dielectric tapes for high frequency applications will allow LTCC technology to expand into many commerical and industrial applications.

## 7. Acknowledgements

The author wishes to acknowledge the help and cooperation received from the engineering and support staff at Scrantom Engineering whose efforts have forwarded LTCC technology to new heights.

## 8. References

1. Green Tape Material System, Design and Layout Guidelines, Dupont Electronic Materials, Research Triangle Park, North Carolina

2. LTCC Data Sheets, Ferro Corporation and Dupont Electronic Materials

3. Horn, Debra S.; Minehan, William T.; Volmering, James E.; Weidner, Ken; Coors Electronic Package Company Chattanooga, TN (1994). Current Technology For Hot Pressed, Co-Fired AlN Electronic Packages

4. Polese, Frank. Sales Information For Polese, Inc.

# DESIGN AND REALIZATION OF HIGH PERFORMANCE CERAMIC HEAT SINKS

E. GAMBARTE, M. TÖPFER, A. PAREDES, M. REILL, M. WEICKHMANN
Technische Universität Berlin, Technologien der Mikroperipherik, Gustav-Meyer-Allee 25, 13355 Berlin, Germany

## 1. Introduction

Due to rising heat dissipation and packaging density on component- , module- and system level, modern applications in the field of High-Power-Modules need new cooling concepts.
Heat sinks with micro channels enable a very efficient heat removal in high power packaging applications. Aluminium nitride is ideal, due to its thermal conductivity and its expansion coefficient matched to silicon for use as substrate material.
Further advantages of aluminium nitride (AlN) are the high electrical isolation resistance, mechanical properties as well as its environmental compatibility.
To optimize multi chip packages a simulation tool was developed which solves the heat conduction equation with nonhomogeneous boundary conditions. The boundary conditions have to be formulated by heat transfer models. In this contribution we describe analytical models for high performance heat sinks. To verify the validity of these models measurements were conducted with realized heat sinks in AlN.

## 2. Modeling of the heat sink

The removable heat is proportional to the available surface concerning convective heat transfer. An increase of the surface can be achieved by using heat sinks, whereas the fins are rectangular in shape and are attached to the spreader surface. As shown in fig. 1 the model consists of one heat source of constant heat flux density $q$ which is mounted on a conductive plate with the cooling fins attached in parallel. Water is forced through the inclosed channels. The design parameters are the channel width $w_c$, channel height $z$, channel length $L$ , fin thickness $w_w$ for given heat flux density $q$ and an allowable pressure loss $\Delta p$ in the channel as well as the material parameters.
In order to calculate and optimize the heat sink the following premises have to be made:

- laminar flow
- constant material properties



W. K. Jones et al. (eds.), MCM C/Mixed Technologies and Thick Film Sensors, 89–100.

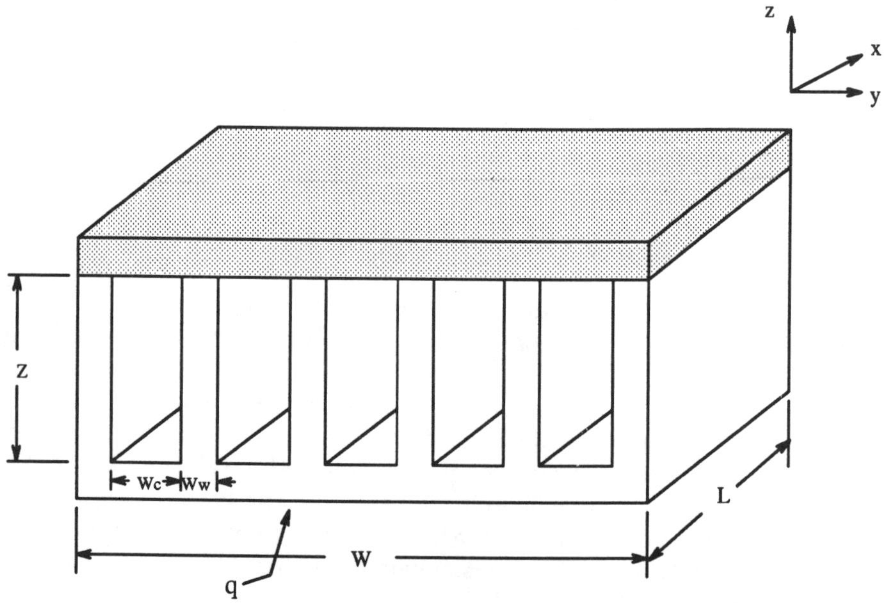


Figure 1: Schematic view of a heat sink

The demand for constant heat-flux-density on the heat sink surface is nearly satisfied, if e.g. the surface is equally packed with heating sources. The capability of the cooler is characterized by using a normalized thermal resistance

$$R^n_{th} = \frac{T_{w,a} - T_{f,e}}{q}. \tag{1}$$

$T_{w,a}$ is the maximum wall temperature, which is reached at the end of the channels ($x = L$) and at the bottom of the fins ($z = 0$). $T_{f,e}$ is the fluid temperature at the inlet region. $q$ describes the flux-density with regard to the surface $LW$. In order to calculate the thermal resistance, we investigate the heat flux within the fins. There is no heat transfer at the top side of the fins ($J(x, Z) = 0$). The heat-flux from the fins towards the fluid is given by the temperature difference between the wall and the fluid. A solution for the thermal resistance is given by Tuckermann [3] :

$$R^n_{th,t} = \frac{(w_c + w_w)\lambda}{k_w w_w \tanh(Z/\lambda)} + L(w_c + w_w)/\varrho c_p f, \tag{2}$$

Whereas

$$\lambda = \sqrt{\frac{k_w w_c w_w}{k_f Nu}} \tag{3}$$

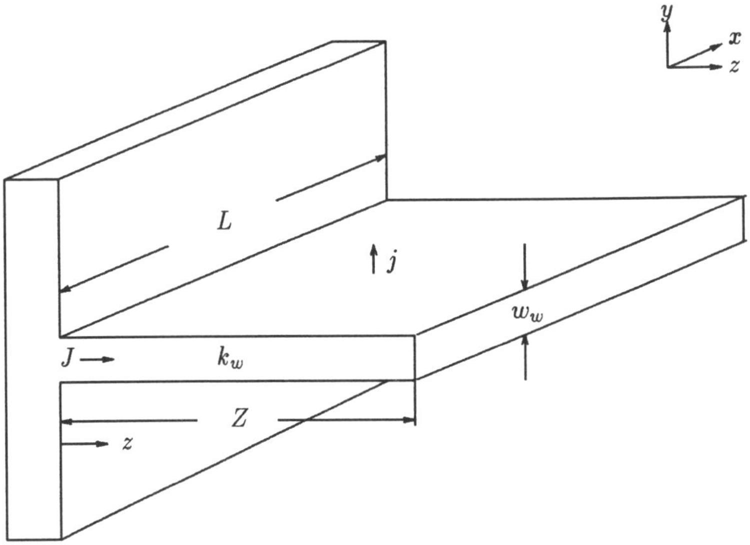


Figure 2: Model of a fin

and $f = v_m w_c Z$ which describes the flow rate through the channel. The geometry is taken from fig. 2 . $k_w$ and $k_f$ are the thermal conductivity of the heat sink material and the fluid, $\varrho$ and $c_p$ are the density and the thermal heat capacity of the fluid respectively. To calculate the velocity $v_m$ we use:

$$v_m = 2\sqrt{\frac{w_c \Delta p}{\xi \varrho L}}, \tag{4}$$

The friction coefficient $\xi$ and the Nusselt-number $Nu$ are given by [2]:

$$\xi = \frac{96}{Re}(1 - 1.3553\zeta + 1.9467\zeta^2 - 1.7012\zeta^3 + 0.9564\zeta^4 - 0.2537\zeta^5) \tag{5}$$

and

$$Nu = 8.235(1 - 2.0421\zeta + 3.0853\zeta^2 - 2.4765\zeta^3 + 1.0578\zeta^4 - 0.1861\zeta^5). \tag{6}$$

whereas $\zeta$ describes the aspect ratio $w_c/z$ .
An alternative solution is proposed by Keyes [1]:

$$R^n_{th,k} = \frac{(w_c + w_w)\lambda^*}{k_w w_w \tanh(Z/\lambda^*)} \tag{7}$$

hereby $\lambda^* = \lambda(1 + L/a)$ and $a = \varrho c_p v_m w_c^2 / k_f Nu$
Up to now we presented two different approaches for modeling heat sinks. In the following we will compare these analytical approaches with FEM simulations to select an appropriate model for optimization of the channel geometry.

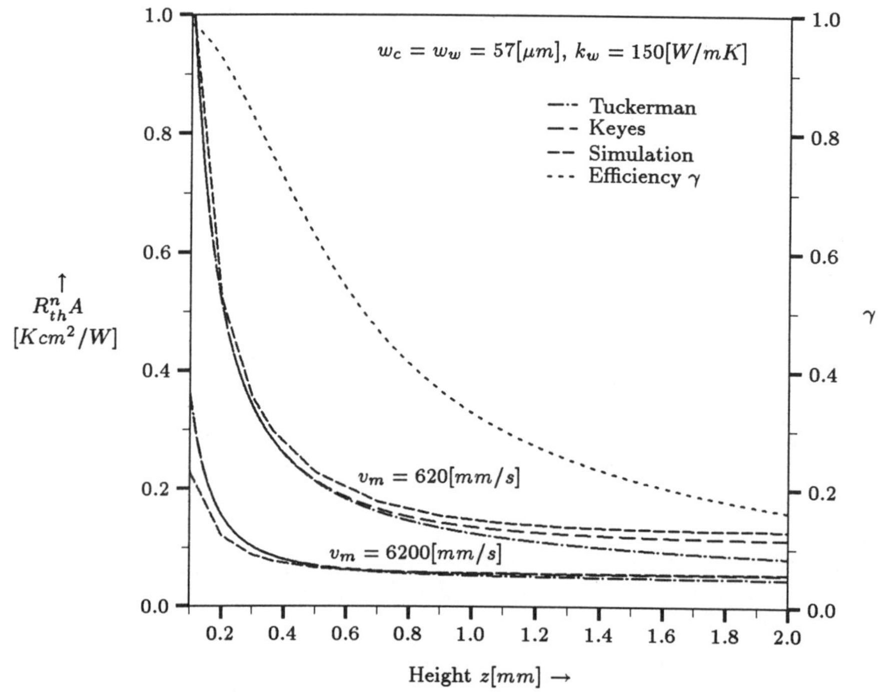


Figure 3: Thermal resistance versus fin height

## 2.1 DISCUSSION OF SIMULATION RESULTS

In a first step the two described analytic procedures were compared with FEM-simulations in order to choose a proper model. The behavior of the models were investigated for different fin efficiencies and flow rates. The fluid heating along the channels can be studied using different flow rates. A FEM-model was used as shown in fig. 2. The thermal conductivity of the ground plate was neglected. The material parameters are assumed to be constant. A given fluid velocity was specified. The energy equation was solved for fully developed flow. The model accounts for the region within the channel, where the temperature profile is not fully developed. The simulation parameters are given in fig. 3.

For computing the thermal resistance one can account for the thermal gradient along the fins in using the fin-efficiency $\gamma = \lambda \tanh(Z/\lambda)/Z$ . In order tc optimize the heat sink or to investigate its geometry, the validity over a wide range of the fin efficiency must be met. These properties are especially satisfied in the model proposed by Keyes. The maximum error is in the range of 9 % whereas applying the model used by Tuckermann the error is calculated tó 30 %. Fig. 4. shows the calculations of a realized heat sink.

It demonstrates that the cooling performance may increased for smaller fin width at equal pressure loss.

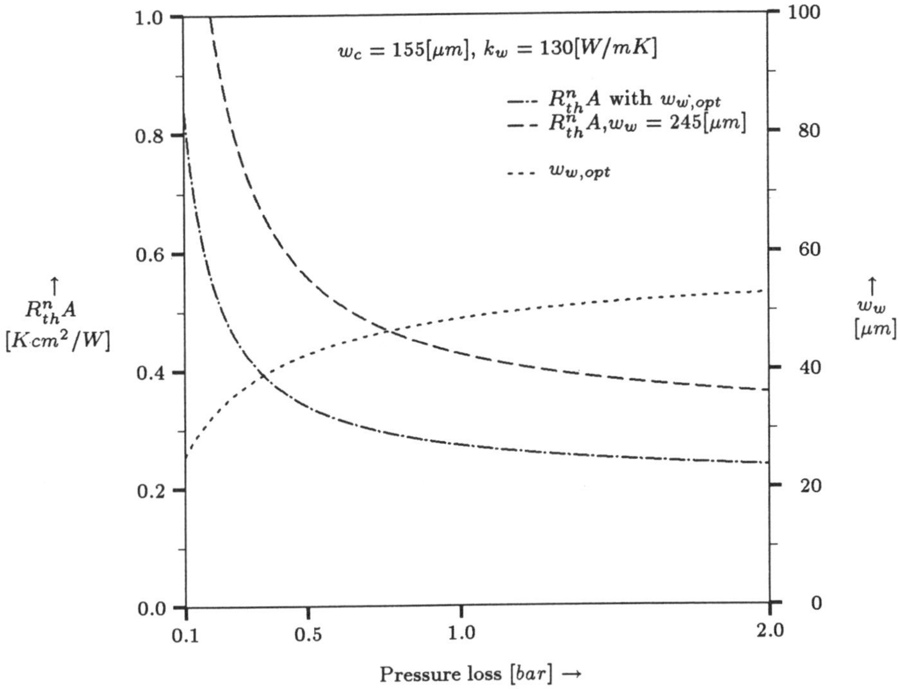


Figure 4: Thermal resistance versus pressure loss

## 3. Laser structuring of the AlN heat sink

Similar structures as fabricated by sawing may be realized with the aid of laser ablation. This technique gives significant advantages for micro structuring with high aspect ratios. The ablation may be effected by different parameters. These parameters should be optimized depending on the material and the structuring. Ablation velocity and gas flux are some of the process parameters. Regarding the laser the parameters such as power, pulse shape, pulse frequency, mode form, divergence, focusing have to be mentioned.

During processing the structured channel walls contaminate and metalize due to ablation. The cooler module is fabricated in AlN ceramic which was manufactured by ANCeram. Its dimensions are 2" × 2" in surface and a thickness of 2 $mm$. The thermal conductivity is 130 $W/mK$ . It has been structured using a Nd:YAG Slab laser. The channels have a pitch of 400 $\mu m$, with a channel width of 155 $\mu m$ and a depth of 700 $\mu m$ . At the inlets and outlets a reservoir of 3 $mm$ in width and 1 $mm$ in depth was added to equalize the pressure and reduce pressure losses. The boundary edges are 3 $mm$ in width to provide substantial area for gluing in order to prevent leakage. The glue used was manufactured by Ciba Geigy.

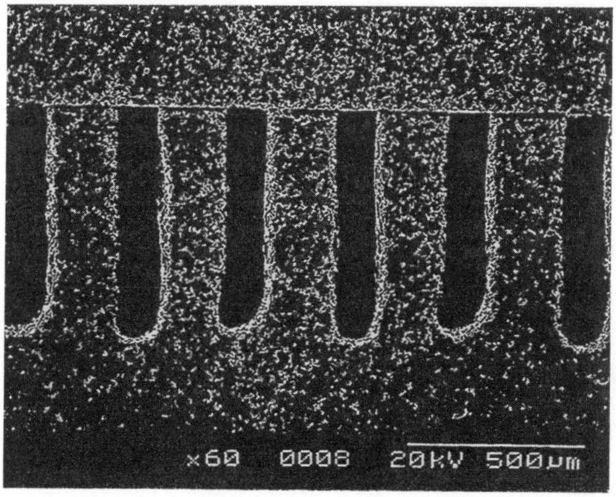


Figure 5: Crossection of the AlN heat sink.

## 4. Measurements

Thermographic measurements were conducted to verify the analytical and numerical approaches. The advantage of this kind of measurement is that the thermal profile of any structure can be detected without being in contact with the samples. The set-up for controlling and measuring of the cooler fluid flow is depicted in fig.

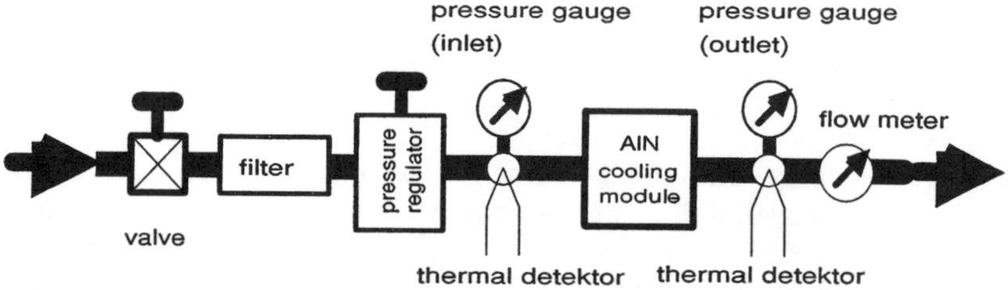


Figure 6: Set-up for controlling and measuring of the cooler fluid flow.

| power dissipation $P_{dis}$ in W | coolant temperature (inlet) in °C | coolant temperature (outlet) in °C | maximum temperature on the substrate in °C | mean temperature on the substrate $\bar{T}$ in °C | thermal resistance $R_{th}$ in K/W |
|---|---|---|---|---|---|
| 54 | 16,2 | 17,1 | 17,5 | 16,4 | 0,024 |
| 108 | 16,2 | 17,7 | 18,5 | 17,3 | 0,021 |
| 161 | 16,3 | 18,3 | 19,8 | 18,5 | 0,022 |
| 215 | 16,4 | 19 | 20,5 | 19 | 0,019 |
| 269 | 16,6 | 19,7 | 21,8 | 20,3 | 0,019 |
| 323 | 16,7 | 20,4 | 23,3 | 21,4 | 0,020 |
| 377 | 16,8 | 21 | 24,5 | 22,5 | 0,020 |
| 431 | 16,9 | 21,6 | 25,7 | 23,5 | 0,020 |
| 484 | 16,9 | 22,1 | 27 | 24,5 | 0,021 |
| 546 | 16,8 | 22,8 | 27,9 | 25,2 | 0,020 |
| 575 | 16,9 | 22,9 | 28,5 | 25,7 | 0,020 |

Figure 7: Results of the temperature measurements on the AlN heat sink.

6. The temperature of the coolant and pressure losses were measured at the inlets and outlets of the module. The fluid flow was measured at the outlet.

In order to minimize the thermal resistance between the heat source and substrate and in order to compare the measured results with the calculations, a heating resistor was realized by sputtering a 0.46 $\mu m$ TiW layer onto the surface of the module. The heating resistor was connected to sputtered gold pads at each end, which guarantees a homogenous current flow through the heater. The heating power was varied from 54 $W$ to 575 $W$ at steps of 50 $W$ at a constant volume fluid flow of 100 $l/h$. As coolant normal tab water was used.

To calibrate the thermographic measurements a calibrated thermo chuck and different thermometers were used. The thermal resistance was calculated considering the maximum temperature of the substrate to the coolant temperature.

Fig 7. summarizes the measured results. Fig 8 and 9 show the thermal profile at power dissipations of 54 $W$ and 575 $W$ respectively whereby the fluid is flowing from the left side toward the right side.

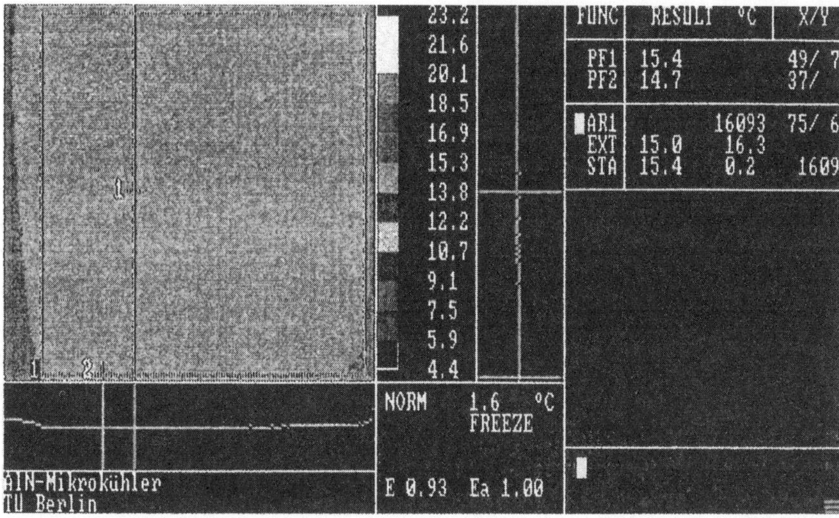


Figure 8: Temperature distribution at a power dissipation of 54 $W$.

In Fig 9 there is an evident temperature profile.
At the outer regions the velocity of the fluid is higher than in the middle region of the cooler. We calculated a higher pressure loss of 6 % at the middle region.
The thermal profile at the connectors of the heating resistance was neglected. Fig 10 shows a nearly linear dependence of the temperature difference between the

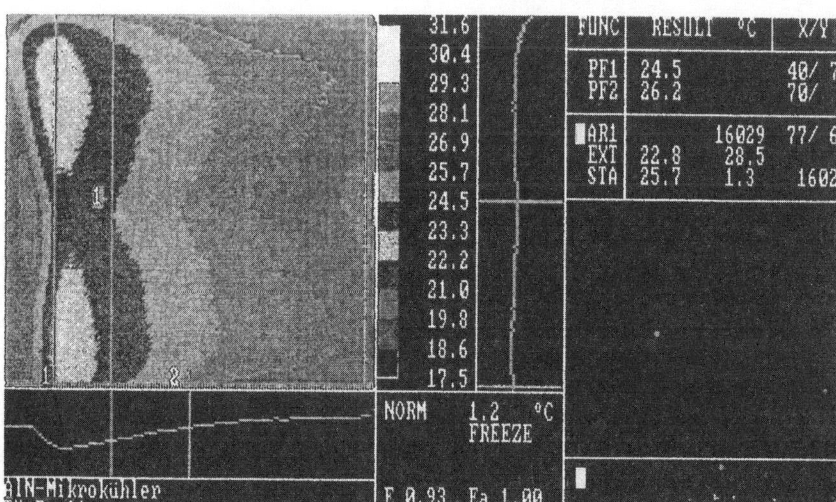


Figure 9: Temperature distribution at a power dissipation of 575 $W$.

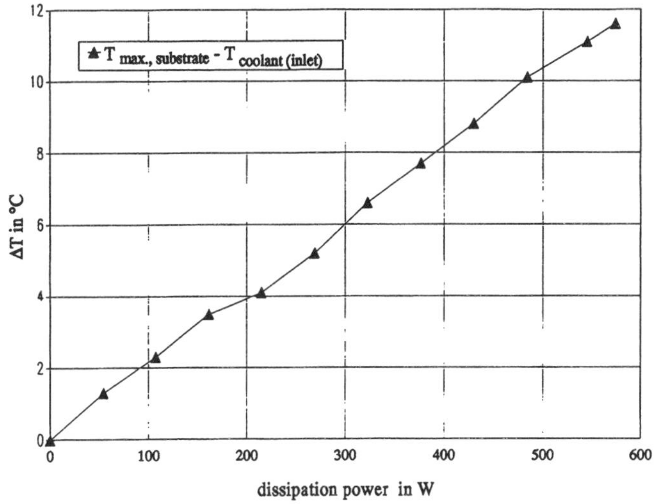


Figure 10: Temperature difference between the maximum temperature on the substrate and the inlet fluid versus power dissipation.

maximum temperature at the substrate and the fluid temperature. This is due to the resolution of the equipment used. Fig 11 shows the thermal resistance of the entire module calculated with the measured data.

The thermal resistance has been predicted within an accuracy of 7 % . The error at low dissipation powers is due to the resolution of the thermograph. Additional

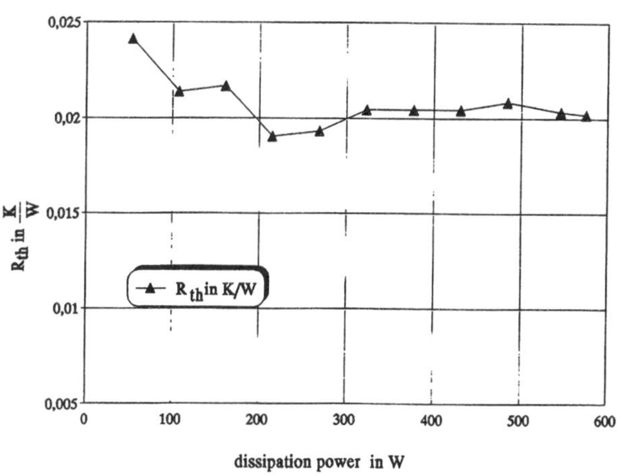


Figure 11: Thermal resistance of the heat sink versus power dissipation.

Figure 12: Example of a multi chip package. The heat flux on the chips (light hatched area) is assumed to be 25 $W$ and 20 $W$. The fluid flow was set to 10 $ml/s$ and the fluid temperature to 20 $^oC$. The temperature field for this package is shown in fig. 13

deviations are caused by temperature depending material properties.

4.1 DISCUSSION OF RESULTS AND SIMULATION

The analytical computations were compared with measurements at a fluid flow of 100 $l/h$ and a pressure loss of 0.6 bar. Calculated and measured thermal resistances are within good agreement. It is shown that the applied analytical models are suitable for designing coolers using the presented technology. Future work is concerned with the characterization of heat sinks by conducting a study of these parameters at different pressure loss and fluid flow. It is planed to realize alternative coolers in order to adapt these methods of design to specific applications.

## 5. Modeling a high power module

In order to predict the behavior of a module a numerical simulation tool was developed which is able to compute thermal distributions of 3-dimensional structures considering following aspects:

- 3-dimensional spreading effects at the interfaces between components and substrates
- thermal coupling by thermal conduction of the substrate
- thermal coupling via fluid

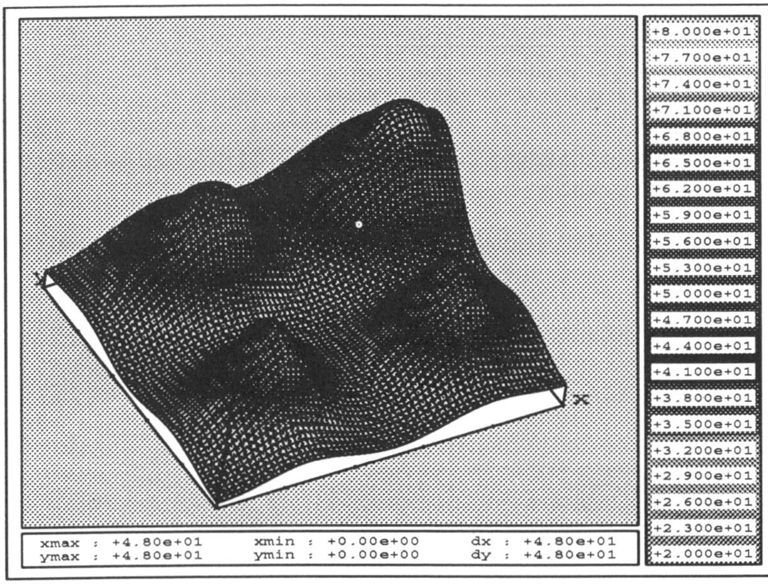


Figure 13: Temperature distribution on the substrate of a multi chip package (ref. fig. 12)

The program *The4Sim* was especially developed at the TU-Berlin for solving thermal problems. It has been adapted to the thermal design of Multi-Chip Modules. The procedure is based on solving the thermal differential equation for stacked rectangular multilayer structures. It allows to consider non-linear boundary conditions. As demonstrated by a 2" × 2" Multi-Chip-Module (fig. 12) the influences of fluid heating on the chip temperatures is shown (fig. 13).

## 6. Summary

In this paper methods for designing micro structured heat sinks were demonstrated. Analytical methods were used to optimize the geometry of the cooler taking into account the entire system. Using numerical simulations the made assumptions for analytical models were verified. On the basis of the obtained optimal geometries heat sinks were fabricated by sawing and laser ablation. The simulations were verified by thermographic measurements. In order to compute the thermal distributions on substrates and components the simulation program *The4Sim* was developed.

**Literatur**

1. R. W. Keyes. Heat transfer in forced convection through fins. *IEEE Transactions on Electron Devices*, ED-31(9):1218 – 1221, Sept. 1984.

2. R.K. Shah and A.L. London. *Laminar Flow Forced Convection in Ducts.* Academic Press, New York, San Francisco, London, 1978.

3. D. B. Tuckerman. *Heat-Transfer Microstructures for Integrated Circuits.* Ph.D. Thesis, Stanford Universtity, 1984.

# ALN COFIRED MCM-C/D

W. K. JONES AND M. A. ZAMPINO
*Florida International University*
*Dept. of Mechanical Engineering*
*Miami, Florida 33199*
*(305) 348-2345 Fax: (305) 348-2649*

## Abstract

A MCM-C/D test cell was designed and fabricated to investigate the compatibility of aluminum nitride with cofired tungsten vias forming a pad/grid array with thin film processing, and the high speed (>2 GHz) electrical performance of deposited TiW-Au / polyimide interconnections and the cofired tungsten vias. The test cell supports a multi-layer interconnect structure designed to allow for characterization of signal propagation properties (e.g. impedance, characteristic wavelength, cross-talk) of various striplines, via chains, and through substrate I/O and ground plane perforation. The design parameters relevant to electrical characterization are reported in detail. The polished A*l*N was found to provide a suitable surface finish (< 5 μin rms) for thin film metallization, however, the cofired tungsten vias maintained a 3-5 micron relief after polishing requiring the use of a polyimide planarization layer. Cross-talk is reported for various stripline configurations on perforated ground planes with 45% and 76% metal area.

Key Words: Aluminum Nitride, MCM-C/D, High Speed Electrical Characterization, Crosstalk

## 1. Introduction

Advanced multichip module applications demand electronic packages with large area mechanical integrity, hermeticity, and thermal management. In addition, high speed electronic designs operating at speeds greater than 2 GHz demand matched impedance interconnects such as microstip and stripline transmission lines requiring ground planes. High speed design also dictate that short electrical paths are desirable facilitating high interconnect densities near 1000 in / in$^2$.



*W. K. Jones et al. (eds.), MCM C/Mixed Technologies and Thick Film Sensors,* 101–111.

In some cases, a combination of MCM fabrication technologies, i.e. (L)aminate, (C)eramic, and Thin-Film (D)eposition, can produce a package with enhanced performance in mechanical integrity, hermeticity, thermal management, and high speed electrical performance. Currently, MCM-C/D technology using ceramic substrate materials and deposited thin films to provide multilayer high density interconnections is being investigated to meet the demands of advanced electronic packaging. The development of cofired multilayered A/N substrates allows for the substrate to contain vertical interconnection through the substrate, ball/pad grid array I/O, power/ground planes, decoupling capacitors, and termination resistors fabricated as part of the substrate.

To evaluate this technology and characterize its performance with MCM-C/D design, a test cell was developed using an A/N substrate with integral cofired tungsten vias, and a multilayer interconnect structure of deposited TiW-Au / polyimide. The test cell was designed to evaluate the high speed electrical performance of the TiW-Au / polyimide interconnect structure and the through substrate I/O as well as, the compatibility of the A/N substrate to thin film processing. The test cell will then be electrically characterized at FIU using a Tektronix IPA-310 TDR (2-20 GHz) and a 6 GHz network analyzer. So as to characterize the dielectric propterties of the A/N material, the dielectric constant of the substrate was determined over a range of frequencies.

In high density, high speed MCM design, the use of thin film resistors for terminating transmission lines is necessary. Fabrication of the resistors directly on the substrate is warranted and hence, the compatibility of the resistors with the A/N substrate is critical. To evaluate this, thin film tantalum nitride resistors were fabricated on A/N and subjected to high temperature stabilization bakes and tested for resistance stability.

## 2. MCM-C/D Test Cell Fabrication

The test cell was fabricated using one inch square A/N substrates with a thickness of 25 mils. Tungsten vias of 20 mil diameter on a 100 mil pitch were cofired into the substrate to provide through substrate interconnection. The substrates were cut from four inch diameter cofired A/N wafers from Coors Electronic Package Company. The wafers were polished by P/M Industries using a propriety polishing process. This produces an A/N surface finish with a surface roughness of less than 25 nm and no sharp discontinuities or grain pullouts [1]. The Coors A/N wafers and the P/M Industries polishing process certainly produce a thin film compatible substrate allowing for direct deposition of conductor metallization and component fabrication such as tantalum nitride resistors.

The addition of the tungsten metallization to create the ball/pad grid array in the A/N wafers adds the feature of through substrate interconnection to a ball/pad grid array. However, the tungsten vias were not completely compatible with the polishing process leaving a tungsten-A/N surface relief of approximately 3-5 μm as characterized on a Tencor Alpha Step 200 profilometer. This creates an incompatibility with direct thin film processing. It is surmised that in a final polishing step, a chemical process is used which is preferential to the A/N, hence, the tungsten vias do not etch back as much as the A/N. This results in the tungsten vias maintaining a surface relief above the A/N. To make the surface compatible with thin film processing, a polyimide planarization layer is necessary. DuPont Pyralin 2721 PD (photodefinable) polyimide was used which was deposited to a thickness of 10 μm for the planarization layers and 12.5 μm for all other layers. Gold was used as the conductor with chrome to aid to adhesion to the polyimide and TiW as a barrrier layer between the gold and the chrome. Subsequently, the gold was electroplated to approximately 3.5 μm. Standard thin film fabrication processes were used in all steps.

### 3. MCM-C/D Electrical Test Cell

The MCM-C/D test cell was designed similar to the test structure developed at the MAYO Foundation as part of the ARPA Advanced Packaging Program. The structure is a set of passive devices which can be used to characterize the electrical performance of MCM-D interconnections at operating in the 2-20 GHz frequency range. Devices included in the test cell include matched impedance transmission lines, via chains, a spiral inductor, a capacitor, and a perforated (meshed) ground plane. All of the test devices in the test cell are designed to be used with 50 Ω impedance co-planar ground-signal-ground three head probes. Electrical characterization is then obtained using time domain reflectometers and network analyzers.

The test cell has patterned metallizations on both sides of the substrate as shown in Figure 1. The front (top) side of the substrate contains two signal layers (M1 and M2) on top of a perforated ground plane (GP) which was either 45% or 76% metal. The signal layers and the two ground planes are shown in Figures 2-5, respectively. The back (bottom) of the substrate contained a single signal layer (M3) and is shown in Figure 6. Vias were patterned in the four polyimide layers to connect the three signal layers.

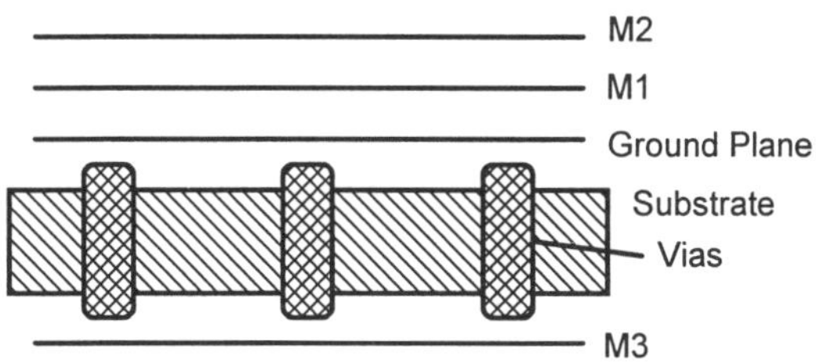


**Figure 1. Schematic showing layers of MCM-C/D test cell.**

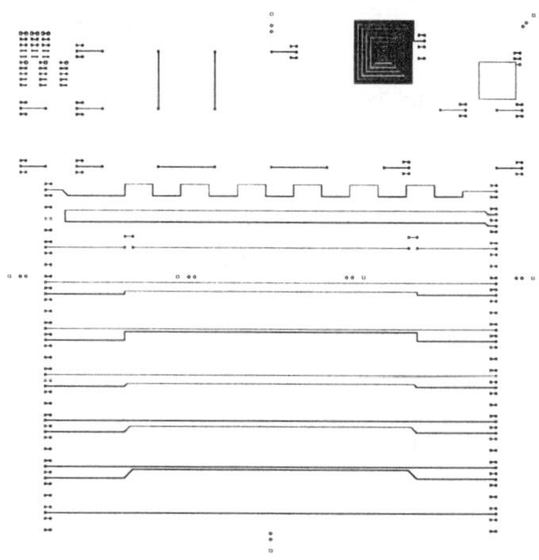

Figure 2. Metallization mask M1.

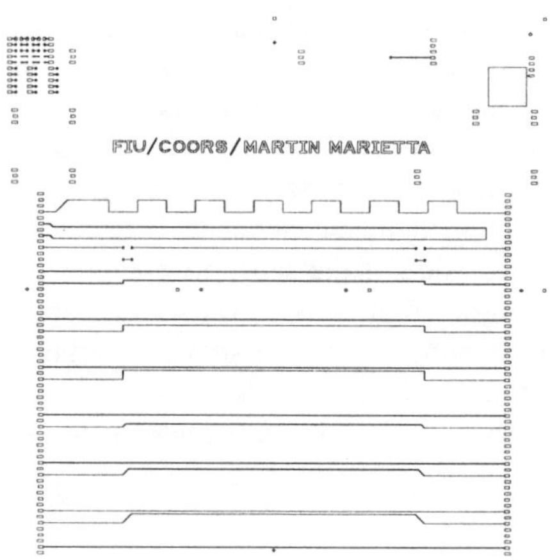


Figure 3. Metallization mask M2.

The front side metallizations contain a collection of 50 Ω matched impedance striplines, via chains, a spiral inductor, and a capacitor. The striplines have a width of approximately 10 μm, hence with polyimide thicknesses of 12.5 and 25 μm, separating the striplines from the ground plane, the width/height ratios for the striplines are 0.8 and 0.4, for layers M1 and M2, respectively. The same stripline set is repeated in both of the signal layers, with the only difference being that the striplines in M1 are buried in dielectric and the striplines in M2 are on the surface of polyimide and exposed to air.

**Figure 4. Ground plane with 45% metal area.**

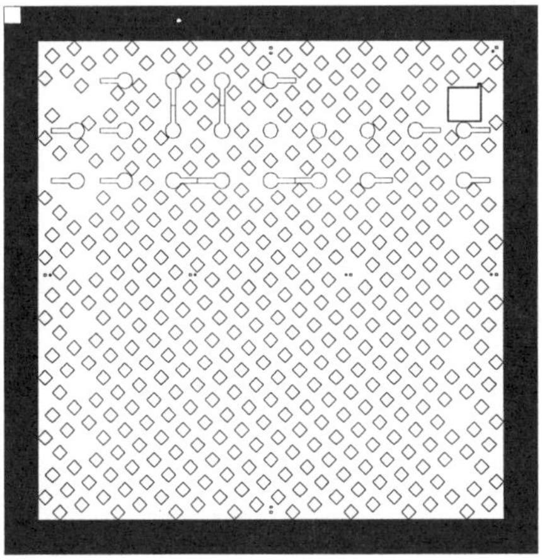

**Figure 5. Ground plane with 76% metal area.**

This configuration allows for a comparison between the electrical performance of buried and exposed striplines.

A variety of stripline patterns are provided so as to investigate the effect of sharp bends in the signal path, in-plane and through plane cross talk, and the effect of vias. From the figures it can be seen that in addition to straight striplines, the test cell also provides striplines which: 1) have four kinks at angles of 45 and 90°, 2) meander back and forth, 3) are daisy chained on the front side only, and 4) daisy chain to the back side metallization using the through substrate vias. For in-plane cross-talk evaluations, stripline sets are available with line spacings (S) of 5, 10, and 15 mils (127, 254, and

381 μm, respectively). For a stripline width (W) of 0.4 mil (10 μm) this gives S/W values of 12.5, 25, and 37.5, with the three spacings provided for lines with kinks of 45° and 90°. For through-plane cross-talk evaluations, any pair of of the striplines can be used with one line on layer M1 and the other on M2. However, for practical purposes the straight transmission lines at the bottom of the stripline pattern set are the most usable.

Since vias are a critical part of high density MCM-D interconnect structures, ten via chains are provided in the upper left corner of the M1 and M2 layers. Via chains with via diameters of 2, 4, and 6 mils (≈ 50, 100, and 150 μm, respectively) are provided to evaluate the fabrication process and for electrical characterization. Finally, the front side metalliztions provides a thin film capacitor and spiral inductor which can be used to determine the properties of the dielectric such as dielectric constant and breakdown voltage.

Test cells were fabricated with either the 45% or 76% metal area perforated ground planes as shown in Figures 4 and 5. Perforated ground planes readily allow for vertical interconnection through the plane and promote adhesion of the polyimide dielectric layers. Successful fabrication of a test cell with a solid ground plane was not possible due to problems with adhesion between the polyimide layers.

The back side metallization (M3) is primarily for use in the evaluation of through substrate signal propagation containing parts of the daisy chained transmission lines which use the tungsten vias in the A*l*N substrate. Referring to Figure 6, the daisy chained transmission line components can be seen in the top one third of the M3 metal pattern. Daisy chained transmission lines are provided using two and six of the tungsten vias, with straight chain segments and segments with 45° and 90° bends. In the middle one third of the M3 pattern are square and circular pads of varying sizes for the evaluation of adhesion. Finally, in the bottom one third of the M3 pattern, are a set of striplines with a line width of 2.6 mil (66 μm) and are designed to give 50 Ω impedance

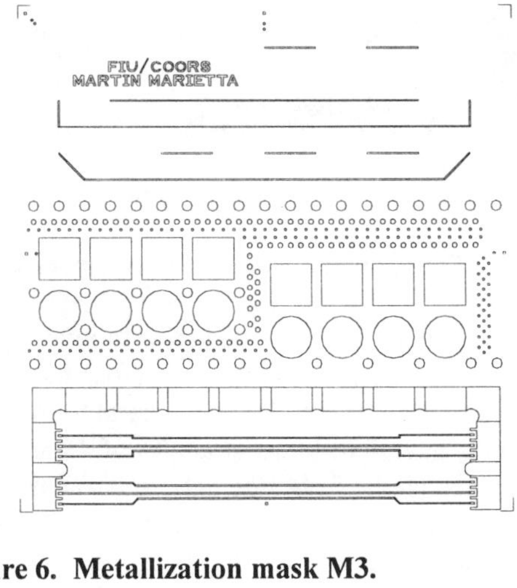


**Figure 6. Metallization mask M3.**

using the substrate as a dielectric. Straight and kinked striplines are available as on the front side metallizations. For in-plane cross talk evaluation, line spacing to line width ratios (S/W) of 5.77 and 3.84 are found.

Through-substrate interconnection uses the substrate as a dielectric medium and the test cell design was created assuming the dielectric constant of the substrate to be 8.0-9.9 @ 1 MHz, as indicated by the manufacturer. For the substrates used, the dielectric constant was measured using an Hewlett Packard Dielectric Tester. Referring to Figure 7, it is seen that the dielectric for the A/N substrates is not constant over the frequency range of 2-10 GHz, but varies within the range of 5.0 to 8.5.

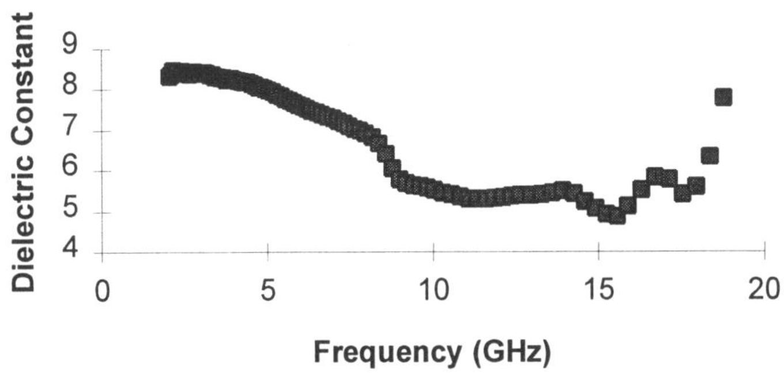


Figure 7. Dielectric constant of A/N substrate material.

## 4. Electrical Characterization of Buried Striplines

Currently in high speed MCM-C/D packages, crosstalk in adjacent transmission lines of 5% or less is considered acceptable. Crosstalk in transmission lines is affected directly by the electrical layout design rules. The primary parameters that a designer can control to affect crosstalk are the line separation, that is, the space between the active and passive transmission lines, and the height above the ground plane of the signal plane. In addition, when a perforated ground plane is used, the percentage of metal area has a direct effect on the crosstalk. Decreasing the metal area of the ground plane will decrease the interconnect capacitance, hence, less of the active transmission line's electric fields can terminate on the ground plane. This results in a higher coupling capacitance between the active and passive transmission lines leading to higher crosstalk.[2]

In MCM-D fabrication, striplines can either be buried in the dielectric or on the surface of the dielectric. Surface striplines have a composite dielectric medium made up of air and polyimide ($\varepsilon_R \approx 3$), whereas for the buried striplines, the dielectric medium is only polyimide. Therefore, for a pair of striplines with the same S/W and W/H design parameters, the buried lines should have a higher crosstalk.

Crosstalk was evaluated in the MCM-C/D test cell using a Tektronix IPA-310 TDR with a 24 ps sampling head, and two Picoprobe 50Ω ground-signal-ground probes (GGB Industries). The test cell was used to evaluate the crosstalk of five buried stripline sets (metallization layer M1) on ground planes of 45% and 76% metal area. The results of the evaluation are shown in Table 1 and follow the expected trends.

Each crosstalk line set is composed of a straight stripline which was used as the active transmission line and a kinked transmission line acting as the passive line. The kinks in the passive line were necessary to bring the lines to a specified line separation. Using the TDR, the crosstalk was determined at the midlength of the passive line where it was maximum.

The 76% ground plane had better crosstalk performance for all the lines with a mean reduction in crosstalk of 59% and for all the line configurations the crosstalk was under 5%. For the 45% ground plane, the crosstalk was acceptable (i.e. < 5%) in all configurations except the closest line separation (S/W = 12.5). Comparison of the data for the passive lines with the 45° and 90° kinks is not conclusive and shows that for the smaller line separation (S/W = 12.5) the crosstalk increased slightly when the sharper kinks were present in the passive line. However, for the larger separation (S/W = 37.5), the opposite occurs with a reduction in the crosstalk for the sharper kinks.

**Table 1. Crosstalk of buried striplines with varying line separation / line width (S/W) for 45% metal area ground plane.**

| Ground Plane | S/W | Kinks | Max. Crosstalk | |
|---|---|---|---|---|
| | | | (dB) | (%) |
| 45% | 12.5 | 45° | -21.8 | 8.8 |
| | 25.0 | 45° | -27.7 | 4.1 |
| | 37.5 | 45° | -33.5 | 2.1 |
| | 12.5 | 90° | -20.5 | 9.4 |
| | 37.5 | 90° | -32.6 | 2.3 |
| 76% | 12.5 | 45° | -27.0 | 4.4 |
| | 25.0 | 45° | -34.8 | 1.8 |
| | 37.5 | 45° | -40.8 | 0.91 |
| | 12.5 | 90° | -26.0 | 5.0 |
| | 37.5 | 90° | -49.5 | 0.33 |

## 5. Thin Film Resistor Compatibility

High speed digital circuits operating above 200 MHz require the use of terminated transmission lines employing either a shunt termination or a series source termination for each signal net. For MCM-C/D designs with high interconnect density (500-1000 nets), the use of discrete termination resistors is too costly and sometimes not logistically possible. Hence, termination resistors need: 1) to be fabricated as an integral part of the MCM-C/D structure to reduce cost, and 2) made as small as possible to reduce parasitic reactances. To meet these needs in MCM-C/D design, thin film resistors (TFR) made from NiCr and $Ta_2N$ can be fabricated directly on the substrate material.

$Ta_2N$ TFR's have the intrinsic properties of a linear temperature coefficient of resistance and a physiochemical inertness. For use with MCM-C/D designs, $Ta_2N$ TFR's have been characterized on Alumina, however, data is limited for $Ta_2N$ TFR's on A*l*N. Therefore, it is of interest to investigate and characterize the performance of $Ta_2N$ TFR's on A*l*N to ensure compliance with MIL-STD-883 and other appropriate standards. For reliable fabrication, the $Ta_2N$ TFR's must be compatible with all the thin film processes used in MCM-D fabrication. Of particular concern is the repeated high temperature cycling required for polyimide curing and the mechanical effect of the polyimide on the resistors performance.

A two inch square wafer of 1000Ω $Ta_2N$ TFR's on A*l*N was manufactured by Mini-Systems Inc. for the evaluation. The wafer was diced into nine square pieces with 80-100 resistors on each test piece and the resistance of all the resistors was measured prior to any testing. Three sets of resistors were baked in air at temperatures of 150 and 300 °C for a period of 100 hours. This procedure follows that prescribed in MIL-STD-883 High Temperature Stabilization Bake. The temperature of 300 °C was chosen because it approximately represents the mean temperature used during polyimide curing. After each test the resistance of each resistor was measured again and the relative change in resistance ($\Delta R/R_0$) was determined for each resistor. Data from resistors which appeared to be initially defective or had a very large relative change in resistance were discarded so as not to falsely skew the results. In all cases, discarded data accounted for less than 3% of the resistors. The remaining data was then used to determine the mean value of $\Delta R/R_0$ for each of the test sets which is shown in Table 2.

For the stabilization bake at 150 °C, the relative change in resistance was 0.011%, which agrees well with the manufacturer's value of 0.03% for $Ta_2N$ thin film resistors on alumina. However, the literature suggests that at temperatures of 250 °C and above, diffusion does occur and the may be some reactions between the $Ta_2N$ structure and the substrate, as well as within the $Ta_2N$ structure itself [3].

The resistors from Group 1 were added to another two reisistor sets which were baked at 300 °C for 100 hours. From the results shown in Table 2, it can be seen that relative changes in resistance was about 17.5%, a figure which is significantly different than that for the lower temperature bake. The actual mechanisms causing this large change are still under investigation, however, it was surmised that diffusion and oxidation was taking place. To further investigate this anomaly, two more groups of resistors were baked at 300 °C for 100 hours. Referring to Table 2, the resistors in Group 4 were baked in nitrogen gas, and the resistors in Group 5 were baked under vacuum.

TABLE 2. Average relative change in resistance of $Ta_2N$ thin film resistors due to high temperature stabilization bake.

| Group | ΔR/R (%) | |
|---|---|---|
| | 150 °C | 300 °C |
| 1 | 0.011 | 19.8 |
| 2 | n/a | 16.8 |
| 3 | n/a | 17.2 |

## 6. Conclusions

A MCM-C/D test structure was fabricated on an A*l*N substrate with a cofired ball/grid array of tungsten to allow for through substrate interconnection. After polishing the tungsten through vias had a surface relief of 3-5μm requiring the use of a planarization layer. Test cells were fabricated successfully with perforated ground planes of 45% and 76% metal areas, however, fabrication with a solid (100%) ground plane was not possible due to adhesion problems. The dielectric constant of the A*l*N substrate material was tested and found to lie between 5.0 and 8.5 in the 2-10 Ghz range.

Crosstalk was evaluated for buried striplines in various configurations using a time domain reflectometer. Acceptable crosstalk performance was obtained on both ground plane designs, however, for the 45% ground plane, crosstalk was too high ( > 5%) for a the transmission line pair with the closest separation (S/W = 12.5). The 76% ground plane had better crosstalk performance than the 45% ground plane with a mean reduction of 59% in crosstalk. Further electrical characterization is being performed using the test cells to determine other parameters (e.g. characteristic impedance, characteristic wavelength, etc.) relevant to high speed electrical design using MCM-C/D technology.

Compatibility of tantalum nitride thin film resistors with the A*l*N substrate was tested by performing stabilization bakes at 150 and 300 °C for 100 hours, following the MIL-STD-883 test procedure. It was found that the resistance stability of the resistors was acceptable for the bake at 150 °C, however, the resistors baked at 300 °C had a shift in resistance of at least 16.2%. Further testing is being performed to better characterize the mechanisms related to the resistance shift.

## 7. Acknowledgments

The authors would like to thank Jack Chicola and Jack Thorton at Martin Marietta for designing and fabricating the test structure. This work was supported by an ARPA grant under the University Research Initiative (URI), Grant No. N00014-91-J-4010. In addition, the authors would like to thank Paul Solan at Mini-Systems, Inc. for supplying the tantalum nitride thin film resistors.

## 8. References

[1] Schardt, C. and Jones, W.K., "Cofired AlN Pad Grid MCM-C/D for High Speed Digital Applications", *Proc. 1993 Int. Symp. on Microelectronics*, ISHM, Dallas, 1993.

[2] Soloman, D., El-Kilani, S., David, F.L., and Lubbers, M.,"Test Vehicle for the Electrical Characterization of MCM-D Interconnects", *Proc. 1993 Int. Symp. on Microelectronics*, ISHM,Dallas,1993.

[3] Au, C.L., Anderson, W.A., Schmitz, D.A., Flassayer, J.C., and Collins, F.M., "Stability of tantalum nitride thin film resistors", *J. Mater. Res.*, Vol. 5, No. 6, pp 1224-1232, 1990.

# Very Fine Line Photoimageable Thick Film Technology Developed at Hibridas , Lithuania

Jurate Minalgiene and Steve Muckett*

'Hibridas' Microelectronics Division , Research Institute of Radio Measurement Engineering ,Kaunas, Lithuania
*Mozaik Technology Ventures Ltd., Farnborough ,UK

## Introduction

The Research Institute of Radio Measurement Engineering at Kaunas was founded in 1958 and became an important research and production facility for the radio and communication requirements of the various military services in the Soviet Union. The Department of Microelectronics , now known as 'Hibridas' was set up during 1972 and specialized in the design and manufacture of microwave devices. Facilities included thick and thin film production equipment as well as metal packaging and hermetic sealing capability.
In the continued drive to improve microwave circuit performance and ease of processing at Hibridas during the early 1980's a project was initiated to study photosensitive vehicle chemistry's that could be used in the formulation of very high resolution thick film conductor materials. A unique vehicle chemistry was developed and the process technology has subsequently been used in production for a range of sophisticated circuits since 1986.
Applied research work is continuing at Hibridas to further develop the capabilities of these photoimageable materials and processing technology for use in microwave, high density interconnect, component and sensor applications .

There is now a renewed interest in photoimageable thick film materials and processing . This is because it is possible to realise thick film features of better than 50µm at normal fired thickness levels using a process which is inherently low cost. In common with the photoetching process the entire area of substrate is printed which means that the print parameters are not critical. The use of a photoimageable vehicle within the thick film paste allows for exposure and developing of the image before firing. Compared to conventional thick film processing the number of additional process steps is small. In the case of the photoimageable vehicle chemistry developed at Hibridas it has been shown that conductor features of 15µm line and space resolution can be achieved at a fired thickness of 8µm. Thick film dielectric and resistor formulations have also been prepared by Hibridas and successfully applied to circuit manufacture using this process. The entire range of photoimageable thick film compositions and process technology is referred to as the "Saule" system within Hibridas.



*W. K. Jones et al. (eds.), MCM C/Mixed Technologies and Thick Film Sensors,* 113–119.

## Process Description

The basic stages of the photoimageable thick film process are described below.
The printing and firing technology used is identical to the conventional thick film process. However , the two additional steps of *exposure* and *development* of the photosensitive thick film material are required prior to firing. Each individual step for processing of the "Saule" photoimageable materials is briefly described below.

Printing
The thick film paste is printed over the total area of substrate onto which the circuit pattern features are to be produced . A nylon mesh screen is employed to apply a wet film of paste ,uniformly thick and free of defects such as pinholes. Only a short time is needed for levelling of these pastes because of their lower viscosity characteristics in comparison to conventional thick film pastes.( The ability to print sharp edged features is not required). The"Saule" photosensitive vehicle chemistry is very tolerant to a wide range of processing environments. No special lighting conditions or atmosphere conditions are needed during the printing or subsequent drying stages. Drying of the printed thick film layers is performed in a box oven or belt oven at 100 degrees Celsius for 10 minutes.

Exposure
The printed and dried thick film layer is exposed to high intensity ultra-violet light through a photomask. The "Saule" system pastes are sensitive to light of wavelength between 320 - 360 nm.During exposure to the intense ultra-violet light polymerisation of the organic molecules within the vehicle occurs. These molecule chains greatly affect the solubility of the vehicle in the locally exposed areas. Exposure times vary with coating thickness ,light intensity and type of 'solids' in the thick film material and are typically 10 seconds for dielectric ,30 seconds for conductor ( Au, Ag, Ag-Pd ,Pt ) and up to 60 seconds for ruthenium based resistor pastes.

Developing
The unexposed regions are removed using a special spraying technique. The polymerised ,exposed regions are not soluble in the developer solution and remain on the substrate. This special spray technique is a high pressure, atomised spray process of less than 10 seconds duration which is optimised to produce very fine line and sharp pattern definition without causing lifting and peeling of the thick film layer from the substrate . The developer medium is aqeous $Na_2CO_3$ solution.

Firing
Standard belt oven firing in air for 1 hour at a peak temperature in the range 850 - 900 degrees Celsius is performed. All the vehicle is volatilised during the high temperature firing to produce thick films with normal sintered structures.

## Compositions and Properties of "Saule" System Materials

All of the "Saule" system of materials are photosensitive thick film pastes which incorporate a vehicle that has the property of polymerisation during exposure to ultra-violet light.The polymerisation sensitivity to intense ultra-violet light for the "Saule" vehicle is very high so that a wide range of powders can be used in the preparation of conductor ,dielectric and resistive compositions without deleterious affect on the exposure times , or resolution characteristics after the developing process. The "Saule" photosensitive vehicle is *not sensitive* to normal, visible light and polymerisation is *not inhibited* by oxygen. This means that all processing prior to exposure can be performed under standard production conditions .The basic properties and character of the "Saule" vehicle chemistry were well researched in the early 1980's.Research into the various powder morphologies and compositions suitable for photosensitive thick film pastes has continued since that time.

Basic properties of conductor, dielectric and resistive compositions are outlined below ;

Conductors

The physical and electrical properties of a photoimageable thick film conductor are determined by the sintered metal-glass composition , because all of the vehicle constituents are burned out cleanly during the high temperature firing process. The result is high conductivity thick film conductors which exhibit good adhesion , bonding and/or soldering characteristics. Table 1 lists the basic characteristics of "Saule" conductors. Depending on the application the conductors can be optimised for good adhesion to a variety of substrate materials. The substrate types most commonly employed at Hibridas are 99% polished alumina , 96% alumina and silicon. The conductor resolutions achievable are widths from 10μm to 100μm , and spacings down to 20μm between parallel tracks. For the very finest lines there will be a trade off against thickness but a fired thickness of 6μm is still achievable at a line width of 10μm. Figure 1 is an SEM photograph for an array of silver conductor tracks 35um wide and 10μm thick on a 99% polished alumina substrate. At high magnification it is evident that the metallisation has sintered to a dense film with vertical edges. For multilayer structures the conductor widths achieved in production are typically a minimum of 50μm.

Dielectrics

The typical properties of the crossover and multilayer insulator composition developed at Hibridas are shown in Table 2. It is a blend of vitreous and recrystallizing glass frits in photosensitive vehicle. A distinct advantage of photoimageable dielectrics over 'printable' types is the lack of pinholes and voids in the fired film which is due to the lower viscosity character of photoimageable pastes. Testing of unencapsulated, fine line crossover structures with the dielectric and silver conductor at 50V bias, 40 C ,95RH for 100 hours has resulted in zero failures. Dissipation factor values better than 0.002 for the frequency range from 1 to 12GHz (important for microwave applications) are achieved .

Resistors

The "Saule" vehicle has been incorporated into several ruthenium oxide resistor compositions . For many applications it is not necessary to achieve very small or very precise resistor geometries , and for this reason the photosensitve resistor paste development has concentrated on low value and high value resistive compositions for use in specific applications which require high precision, ie) thermistor chips and microwave filters. The fired thickness of the photosensitive resistor pastes is in the range of 4µm to 10µm ( depending on the requirement) which is much thinner than for conventional screen printable resistor pastes. Perfect resistor geometries are possible at dimensions of 100µm x 100µm and resistor tracks of 50µm width have also been produced. Figure 2 illustrates a 2mm x 1mm high value resistor chip ( 400 Megaohms) that has been produced using the photoimaging process. The resistor track is 50µm wide and 3.88mm long. Such small resistor geometries could cause problems due to metal diffusion at the conductor interface. For this reason careful consideration must be given to conductor compatibility and resistor design . Using the "Saule" process it is also possible to produce resistors from conductor pastes such as platinum and palladium-silver in a relatively small area of substrate, by developing and firing a meander track of high precision . Such geometries would not be possible using conventional thick film processing.

## Applications

During the last 7 years photoimageable thick film materials have been used at Hibridas for several production applications. These are summarised below ;

Microwave Products

Photoimageable thick film technology is currently employed in the construction of microwave integrated circuits and substitute for thin film in devices such as filters, modulators ,converters ,delay lines ,etc. One example is the frequency multiplier -signal modulator ( output signal 0.1 - 18GHz) . On the top side of the 99% polished alumina substrate are printed pure silver microstrip segments with line widths of 50µm, and resistors. On the back side are printed silver ground plane , dielectric isolation layers and a palladium-silver conductor layer for low frequency connections. The circuit dimensions are 60 x 48mm and it contains 25 chip capacitors ,35 wire bonded diodes ,8 chip transformers and 12 chip inductors. Connections between conductor layers are through vias in dielectric and laser drilled holes in ceramic.

Multilayers

One production example of a photoimaged thick film multilayer is a circuit of dimensions 24 x 15mm with 4 conductor layers. The conductor tracks are silver (inners layers) and a top conductor of palladium-silver for optimum solder aged adhesion. The finest line and space features are 50µm. The three isolator layers are each formed from 2 prints of dielectric and the smallest vias are 75µm. The via connections are made during the screen printing of conductive layers.

Sensor Components

A wide range of thermistor chips and other component elements have been developed and produced at Hibridas. Two such examples are i) a 1000 ohm thermistor chip of dimensions 2 x 2mm with Pd-Ag terminations made with 2 ohm per sq. resistor paste and with a TCR of +4000ppm. The resistor width is 50µm and length is 500 squares. ii) a high value (1Gigaohm) resistor chip of dimensions 2 x 4mm with Pd-Ag terminations made with a 5 Mohm per sq. paste.

Special heaters and temperature sensing elements have also been manufactured with photoimageable platinum paste with track widths of 20 to 50 micrometres.

Solar Cells

Special photoimageable silver pastes for use on silicon substrates have been developed but are not used in production at Hibridas. The possibility of producing conductor tracks 20µm wide and 10µm thick is achievable and offers opportunities to further improve the efficiency of thick film solar cells by increasing the area of silicon exposed to sunlight , without reducing electrical conductivity of the silver tracks.

## Conclusions

The paper has described how the photoimageable thick film process developed at Hibridas can be used to produce very high precision and fine line circuit structures. The chemistry of the photosensitive "Saule" vehicle developed at Hibridas is unique with high polymerisation sensitivity to intense ultra-violet light. The technology is inherently low cost and has been successfully employed for more than 7 years in a wide range of applications such as microwave integrated circuits, multilayer circuits and sensor components .

| PHOTOSENSITIVE CONDUCTORS | |
|---|---|
| **METAL** | Gold<br>Silver<br>Silver-Palladium<br>Platinum |
| **FIRED THICKNESS** | 6-10 $\mu m$ |
| **RESOLUTION:** | |
| **ON CERAMIC SUBSTRATES** | |
| Line | 15 $\mu m$ |
| Space | 20 $\mu m$ |
| **ON DIELECTRIC ISOLATORS** | |
| Line | 50 $\mu m$ |
| Space | 50 $\mu m$ |
| **RESISTIVITY:** | |
| Gold Conductor | $<0.007\ \Omega/sq$ |
| Silver Conductor | $<0.005\ \Omega/sq$ |
| Silver-Palladium | $0.015 - 0.05\ \Omega/sq$ |
| Platinum Conductor | $<0.3\ \Omega/sq$ |

**Table 1**
Basic properties of photosensitive conductors developed at 'Hibridas

| PHOTOSENSITIVE DIELECTRICS | |
|---|---|
| **FIRED THICKNESS:** | |
| Two Layers | 25-30 $\mu m$ |
| Three Layers | 35-40 $\mu m$ |
| **DISSIPATION FACTOR** | <0.002 |
| **DIELECTRIC CONSTANT** | 8.0 |
| **BREAKDOWN VOLTAGE** | >15 V/$\mu m$ |
| **VIA RESOLUTION** | 75 $\mu m$ |

**Table 2**
Basic properties of photosensitive dielectric developed at 'Hibridas'

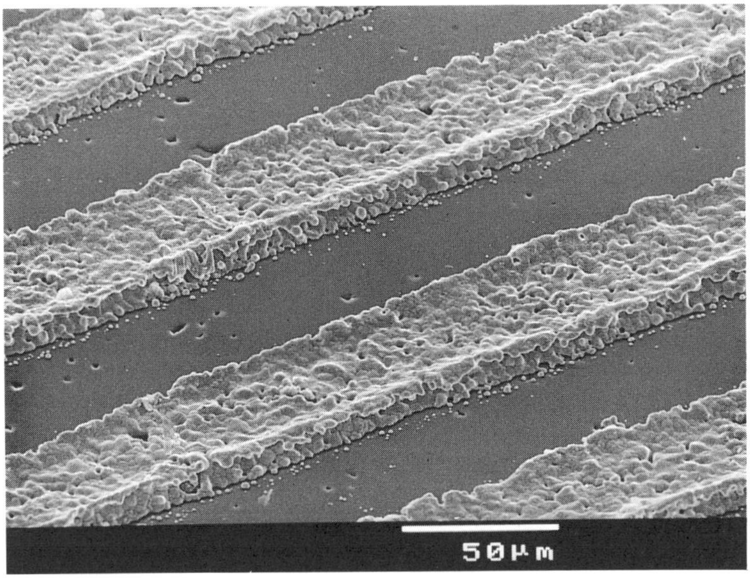


Figure 1. SEM for an array of silver conductor tracks on a polished 99% alumina ceramic substrate

Figure 2. High value (400 Megaohm) resistor chips produced with photosensitive thick film materials at 'Hibridas'.

# HIGH PERFORMANCE SOLID STATE MASS MEMORY MODULES

A. COELLO-VERA, P.OUDART,
*Alcatel Espace,*
*26 Av. J.F. Champollion, 31037 Toulouse CEDEX, France.*
P. D'ANDREA,
*Montpellier Technologies, IBM France.*
*Rue de la Vieille Poste, B.P. 1021, 34006 Montpellier.CEDEX, France.*
G. CYSAREK
*Sindelfingen Elektrokeramics, IBM Deutschland.*
*Postfach 266, D-71044 Sindelfingen, Germany.*

## 1. Introduction

Solid State mass memory systems with memory capacities up to 1 Terabit pose formidable packaging challenges. There are thousands of components to be assembled and, for Aerospace applications where the system is aboard an airplane or satellite, the equipment has to be very light and occupy the smallest volume possible. Existing applications use Magnetic Recorders for mass storage. The weight and volume occupied by these solutions limit the amount of memory available on-board. System designers are therefore very limited in their possibilities for making full use of the data gathering/processing potential of satellites and airplanes. MCM technology has the potential to meet these objectives.

We will describe a Solid State Mass Memory Application where the building block is a Multichip Module using High Temperature Co-fired Technology. The maturity and proven reliability of the MCM-C Technology make the resulting Mass Memory System very competitive with existing Systems based on PCB, Hybrid or Magnetic Storage Technologies. This system is open and can include future evolution in Memory Chip Technology.

## 2. MCM-C Substrate Technology.

The main objectives which drove the technology choices were : Performance, reliability and Price. Price has to be understood at application level, which implies such parameters as volume and weight.

In order to meet these requirements, it was decided to capitalize on the MCM technology IBM had used for years and which combines reliability, quality and volume production price advantage.



*W. K. Jones et al. (eds.), MCM C/Mixed Technologies and Thick Film Sensors,* 121–130.

The mass memory module is built on an MLC substrate for which IBM has 15 years of experience.

The technology is a high temperature cofired ceramic (HTCC) whose characteristics appeared in *Figure 1.*

The multilayer structure allow to take full advantage of the C4 chip interconnect, for packaging density as well as for electrical performance as it will be described in paragraph 3..

## Substrate Specifications

| | HTCC | Glass Cer. | Thin Films | LTCC |
|---|---|---|---|---|
| Conductor material | moly | copper | copper | silver |
| Dielectric material | alumina | cordierite | polyimid | al/glass |
| Dielectric constant | 9.5 | 5.2 | 3.5 | 7.8 |
| Thermal Conductivity (W/m/°C) | 20-30 | 5.0 | 0.09 | 3.0 |
| TCE (ppm/°C) | 6.5 | 3.0 | n/a | 5.5 |
| layer thickness (mm) | 0.15 | 0.1 | 0.006 | 0.1 |
| min via pitch (mm) | 0.225 | 0.225 | 0.05 | 0.225 |
| min via diameter (mm) | 0.08 | 0.08 | 0.01 | 0.08 |
| line pitch (mm)al | 0.45 | 0.45 | 0.05 | 0.45 |
| line width (mm)al | 0.1 | 0.075 | 0.015 | 0.075 |
| line capacitance (pF/cm) | 2.4 | 1.3 | 1.5 | 1.8 |
| line resistance (ohm/cm) | 0.32 | 0.22 | 2.3 | 0.2 |
| line impedance (ohms) | 50 | 60 | 50 | 60 |
| line delay (ps/cm) | 118 | 80 | 66 | 95 |

*Figure 1* Co-fired Substrate Specifications

The main characteristics for the MCM-C substrate are:

- TCE close to silicon
- Supports high packaging density.
- Good thermal conductivity.
- High reliability.
- Tolerates high process temperature.
- Large variety of packaging alternatives.
- Low dielectric loss.

### 3. DRAM Flip Chip Bonding.

One of the key elements of the IBM MCM technology is the chip to substrate interconnection, using a flip-chip concept. The specific process being used is the C4

(Controlled Collapse Chip Connection) process for which there are almost 30 years of experience in terms of process control as well as reliability data.

The structure of the C4 interconnection appears in *Figure 2.*

By structure the C4 technology gives performance advantages : chip to substrate is at its minimum (80 to 100 μm). In addition, the array pattern allows for power and ground pads to be interleaved with signal pads, thus improving drastically inductance characteristics as well as noise performance.

The technology is particularly suited for MCM applications. It allows for easy repairability during the module assembly process as no adhesive is used for chip attachment. In addition, the C4 pads permit bare die testing and burn-in prior to chip assembly on the module.

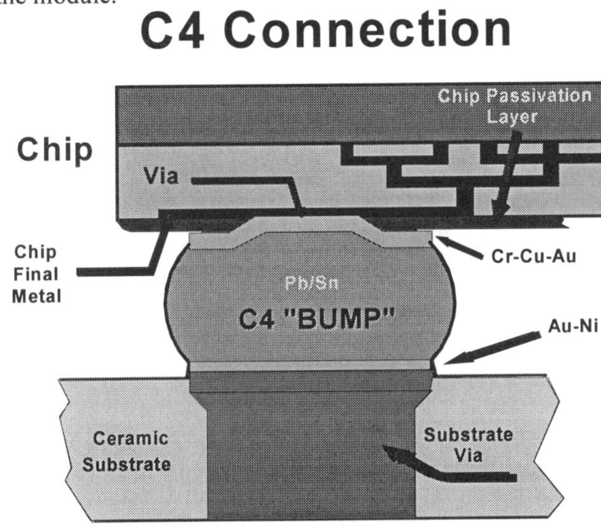


*Figure 2.* Structure of the C4 connection.

To take advantage of these characteristics, a C4 version of the IBM 16 Meg DRAM was developped for this solid state mass memory project. The main characteristics of this DRAM are listed in *Figure 3.*

## LUNA-C 16 Mb DRAM features

- Technology : 0.5 µ
- 4 Mbits x 4 - Data Selects Enable/Disable I/Os
- Single 3.6V+/-5% power supply
- High Performance
  - RAS access time 50ns/70ns (max)
- Std Retention
  - Standby (TTL inputs) 2.0ma (max)
  - Standby (CMOS inputs) 1.0ma (max)
  - Active (max) 210ma (110ns Cyc, 11/11 Add.)
- 2048 refresh cycles/32ms (11/11 Add.)
- 4096 refresh cycles/64ms (12/10 Add.)
- Fast Page Mode, Toggle Mode, Data Selects, Output Enable, On-Chip ECC, Soft Error Scrub, Internal Row Address Counter
- Input/Output: LVTTL for 3.6v supply
  Tristate
- SER: <0.1% per KPOH per chips
  - Actual SER will be too low to measure
- Long Retention Part
  - 200µA Standby Cmos levels
  - 12/10 Addressing
  - 256ms And STR
  - 340µA Keep Alive Current
- ECC
  - DED/SEC Error Checking and Correcting
  - Hamming code (odd-weight)
  - Implemented in each quadrant
  - Improves reliability on both hard and soft errors.

| | -50 | -70 |
|---|---|---|
| $t_{RAC}$ RAS Access Time | 50ns | 70ns |
| $t_{CAC}$ CAS Access Time | 10ns | 12ns |
| $t_{TAC}$ TOG Access Time | 10ns | 12ns |
| $t_{AA}$ Column Address Access Time | 18ns | 20ns |
| $t_{RC}$ Cycle Time | 110ns | 130ns |
| $t_{PC}$ Fast Page Mode Cycle Time | 22ns | 24ns |

*Figure 3*

The main steps of the C4 process are described in *Figure 4*. They start after the last passivation layer of the wafer process. Openings are photoetched to allow for contact to the last level of metal interconnect. Then a metal mask is used to vacuum desposit a structure of Cr, Cu and Au which will insure the proper adhesion to the wafer and allows for adequate intermetallics structure with the lead-tin pad to be vacuum deposited next.

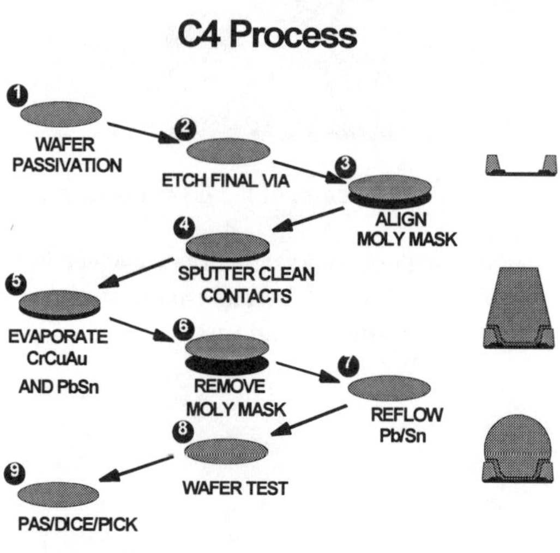


*Figure 4*. Main steps for the C4 process

After the lead-tin deposition the wafer goes through a reflow process which, by surface tension, will allow the pads to reach their rounded shape.

Once the dices are properly screened for performance and reliability, they are ready for bonding to the substrate, which can be an MLC structure, as described below.

Chip to substrate attachment is obtained through a single reflow operation for all the chips contained in the MCM module.

Regarding the packaging density, the C4 bumped dices are directly soldered to the MLC substrate which is used for chip to chip interconnect as well as for chip packaging. A hermetic sealing is provided on the top surface through brazing and seam sealing.

## 4. MCM-C Module Design and Manufacturing.

In order to improve the packaging density of the solid state mass memory module a double sided structure is to be used thus providing a 320 Mbit module on less than 25 cm² with associated logic control. A cut-away view of the module is shown in *Figure 5* and the finished module in *Figure 6*.

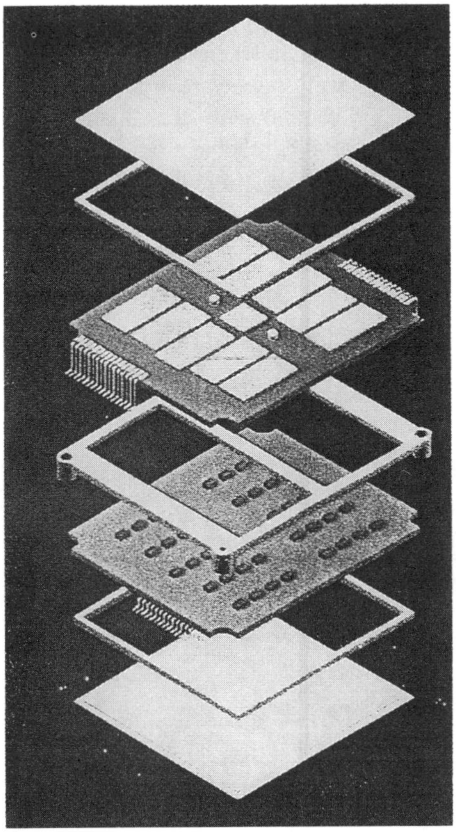

*Figure 5*. Exploded view of the MCM-C Module.

*Figure 6.* Closed MCM-C Module.

The multilayer ceramic substrate can match the electrical performance requirements by adequately designing and interleaving the signal planes as well as the ground and power planes. Through proper electrical simulations (it was done with ASTAP tool) it is possible to accuratly project the noise characteristics of the module : this is a key characteristic in application where high number of simultaneous switching can occur. Critical path delay, as well, can be projected.

This process, in conjonction with the quality system which has been implemented, has proven to give 100 % reliable product : during the 15 years of volume production no defect has ever been identified in the field.

## 5. System Integration

The modules are mounted onto a double sided EUROCARD as shown in *Figure 7.* The card is made of Polyimide-glass and has up to 14 layers. The capacity of the card can be adjusted up to 5 Gbits.

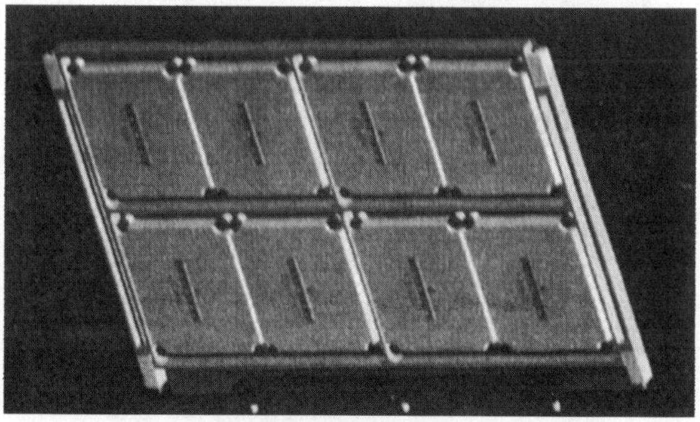

*Figure 7.* Memory Card

Each card has built-in redundancy in case of failure of a module in the card. The card inputs and outputs can be either series or parallel depending on the data rates that have to be handled. The double sided EUROCARDS are next assembled to form the final Mass Memory Equipment as shown in *Figure 8.*

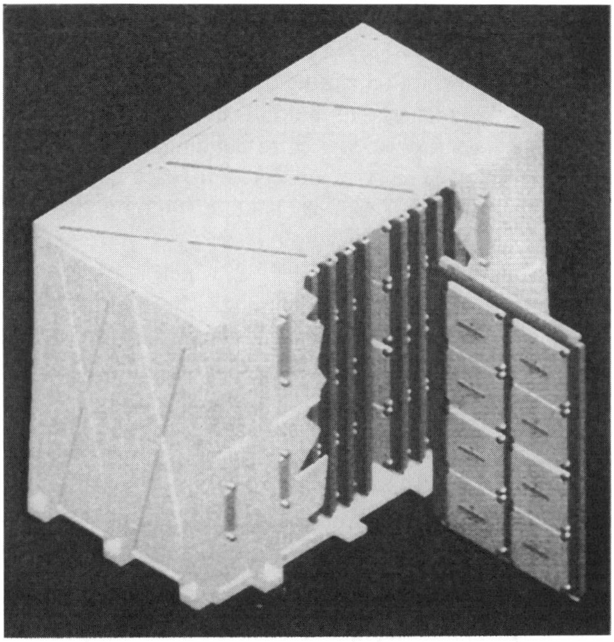

*Figure 8.* Solid State Mass Memory (SSMM) Equipment

This equipment has a capacity of 30 Gbits, weighs 18 Kg and occupies a volume of 15 liters. The estimated power consumption in the standby mode is 20W and during operation (read/write) 70W. The electrical block diagram is shown in *Figure 9.*

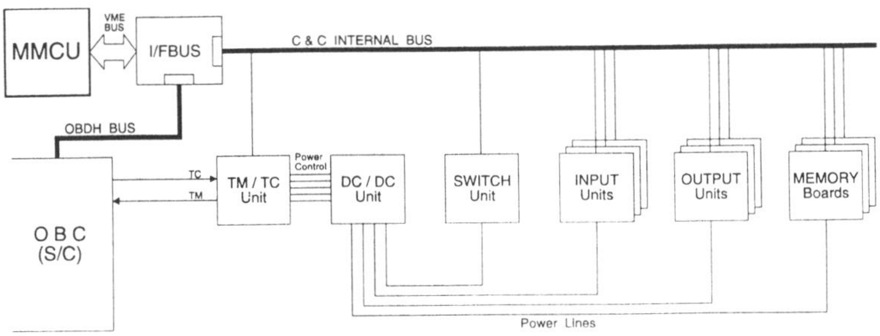


*Figure 9* . Block Diagram of SSMM Equipment.

The Mass Memory Control Unit (MMCU) is the management unit which ensures the management of the Solid State Mass Memory (SSMM) by an external user and it is the unit that controls the different parts which constitute the SSMM equipment.

The TM/TC Unit is the node used to receive external user telecommands and transmit the required SSMM status. In addition the TM/TC controls the power switches of the other SSMM units: Input, Output, Switch and Memory Boards.

The Switch Unit links in a physical way all the units involved in a recording, reading or reading/recording operation, during the time it lasts.

The Input Units are in charge of multiplexing the external Input Channels and transmitting the internal data flow to the corresponding Memory Board through the Switch Unit. One internal Channel is connected to each Input Unit. The SSMM Equipment has eight Input Units and each one has a total throughput of 200 Mbits/s as a maximum.

The Output Units are the specific devices which are used to restore the data from the Memory Boards and drive them through the Switch Unit to the SSMM external output channels. The SSMM Equipment contains up to 3 Output Units , each one containing up to 8 output channels. The maximum throughput is also 200 Mbits/s.

## 6. Future Trends.

The system presented here uses 16 Mbit DRAM chips. It has been designed to evolve with the technology. This evolution may come from two fronts: higher density chips (64 and 256 Mbits) and memory stacks. The memory stacks (3D integration) allow, for a given memory chip, the highest memory density (in Gbits/liter or in Gbits/Kg.). Or, alternatively, it is possible to envision a 256 Mbits stack, using 16 Mbits memory chips, having the same footprint as the future 256 Mbits chip.

ALCATEL has indeed developed such a 256 Mbits stack using sixteen 16 Mbit DRAM LUNA chips from IBM. The stack measures 19.0 mm x 19.5 mm x 11 mm which gives a total volume of 4.0 cm3 (0.25 cubic inches) which represents a memory volume density of 64 Gbits/liter. This 256 Mbits stack is part of the research project TRIMOD. TRIMOD is part of the R&D Programme on IT (Information Technologies) of the DG III of the European Union. This R&D Programme on IT was formerly known as ESPRIT. The 256 Mbits stack, designed by ALCATEL, has been built by THOMSON CSF/DOI of Colombes, France, using the 3D stacking technology being developed in TRIMOD [1]. This stack is shown in *Figure 10.*

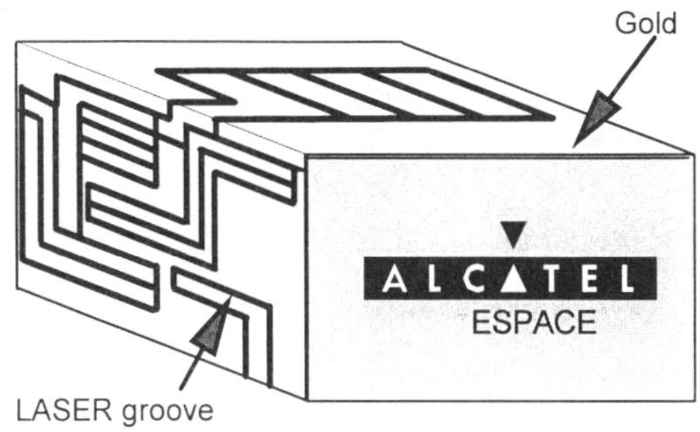


*Figure 10.* 256 Mbits Memory Stack

The THOMSON CSF stacking technology is a true Vertical Multichip Module (MCM-V) Technology. This means that with this technology it is possible to stack chips of different sizes and nature (in addition to assorted passive components). For the SSMM it would therefore be possible to stack together the memory chips and the control ASICs. This is shown in *Figure 11.*

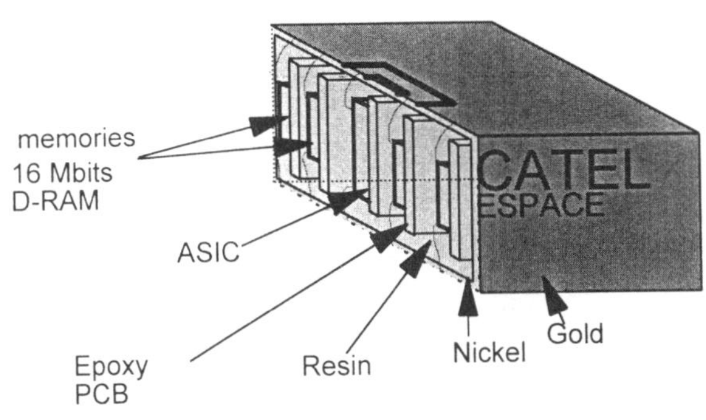


*Figure 11.* Cross-section of Memory Stack including the control ASIC.

The resulting cubes will still be assembled onto a co-fired multilayer substrate. It can be done par exemple using BGAs (Ball Grid Arrays) or simply by wire-bonds. This is illustrated in *Figure 12.*

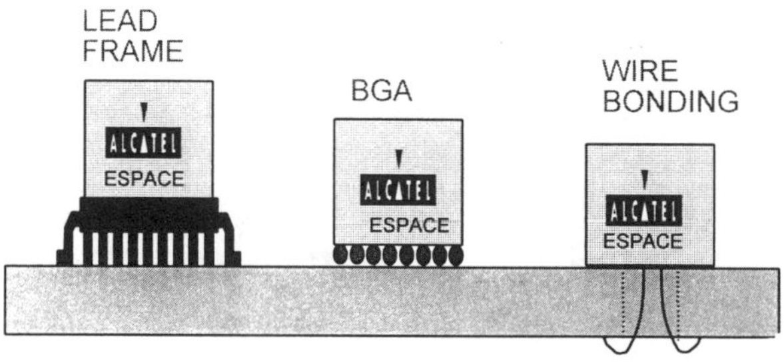


*Figure 12.* Different mounting possibilities for the 3D Stacks.

The resulting SSMM will have a significant reduction in weight and volume. For the 30 Gbits SSMM equipment, the estimated gain is shown in *Table 1:*

*Table 1* Estimated imrpovement using 3D stacks for a 30Gbits SSMM Equipment

| | Weight | Volume | Density Gbits/liter | Density Gbits/Kg. |
|---|---|---|---|---|
| Current MCM-C Technology | 18 Kg | 15 Liters | 2 | 1.6 |
| Future MCM-C using 3D Stacks | 4 Kg | 5 Liters | 6 | 7.5 |

## 7. Conclusion

A very dense solid state mass memory system has been presented. Because of the utilization of MCM-C technology, the system is very reliable and ready for today's needs. It has been designed keeping in mind technology evolution. It will thereby be able to accomodate higher density memory chips as well as 3D stacks. As the technology will allow higher density for the same weight/volume/price, system designers will readily make use of the extended memory capability.

## 8. References

1. C.Val and M.Leroy, (1991) The 3D interconnection - applications for mass memories and microprocessors, *ISHM '91 Proceedings*, 62-67.

# MASS MEMORY PACKAGING FOR SPACE APPLICATIONS

J.A. DE GIVRY
*MATRA MARCONI SPACE*
*Direction des Equipments et Technologies*
*37, Avenue Louis Breguet - 78146 VELIZY (FRANCE)*

## 1. Mass memories and the space context

### 1.1. SPACE NEEDS

A large market is opening up for earth observation and meteorogical satellites. The recording of datas on board is mandatory, for the main reason that the satellite is not always in visibility of earth stations, and so cannot transmit its informations in real time. Optical and radar instruments are now generating high output data rates ten times higher than the maximum speed of the link between the spacecraft and the ground. The required capacity for year 1998 is around 100 Gbit, end of life, and the input data rate is 400 Mbit/s minimum.

### 1.2. SPACE CONSTRAINTS

The main specific difficulty is radiation level, with three parameters : Total dose, Singe Event Upset, Latch-up. The choise of memory chips is tightly linked with radiation tolerance performances. Other concerns are high level of vibrations during launch, and good thermal dissipation, the equipment life being in inverse ratio to temperature. (A seven year mission is typicaly required).
Power must be lower than 2W/Gbit, to save weight of solar generators. The weight of the memory stack must be less than 1 Kg/Gbit. And cost must be competitive in comparison with tape recorders used as yet.



*W. K. Jones et al. (eds.), MCM C/Mixed Technologies and Thick Film Sensors,* 131–138.

### 1.3. SOLID STATE VERSUS TAPE RECORDERS

Solid state recorders can compete favorably with tape recorders because the latter are :

- Unreliable in space environment, for long life missions. Hermeticity of large equipment is difficult to maintain along years : Tape and mechanical moving parts need to work under gas pressure to be reliable.
- Not flexible, as they need a sequential access, and are not easily adaptable to various missions.
- Noisy, as they generate a mechanical vibration, disturbing the stability of instruments during image capture.

Thanks to the semiconductor density progress, since the 1 Mbit chip availability, it has been demonstrated that SSR were competitive, avoiding all the drawbacks listed above, and offering a set of other advantages :

- High speed data rate
- 50 to 70% weight saving
- Lower cost in high volume production

Nevertheless it is not obvious to package 1000 to 10,000 die and to design a redundant and fault-tolerant memory equipment. A careful selection of packaging processes is a key issue to achieve this goal.

## 2. Solid state packaging review

### 2.1. GENERAL PACKAGING SOLUTIONS

A detailed analysis of the different packaging options has been done. Table 1 shows the available solutions. Two main famillies are related to chip presentation : packaged or bare die. LCCC style of package were used in the past, with low capacity chips. TSOP plastic packages present a better packaging efficiency, but they are not space qualified, at least in Europe. Same situation for chip-on-board processes.

Memory wafers are not easily available, and manufacturing yield control is difficult. That is why the assembly of memory chips ans associated circuitry is frequently based on hermetic modules with a MCM substrate, or a conventional thick film substrate, the number of inputs being relatively low. The hermetic package is metallic or ceramic.

These modules are generally mounted as flat-packs. But "Single-in-line" vertical packages are also used. Chips are sometimes stacked in 3 dimensions to improve performances.

A printed circuit mother board allows the interconnection between modules.

TABLE 1. SSR Packaging - A review of solutions

| PACKAGED DIE | | | BARE DIE | | | WAFER |
|---|---|---|---|---|---|---|
| Large PCB | MODULES | | CHIP ON BOARD | MULTI CHIP MODULES | | MODULES |
| TSOP or LCCC | TSOP or LCCC | TSOP or LCCC | Wire bonding or TAB | Wire bonding or TAB or Flip chip | Wire bonding or TAB or Flip chip | Wire bonding |
| (Double-sided SMT) | PCB daughter boards | Ceramic daughter boards | Single-sided PCB + Glob Top Gel | Metallic packages | | Metallic packages |
| -- | SIL | SIL or Flat Pack | -- | Flat Pack | SIL or Flat Pack | Flat Pack |
| 2D/3D | 2D/3D | 2D/3D | | 2D/3D | 2D/3D | 2D/3D |
| | PCB Mother board (Single or double-sided) | | | PCB Mother board (Single or double-sided) | | |

TSOP : Thin Small Outline Pastic packages
LCCC : Leadless Ceramic Chip Carriers
SMT : Surface Mount Technology
SIL : "Single-In-Line" package
3D : Three dimentional assembly

## 2.2. THREE DIMENTIONAL PACKAGING

Considerable research and development work have been undertaken on 3D processes. Mass Memories, as they gather thousands of repetitive and relatively simple structures, represent a major application of 3D.
Here also, 3D processes can be splitted in 3 categories, depending on the chip style :

| | |
|---|---|
| - Packaged Die : | - Standard or custom package |
| - Bare Die : | - Sacked standard chips |
| | - Stacked custom chips |
| | - Stacked multichip modules |
| - Wafer : | - Stackable custom wafers |

The use of bare die is presently predominent, with memory chip "cubes" (T.I., IBM, Thomson-CSF, ...) or stacked MCM..

# 3. Description of Matra Marconi Space packaging

In order to adapt easily the hardware to mission needs and to chip evolutions, a modular architecture was specified. The modular concept is based on :

## 3.1. MEMORY COMPONENTS

From 1 Mbit SRAM to 16 Mbit DRAM, with extension possibility to 64 Mbit DRAM.

## 3.2. MCM

This is a level 1 assembly, with 2, 3 or 4 memory dice, so ranging from 2 to 64 Mbit global capacity. When 3 or 4 dice are used, 3D stacking of chips is used. Depending of the bonding pad topology on chips, different processes are used.
Fig 1 shows the "crossed pairs" concept ($3D^{\#}$ MMS patent).
Fig 2 shows the "Tiles" concept ($3D^{T}$ MMS patent), using 4 rerouted unwedged dice, with relay pads for "chip select" outputs.
Fig 3 shows the Flip-Chip concept ($3D^{FC}$ MMS patent) where dice are turned upside down on a silicon MCM. A second MCM is mounted on top of the dice of the first MCM. Conventional wire bonding is used to interconnect MCM. All materials being Silicon, reliability is not affected by thermal mismatch.
For all three cases, MCM substrates, which support the two layers of dice, can be made with thick film on alumina (Dual side with metallized through holes, or multilayer). Fine line are mandatory (100 µm width, and 75 µm between lines). Wire bond pull tests and outgassing tests are more consistent than other MCM techniques tested with organic or Silicon MCM. High purity adhesives are used for stacking.
MCM are modified when a new type of component is used, but interfaces do not vary and next level assembly is preserved.

*Figure 1. MMS 3D# concept*

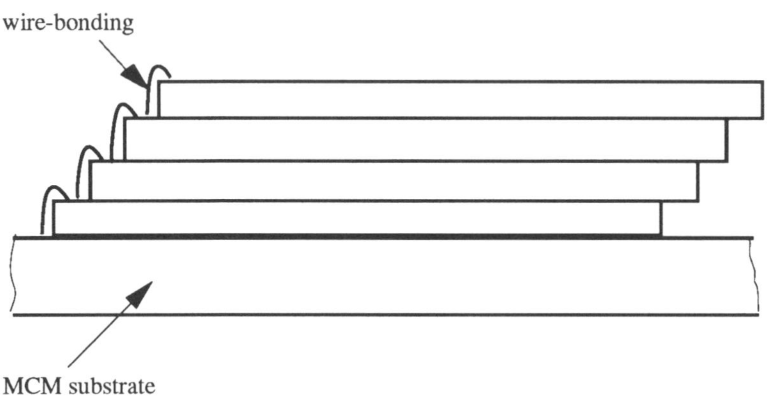


*Figure 2. MMS 3D$^T$ concept*

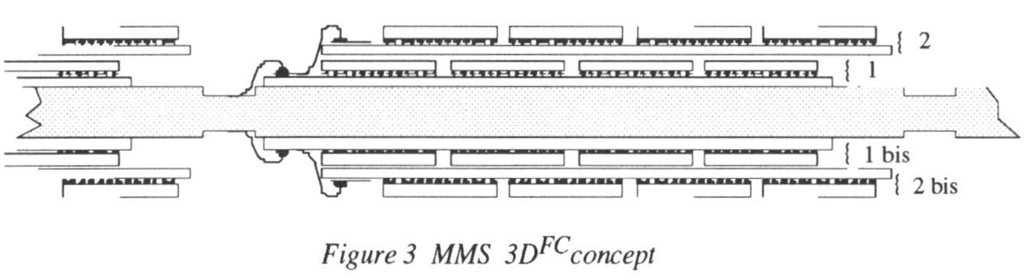


*Figure 3 MMS 3D$^{FC}$ concept*

3.3. MODULES

A dual cavity package includes 8 MCM. Its capacity range is 16 Mbit with 1 Mbit SRAM up to 512 Mbit with 16 Mbit DRAM. High temperature cofired MCM-C was selected, in order to optimize weight, interconnect easily the two cavities, and simplify output lead attach (No glass feed through).
Fig 4 shows one side of the bare module, with its Kovar frame and its cavity, which length is 80 mm.
The substrate includes 8 conductive layers. Line width is 200 µm, and pitch 381 µm. Conductors are gold plated. NXERAM (Bollene, France), manufacturer of this package, has now in development a 254 µm pitch technology. This explains the high production yield they have experienced on 381 µm pitch. Lids are seam welded Kovar plates.
Fig 5 shows the module populated with MCMs and buffer chips.

3.4. THE MEMORY STACK

The assembly of 64 modules is made on a mother board within a metallic frame (Slice).
Fig 6 shows such a slice, half populated with 32 modules. "Single in line" modules are monted with SMT specific connectors developped by Amphenol-Socapex or with "Spring-Chip" (MMS patent). "Digi-clip" can also be used for prototypes. Both extremities of the modules are tightly fastened to the frame with an expansion concept. Mechanical and thermal performances are very high.
On the other side of the mother board, control functions (ASICs and anti-latch-up hybrids) are SMT mounted, close to the modules.
Several slices (up to eight) are stacked with tie-rods and interconnected by a multilayer board on a base plate. Input/output and DC/DC converter slices are added on both extremities of the memory stack.

## 4. Performances

The following table (Table 2) gives the performance range that such a Mass Memory concept permits to achieve, depending on the type of the memory chip and on the 2D or 3D assembly. Results are given for the module level and for the slice equiped with an interconnection base plate (Stack level). The lowest capacity, with 2D 1 Mbit SRAM, is useful when a few Gbit buffer memory is needed. The largest capacity, with 3D 16 Mbit DRAM, allows for example the design of a 100 Gbit SSR within a 12 Kg stack. This is an order of magnitude lower than a tape recorder system of the same capacity.

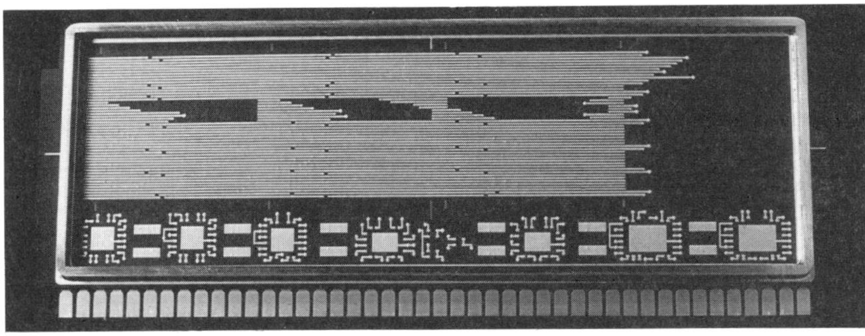

*Figure 4. Bare co-fired module (Recto side)*

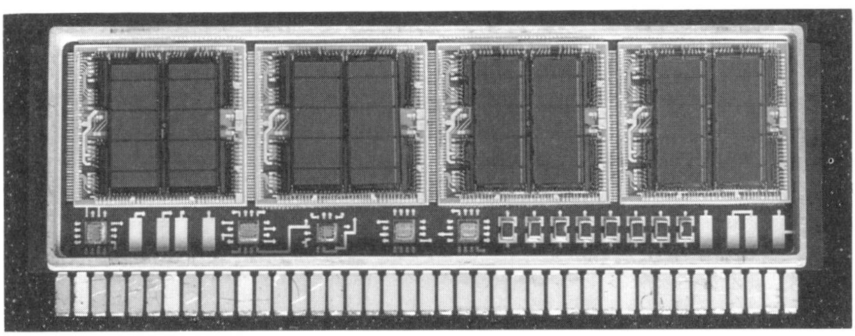

*Figure 5. MCM assembled in the module (Recto side)*

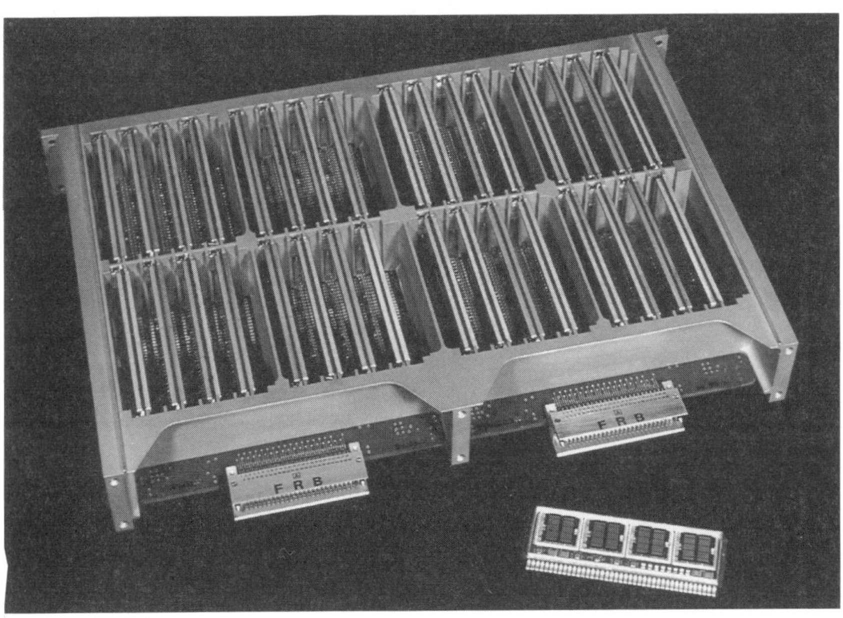

*Figure 6. Memory slice (up to 33 Gbit)*

TABLE 2. Density perfomances

| CHARACTERISTICS | | MINIMUM --> Buffers (1 Mbit DRAM/2D) | MAXIMUM --> SSR (16 Mbit DRAM/2D$^T$) |
|---|---|---|---|
| Weight density | Module Level | 0.53 | 15.2 |
| Gbit/Kg | Stack Level | 0.35 | 11.3 |
| Volumic density | Module Level | 0.89 | 28 |
| Gbit/dm$^3$ | Stack Level | 0.48 | 15.3 |

## 5. Main advantages

Advantages of the packaging described above are the following :

- Competitive density, especially at equipment level.
- Most processes are already space qualified.
- New processes have been successfully evaluated .
- Flexible concept : the same module accepts most die types (even future 64 Mbit chips). Modularity allows an easy adaptation to capacity needs.
- Mechanical stiffness and high heat transfer are achieved with the slice frame.
- Test and burn-in at MCM level permit a high manufacturing yield and a good production cost control.

## 6. Conclusion

MCM-C offer an adapted solution to mass memory packaging for space applications. Ceramic substrate processes are mixed to reach the needed density and reliability :

- Fine line thick film MCM on alumina.
- Double sided cofired ceramic package.
- Simple 3D concepts and modular packaging offer a very efficient and flexible answer to ever evolving capacity needs. Adaptation to new generations of memory chips without expensive non-recurring engineering is also a major advantage of the technology developped by Matra Marconi Space.

# ADVANCED MULTICHIP MODULES FOR TELECOM APPLICATIONS

K. Lösch
*Alcatel SEL Research Center*
*Lorenzstraße 10, D-70435 Stuttgart, Germany*

H.M. Rappold
*Alcatel SEL Components Group*
*Nimrodstraße 9, D-90441 Nuremberg, Germany*

Interconnection and packaging technologies are becoming more and more important for a successful implementation of new telecommunication systems. Advanced assembly methods like multichip modules and flipchip technology offer competitive advantages such as better performance, smaller equipment size or cost savings on system level.

## 1. Introduction

Telecom equipment manufacturers are faced with a continuously growing demand for higher system performance in combination with smaller size and less power consumption. The rapid progress of silicon integrated circuit technologies makes it possible to increase the operation frequencies and to include more and more functionality on one chip. Standard electronic assemblies and packages impose a serious penalty on the speed of chips due to the parasitic inductance and capacitance in the chip-to-chip interconnection path. Thus many systems are limited by interconnection and not by chip performance, often called "interconnection bottleneck". With the multichip module (MCM) approach, where a set of bare chips is mounted on an interconnect carrier, many of these penalties can be avoided. However a very wide range of performance parameters has to be covered in advanced telecom systems: for line transmission very high data rates up to 10 Gbit/s (40 Gbit/s in near future) with a low number of channels are typical, whereas broadband switching requires many channels at data rates of 622 Mbit/s (later 2.4 Gbit/s) with a very complex interconnection scheme; finally in radio communication analogue signals up to 60 GHz are applied.

The specific advantages of MCMs for Telecom systems are: higher density, better signal performance, lower cost on system level and, what is becoming more and more important, shorter time to market.



*W. K. Jones et al. (eds.), MCM C/Mixed Technologies and Thick Film Sensors,* 139–146.

## 2. The Alcatel Approach

In order to stimulate the use of MCMs we have set-up an Alcatel-wide collaboration within the frame work of the Multichip Module Competence Center. In this Competence Center (CC) there are contributions from research teams in Belgium, France and Germany, and from industrial units in Belgium , Germany and United States with well defined work split. The chip manufacturer Alcatel MIETEC is contributing to testing and Known-Good-Die (KGD [2]) issues. In addition to this close collaboration between "technology providers" we have regular meetings and discussions with potential in-house users of MCMs (User Group). On a yearly basis, results and future program of the Competence Center are presented to an Advisory Board, where all Alcatel product groups are represented.

As an example, the approach of the Alcatel SEL Components Group as a MCM Supplier will be described in more detail. It turned out, that an "integrated" approach is necessary in order to become a competent partner for our internal and also for our external customers. This covers top level design, cost modelling, thermal and electrical simulation, mask layout (Cadence system with interface to all major CAD systems), substrate manufacturing (in the case of MCM-C), substrate characterisation and trimming, component assembly, packaging and final testing as well as quality and reliability aspects. For all substrate types approved suppliers exist (internally and externally). A special framework on KGD delivery with our key chip suppliers has been set up via Alcatel's central purchasing. MCMs were realised both for external customers (e.g. automotive motor control circuit, sensor circuits) and for internal applications.

In order to provide the optimum solution for the user, all MCM-types are thoroughly compared with each other and also with alternative solutions (e.g. monolithic integration) using a cost model based on own experience and price information from external sources. For this reason a spread sheet has been defined, where important topics such as technical performance, manufacturing requirements, availability of components (KGD), expected costs and marketing aspects (each with certain sub-topics) are listed and weighted by the customer or system designer according to the relative importance in his specific application (weight factor WF). Then the MCM experts suggest different MCM solutions and compare them with other alternatives (standard packages, direct chip attach to the board and higher degree of monolithic integration) by a ranking figure R for each of the topics, which describes how good the quantitative customer requirement is met by the specific solution. The total ranking of a solution is the sum of all products R * WF. Thus both aspects, the customer expectations and the technical expertise are combined in one figure.

Currently all three MCM types are investigated at Alcatel. The main technical parameters are listed in Table 1 for comparison. The interconnection density, which can be achieved by a specific MCM-type, is mainly given by the line width and the land size, which is the area around a via not available for routing. The large difference in the via density (number of connections between different conducting planes per square centimeter) is quite evident.

Table 1: Typical technical parameters of interconnection carrier types (for definition see [1]

| | MCM-L | MCM-C | MCM-D |
|---|---|---|---|
| Line width [µm] | 100 | 100 | 50 |
| Wiring density per layer [cm/cm²] | 50 | 50 | 100 |
| Land diameter [µm] | 500 | 200 | 50 |
| Via density [1/cm²] | 60 | 120 | 5000 |

## 3. Selected examples from various Telecom fields

For one example - the dual processor MCM - we have realised size- and pin-compatible versions as MCM-C (thickfilm screen printed, diffusion patterning for vias), as MCM-D (external source: Dyconex) and as MCM-L (FR4 material, for details see [3]). The MCM approach was selected to combine the functions of two printed boards on one. Each MCM comprises 2 digital signal processors (die size 7.9 x 7.9 mm², P = 1 W) and 12 static RAMs (10.2 x 3.5 mm², P = 1.05 W) as bare dies with a total of 672 internal wire bond loops and 22 surface mount capacitors. Connection to the PCB is provided by 4 connectors with 192 pins in total. Present clock frequency is 40 MHz. For next generation 80 MHz are envisaged. Due to the large via diameter and the corresponding land size, routing was very difficult and time consuming for MCM-L and some longer paths could not be avoided, whereas the other two solutions made no difficulties and had enough spare area for more complex routing. For MCM-C routing and thermal simulations led to a design with 4 metal and 3 dielectric layers on $Al_2O_3$ ceramic (63.5 x 63.5 mm²). A line width of 120 µm (pitch of 250 µm) and a via diameter of 200 µm was used.

Fig. 1 shows the MCM realised at Alcatel SEL.

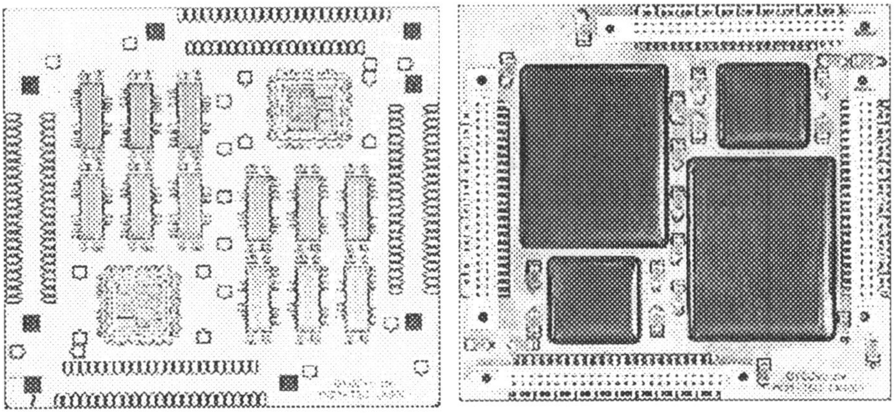

*Figure 1: Dual processor MCM before and after local encapsulation*

## 3.1. APPLICATION IN TRANSMISSION

MCM-D technology was studied for a new synchronous transmission system (Alcatel Multiport 1641X). The central part of the Multiport is the matrix which is able to cross connect virtual containers and administrative units. As a first step the basic switching chips (BSCs) were selected for construction of multichip modules. Prototypes have been prepared using a three metal layer interconnect on $Al_2O_3$ placed in a covar housing with 300 feed-throughs (Fig. 2). The main fabrication steps of the carrier are spin coating of a polyimid dielectric layer (with low dielectric constant $\leq 3$) and sputter deposition of metals. Lines and vias are formed by standard photolithography and dry etching (line width 50 µm, via diameter 60 µm). Electroplating follows to increase the metal thickness to about 5µm [4,5]. A formerly used baby board was no longer required, all functions fitted now onto the PCB as shown Fig. 3.

*Figure 2: MCM (Type D) for crossconnect application*

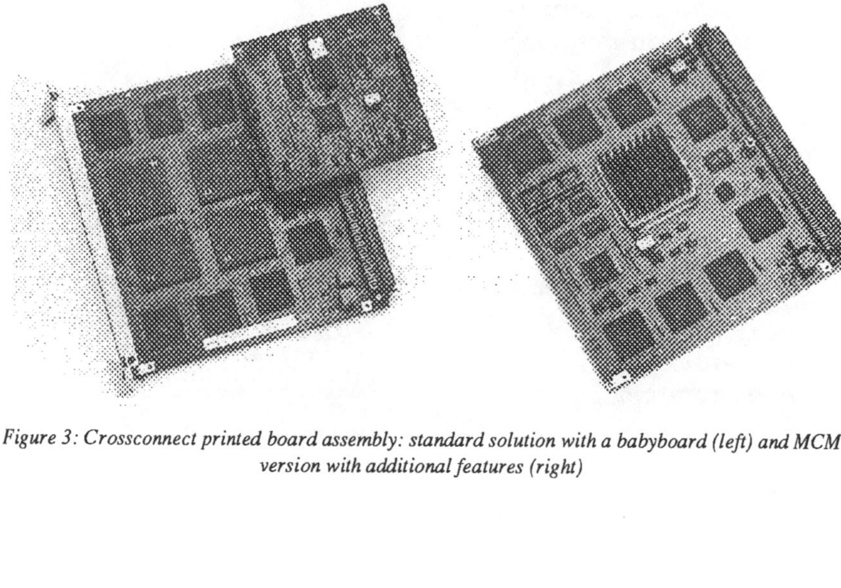

*Figure 3: Crossconnect printed board assembly: standard solution with a babyboard (left) and MCM version with additional features (right)*

## 3.2. APPLICATION IN SWITCHING

Great complexity combined with high data rates are found in modern switching system applying the ATM principle (Asynchronous Transfer Mode, a type of packet switching). A typical printed board assembly handles 4 channels each with 622 Mbit/s. Impedance controlled lines are needed to connect the ICs. Power dissipation of the switching ICs is at the limit of semiconductor technology (up to 6 W/chip). A study on the use of MCM technology has been made both for technical reasons and cost improvement. The outcome is: Maximum benefit of MCM technology can be gained, if all critical chips are concentrated in one MCM with 3600 internal connections and only 360 external I/O's. An impedance controlled thin film multilayer carrier (4 planes) is needed to handle all the internal lines on a reasonable overall size of the MCM (about 50 x 100 mm$^2$). Testing of such complex MCM at operating frequencies is a challenge for test equipment manufactures. The MCM solution offers slight cost advantages on system level, because much simpler printed circuit board and less board testing can be used, provided that the naked dies are available for 70% of the cost of packaged components.

## 3.3. APPLICATIONS IN RADIO TRANSMISSION

Front-ends of millimeter wave radio transmission equipment have to combine excellent high frequency characteristics (extremely low loss) with compact size for shielding and cost reasons [6]. The use of impedance controlled multilayer carriers with low dielectric constant ($\varepsilon \approx 3$) instead of strip lines on alumina ceramics ($\varepsilon \approx 10$) gives the potential of size reduction by more than factor of ten (see Fig. 4).

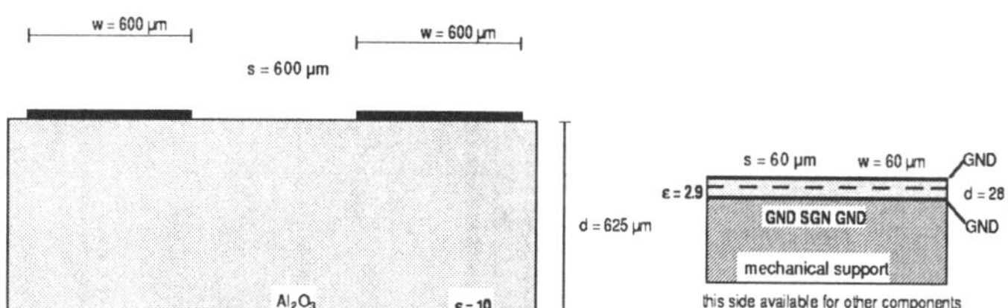


*Figure 4: Comparison of Rf carriers: state of the art strip lines (left) and advanced polyimide multilayer technology (right) allowing an increase of line density up to a factor of 10 and reduced crosstalk*

In order to find the limits a high frequency test MCM has been designed and realised using a thick layer of polyimide as dielectric medium. Excellent properties of the test structure were found, provided that the thickness and dielectric constant of the polyimide layer are carefully controlled (thickness: 28 µm ± 1µm). S-parameter measurements with a cascade wafer prober are shown in Fig. 5. It is intended to realise a front end MCM for 38 GHz radio transmission, comprised of flipchip bonded GaAs MMICs, impedance controlled thinfilm interconnect and a planar antenna.

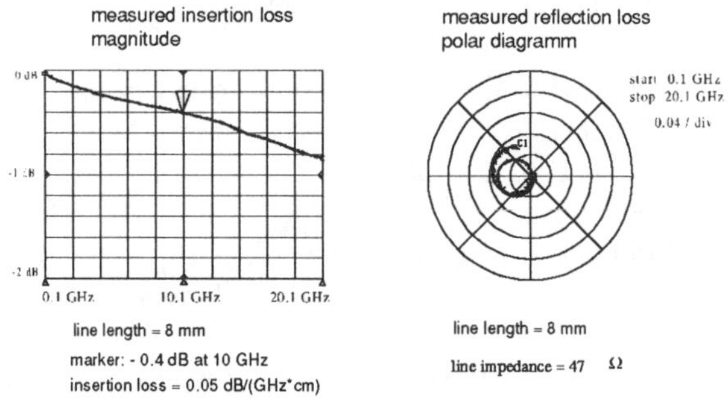


*Figure 5: High frequency properties of advanced MCM-D carrier with polyimide dielectrics: line attenuation (left) and reflection (right)*

## 4. Flipchip Technology

In the Alcatel plants wire bonding is used as standard method, however, increasing demands on system performance, miniaturisation and cost reduction require the investigations of alternative methods. Due to shortest signal paths flipchip interconnections show the best electrical performance of all chip interconnection technologies. In addition flipchip technology allows mounting of very high pincount ICs by use of area configurations, which enables significant reduction of on-chip wiring.

Silicon chips have to be bumped at wafer level using PbSn solder alloy. At Alcatel a special micro galvanic process has been elaborated which is compatible with standard bond pad sizes and pitches used for wire bonding. Thus chip redesign can be avoided. As most GaAs MMICs have gold metalisation bumps of Au are appropriate. As specific advantage, the Alcatel technology allows to bump the interconnect carrier instead of the chip. Thus no wafer processing of the GaAs components is required. Components from different suppliers can be used off-the-shelf [7].

Flipchip bonding can be done in the thermocompression, the eutectic and the solder reflow mode. It starts with aligning the semiconductor device in face-down position to the footprints of the MCM-substrate by use of a split field camera. When alignment is done, the camera is pulled out and the bond process itself takes place.

In the case of solder bumps the bond shape is optimised by z-axis control. Fig. 6 shows a Si-chip on a Si-substrate. The advantage of this kind of solder joint is, that the stresses caused by mismatch of the thermal expansion coefficients are removed from the interfaces of the chip and substrate metalisation. The stress is concentrated on the waist of the bulk solder alloy, where strains can easily be accommodated by creep. Flipchip bonding onto a standard thickfilm circuit from our plant has been used to realise prototypes of a transmit/receive module for optical transmission., which is currently in reliability tests for qualification.

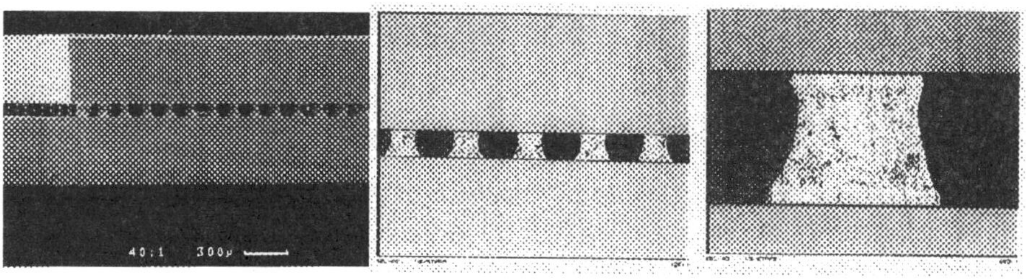

*Figure 6: Flipchip connections (Pb/Sn(37/63), pitch 170 µm, height 70 µm) with waiste: SEM photo (left), cross sections (middle and right)*

For III/V-semiconductors, where gold is the preferred material for the flipchip interconnects, bonding is performed in the thermocompression mode. The chip positioning accuracy achieved is ± 1µm in x/y-direction, which facilitates optical coupling of discrete and integrated optoelectronic devices. Millimetre wave components have been flipchip assembled. The excellent electrical performance of flipchip interconnects is shown in Figure 7 in comparison with standard die attachment and wire bonding [7].

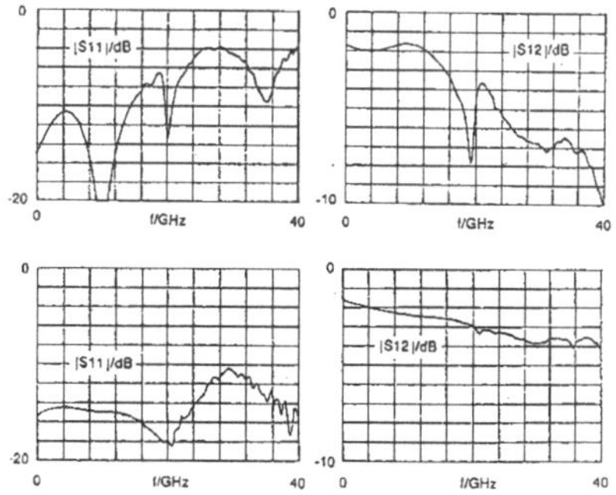


*Figure 7: Rf characteristics of wire bonded (top) and flipchip bonded GaAs chip (bottom)*

## 5. Conclusion

MCM technologies and "Known good die" availability have reached a status, which allows the introduction of MCMs into telecom products at a very attractive and competitive cost level. It has been shown that due to the wide span of required performance and cost targets of telecom systems specific MCM solutions have to be elaborated depending on the application areas, ranging from low cost MCMs up to MCMs with optical interfaces in the near future.

## 6. Acknowledgement

The authors would like to thank all of their colleagues in Antwerp, Nuremberg, Richardson and Stuttgart, who have been involved in the activities reviewed in this article. Financial support in part for selected activities is gratefully acknowledged to the German Ministry for Research and Technology (BMFT).

## 7. References:

1. Guidelines for Multichip Module Technology Utilisation; edited by The Institute for Interconnecting and Packaging Electronic Circuits, 7380 North Lincoln Avenue, Lincolnwood, IL-60646-1705 ref. no. IPC-MC-790, August 1992
2. Martin S., Gage, D., Powell, T. and Slay, B. (1993) A practical approach to producing known-good die, in *Proc. of the 1993 International Conference on Multichip Modules*, edited by International Society for Hybrid Microelectronics (ISHM), P.O. Box 2698, Reston, AV 22090-2698, p. 139
3. Ackaert, A., Van Dam, C., Lamers, G. and Allaert, K. (1993) The use of MCM-L manufacturing and assembly technology for broadband application, *ibid.* p. 136
4. Baumgärtner, A., Richter, H., Springer, J., Meier, U. and Heck, W. (1993) Advanced Multichip Module Technologies, *Proc. of the 9th European Hybrid Microelectronic Conference ISHM*, Nice, June 2-4
5. Florjancic, M., Richter, H. and Springer, J. (1992) Multichip Modules for High Bitrate Telecom Applications in H. Reichl (ed.) *Micro System Technologies '92*, VDE-Verlag GmbH Berlin, Offenbach, p. 457 - 470, /9/
6. Meier, U. and Müller, H. (1993) Substrate integrated transition between planar millimetre wave antenna and hermetically sealed integrated circuits, *Electronics Letters* ***29***, 2205-2206
7. Richter, H., Baumgärtner, A., Baumann, G. and Ferling, D. (1994) Flip chip attach of GaAs-devices and application to millimeter wave transmission systems, in H. Reichl (ed), *Microsystems Technologies '94*, VDE-Verlag GmbH Berlin, Offenbach, to be published and Microsystems '94, Berlin, 19.-21.10.1994

# TELECOM APPLICATIONS OF MCM TECHNOLOGY

JOSE LUIS CONESA
PIERRE PLAZA
*Telefónica I+D*
*C/ Emilio Vargas 6*
*28043 Madrid (Spain)*

## Abstract

Telecom systems are seen as one of the major application areas for MCM technology. The most interesting application fields like portable terminals, and ATM networks are presented and discussed, as well as an overview on those aspects that are still to be improved to allow extensive use of MCM technology within the mentioned fields.

## 1. Introduction

Present telecom systems are extensively using the most advance semiconductor processes to reduce cost, increase system performance, and reduce weight size and power. However, the evolution in silicon technology has been much faster than in the packaging and interconnect area. In the present situation, a new interconnect technology is required to take benefit at the system level of present semiconductor capabilities. This paper will review the main application areas of MCM technology from a telecom perspective. The focus will be placed on the benefits that this technology is expected to bring into the systems, and the key aspects to be solved for this technology to become a major player in the telecom sector.

## 2. Telecom Networks

A telecom network can be divided into three main elements: the first one is the access network, were all customers interface to the telecom network itself. The second element is composed of the switching nodes, where all the traffic is routed. And the third element is the transit network, that performs the high capacity signal transmission between the switching systems. In all the three areas pointed above MCM can play a major role.



*W. K. Jones et al. (eds.), MCM C/Mixed Technologies and Thick Film Sensors,* 147–152.

In the access network there is a strong push towards portability. The usage of portable phones, in the different flavors (cordless, cellular, etc.), has seen an exponential growth in the past years. This trend is not however unique to the telecom market, since it is present in almost all electronics sectors like computers (laptops and notebooks), and consumer electronics (camcorders, portable CD players, games, etc.). Portable telecom equipment, like any other portable system has very strict requirements for space, weight and power, making them an ideal field for the application of MCMs.

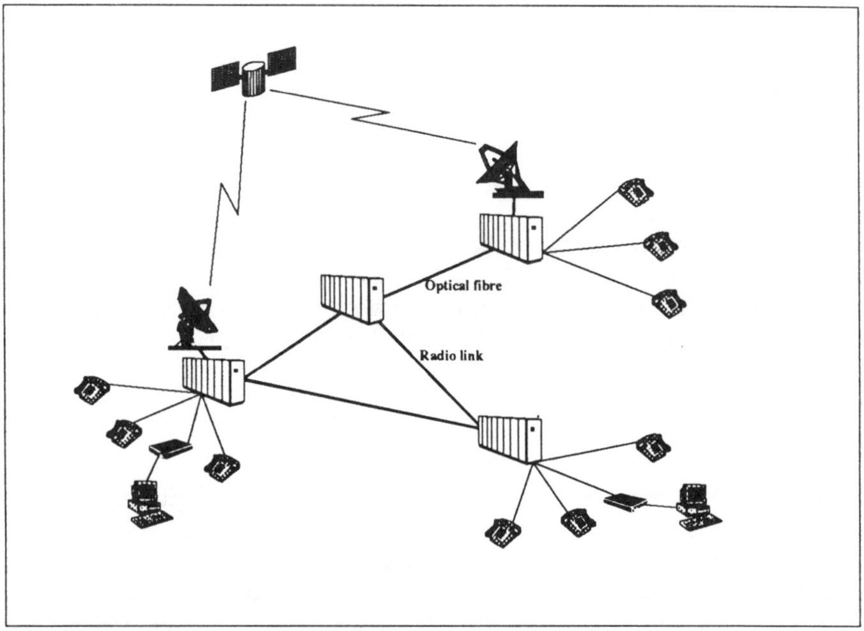


Figure 1: Simplified telecom network

The second application area for MCMs is the broadband integrated network. This network is being designed to integrate the multimedia communications needs. The requirements in terms of chip complexity and speed, are pushing interconnect technology to its present limits. In this area the main advantage that MCM can provide is the increment in systems performance. This increment in performance is also interesting for the transport network, since most of the transmission systems require high speed electronics.

### 3. Portable Terminals

The necessity of being able to access the communication networks independently of the user location is provoking a increased interest on portable systems. In all portable systems the information is transmitted via radio. The band depends on the system itself but due to the congestion in the radio spectrum they tend to be located around or above 1 GHz.

Probably the most extended portable systems are the cordless telephones. Those telephones are connected via radio to a base station connected to the user phone wires. For any other network user, and for the network itself, those phones work just as any other terminal, and the maximum distance from the base station does not exceed 50 or 100 meters. An evolution of this system is the DECT (Digital European Cordless Telephone), as opposed to the domestic ones, this specification is intended for use in medium industries and buildings. DECT allows a quite dense packing of the communications as well as data transmission.

Another portable systems are cellular phones. Those phones can communicate with any of the base stations available in a complete network. Base stations are usually located in many buildings inside towns, and along main roads. With this system the user can communicate from any location within a certain coverage. The network performs all the functions to guaranty that the link is not lost when the communications switches from one base station to the next while the user is moving. The network also locates and directs the calls to the cell where the user is located. There are several specifications for such systems, and most of them cover just one country. A new standard is GSM, that can work in all Europe independently of the country were the user is located (needless to mention that accounting and billing in such system is quite complicated). The evolution of cellular systems is going towards smaller cells (of around 500 meters) that allow a very high communications density. An even more revolutionary version of the same service is being addressed by Motorola with its Iridium project. In this case the portable phones will communicate directly with a satellite instead of a base station. To allow the system to work in the whole world, the network will include 60 low orbit satellites.

There are other portable communication systems besides phones that are already in production. Although not normally counted as part of a telecom network, GPS (Global Positioning System) receivers share the same basic technology as portable phones. GPS calculates the user position by receiving a signal from several satellites at the same time.

All the terminals of the systems mentioned above share the same requirements as the consumer electronics. The price is a critical factor, and the market windows are quite narrow, making time to market a key factor. The trend in those systems is to reduce size and weight, and increase operation time on one charge and functionality. This means that low power and high integration is a must. The electronics involved in such systems are quite complex. they usually include the high speed electronics for the radiofrequency transmitter and receiver, a number of ASICs to perform the required DSP functions, and a microcontroller that takes care of the user interface, the protocol with the base station etc.

The benefits that MCM technology can provide to those systems are quite clear from the previous description: namely a reduced size and weight. As mentioned previously, the key aspect for MCMs to take over in this field is the price. Among the different types of modules, MCM-L has already been demonstrated as suitable technology for consumer applications like watches and calculators, so it can be the best candidate at least in the near future.

## 4. Broadband networks

Telecom networks are going through a rapid evolution. Some years ago the services offered by those networks were limited to what is known as POTS (Plain Old Telephone Services), that included basically voice transmission. Some terminals have been designed to used the same interface as voice, to complement the communications possibilities. Examples of those systems can be fax machines or modems, since those terminals use the same interface as human voice, its existence is irrelevant to the network itself.

Up until now, every telecom service requires the installation of a specialized network to support it. Voice is supported by circuit switching equipment, computer communications has dedicated protocols like X.25 and are routed through packet switches, and video signals are broadcasted through satellite and radio or distributed trough cable. This network architecture was chosen since the requirements for each service are very different from each other. However this implementation is quite inefficient since one of the services might have a congested network, while other network is idle. The idea of providing an integrated network for all services has been quite appealing for a long time.

The transfer mode that is agreed by the experts as the most promising solution is ATM (Asynchronous Transfer Mode). In ATM data from any service (voice, video, computer data, etc.) is divided in very small pieces called cells. Cells are transmitted through the network, and regrouped back together at the receiver end. ATM networks are quite flexible in terms of the services that can be transported on them, since the bandwidth can be allocated as needed, providing a good platform to integrate multimedia communications. As a consequence of the flexible bandwidth allocation, ATM systems can take advantage of statistical gain. If a large number of users are connected to a single high speed route, it is very unlikely that all of them will require the maximum bandwidth at the same time, so the route does not have to be sized to accept the full bandwidth from all user at the same time.

Since ATM is quite flexible it can be applied not only to the long range network services, but also to LANs ( Local Area Networks) and WANs (Wide Area Networks). In fact, ATM is gaining a lot of interest as a backbone technology to interconnect already existing Ethernet segments or Token ring LANs within medium and big companies facilities. This solution not only can provide a fast communication channel, but will also ensure future compatibility with the public broadband telecom network. On the other hand ATM will pave the way for new services like videoconferencing, video on demand, etc.

Although ATM does not impose any restriction on data rates it is usually associated with high speed communications. ATM can work at low data rates, but statistical gain can only be achieved when the data from a large number of users is aggregated. Another reason for the usage of high speeds is that ATM allows to integrate not only voice and data, but also multimedia, and those signals usually require a high bandwidth. Typical speeds for ATM are 51 Mbit/s in the low end, 155 Mbit/s as typical end user access interface and then 622 Mbit/s, 1.2 Gbit/s and 2.4 Gbit/s internally inside the network. To be able to reach those speeds, the switching of cells is done in hardware instead of by software (as it is common in conventional digital switches of voice or data).

Both transmission and switching equipment in ATM need to use the latest semiconductor technology to be able to obtain the speed requirements. To reduce those speed requirements, usually data is treated in parallel buses, that are only converted to serial to transmit the information through the optical fibers. Even in this case, is quite normal the use of submicron CMOS, BiCMOS and ECL circuits, that operate at several hundreds of MHz. A typical ATM switch is composed of a large number of very complex ASICs that perform all the switching, routing, and monitoring functions.

The typical integrated circuits for ATM applications are in the range of 100 to 300 pins, with up to 700 K transistors, and operating frequencies up to 150 Mhz for CMOS, about 300MHz for BiCMOS and up to 2.5 GHz for pure bipolar (in this the transistor count should be limited to about 50K). The high number of pins and high power dissipation force the use of rather big packages (like CPGAs) for the CMOS and BiCMOS devices. Those big packages tend to have quite high pin inductance, and a very poor area efficiency (typically less than 10%).

MCM technology can help obtain at the system level the benefits of the silicon high speed potential, allowing a higher system performance by reducing I/O capacitance and inductance. And also allowing to confine the most complex parts of the system in a module, thus reducing the required complexity of the PCB, and at the same time simplifying the reusability of the technology.

## 5. MCMs from a system perspective

As has been explained above, several telecom systems can obtain high benefits from the use of MCMs interconnect technology. There are however some areas that might be slowing down the usage of this technology.

The adoption of the MCM technology resembles the introduction of ASICs in the systems. At the beginning system designers view ASICs as an "exotic" solution, and were concern about the development cycles, and overwhelmed by a large number of technicalities and options that didn't really understand. The situation now with MCMs is somehow similar. Most system designers are faced with a huge number of technical solution within the MCM technology: there are different materials for the substrate ranking from FR4 to silicon, then there are a number of possibilities for first and second level interconnect, module mechanical constructions present even more options... And all decisions taken can drastically affect the price and performance of the system.

From the design methodology, MCMs also present some new challenges to system designers. Preparing the module test can be difficult depending on the selected options for the design. Structured testing techniques as boundary scan can be useful but, what about the chips that do not provide it?. Although functional simulation of the module is a must it cannot be exhaustive, so problems might be found that force to test inside the module and maybe to change some connections, in a MCM that is more difficult than in a board. On the other hand if the microprocessor of the system is included in a MCM with some more chips, using an emulator for the software debugging is not possible. Some questions may arise like: should I make a breadboard with packaged chips before committing to the final MCM? (and run it possibly at a lower speed), is this going to cause a longer development cycle for my system?

Apart from the purely technical aspects, MCM present some new challenges for the procurement point of view. Few companies have been involved previously in the purchase of bare dies, and is not clear how the interface with the vendor should be setup. Testing the naked dies at speed it not easy, neither it is the burn-in. This is worse for the commercial circuits, since there are still few vendors that provide a naked chip option. All those aspects are simpler to solve for the vertically integrated companies, that develop their own chips, their own MCMs and boards, to be mounted in their systems.

The fact that MCM will have a significant impact in the future systems is absolutely clear. However it is not so clear in which form this technology will be introduced. Some of the advances offered by MCM developments might hit the market sooner than others helping in the introduction of the technology, this seems to be the case for example for BGA. Interim solutions should be found that enable system houses to get access to this new technology minimizing the entry risks. Using again the ASIC model, maybe the solution is a "semicustom" MCM technology that can help develop the market. This solution could be the use of BGA subcarriers to alleviate test problems, or it could be based on the development of "standard" MCM packages so that only interconnect must be defined by the user.

# DESIGN AND REALIZATION OF A MULTICHIP MODULE AS A MOTION ESTIMATOR FOR HDTV-APPLICATIONS

S. FAZELPOUR*, B. BÖLIKE**, R. BRODOWSKI*, R. DÜMCKE*, H. REICHL*
*FhG-IZM, Gustav-Meyer-Allee 25, 13355 Berlin, Germany, Fax: +4930 46403111
**Deutsche Bundespost Telekom, Forschungs- und Technologiezentrum Forschungsgruppe FZ 143, "Neu Fernsehübertragungssysteme",
Postfach 8, 12473 Berlin, Germany, Fax: +4930 67082179

**KEYWORDS/ABSTRACT:** multichip module/chip-and-wire/polymer laminate technology/ground bounce/cross talk/thermal simulation/motion estimator/hdtv

In this paper the design and realization of a multichip module as a motion estimator for HDTV applications will be shown. The substrate technology used is MCM-L with polyimide as its dielectric material. The module has an area of 110 x 125 $mm^2$. There are 32 bare chips mounted on this module, three of which have approximately 300 pins with a pitch of 120 μm. The chips are wire bonded on the module. The module will be mounted face down on the motherboard. The first prototypes proved to be functional.

## 1. Introduction:

Multichip module (MCM) technology provides to the system designer an alternative to realize more compact electronic systems with better electrical performance, higher reliability and better cost/performance ratio than with conventional technologies. To realize a MCM one requires /1/, /2/, /3/:

a) supply of fully tested bare dies
b) choose a cost effective and appropriate

I) chip contact method
II) substrate technology
III) connector

c) careful electrical design
d) well planned heat removal strategy
e) choosing materials with matched thermal expansion coefficient (α)
f) a test friendly module design

Because the supply of Known Good Dies (KGDs) from chip manufacturers at least for ASICs is pretty rare typically the first step is not an easy one. In our case there are five different types of chips placed on the module, three of which are ASICs and we were confronted with the chip testing problem.
The question of the most appropriate chip contact method can be answered depending on the specific application of the module. Usually for high frequency environments flip chip is the matter of choice. If burn-in test for unpackaged chips is needed TAB offers advantages for it. If these special requirements are not present, chip-and-wire can be used as a simple and well established method. That is why we decided here for it.



*W. K. Jones et al. (eds.), MCM C/Mixed Technologies and Thick Film Sensors,* 153–167.

One important parameter influencing the performance and cost of multichip modules is the choice of substrate technology.
MCM-L typically results in a cheaper module compared with other technologies, but it is usually not the most compact choice specially because of its poor via design rules. With the aid of non mechanical via processing methods MCM-L has narrowed significantly its gap with MCM-D regarding the packing density. The MCM-L technology used in this work is the Dycostrate™ technology (a trademark of DYCONEX). It will be shown that for our complex module design, considering all the electrical requirements, two signal layers and two power, ground layers were sufficient to generate the layout of the module. In all other technologies considered, a minimum of four layers was also required.
Optimum electrical design specially for modules with critical lines or EMC sensitive systems is still a tedious work. With clock frequencies steadily increasing, performing complex two and three dimensional field simulations become a natural part of electrical simulations /4/. A popular approach, which is also followed here, is first to calculate the RCGL matrix of line parameters from the field simulations. In the next step, from these data lumped elements for modelling of interconnects can be gained. A network simulator in time domain can then show the quality of signal transmission between the driver and receiver chips /5/.
For making a good thermal design powerful software tools are available on the market /6/. But the problem is usually to have access to the real values of the power dissipation of the chips in order to avoid the overdesigning of the heat sink of the module. The maximum values delivered by the IC manufacturers are often only valid for extreme conditions of operation.
The issue of mechanical design and the selection of $\alpha$-compatible materials for the assembly of the module must specially be addressed in the military and harsh environment applications. This was not the case for this prototype so that costly and tedious precautions were avoided here.
Regarding the testability of the module one must keep in mind that the more complex a module becomes the more cumbersome the test procedure will be. Alone because of the number of the test points on our module of about 2500 in a module area of 110 x 125 $mm^2$ a complex test vehicle had to be developed for our module.

## 2. Principle of the motion estimator module

Figure 1 shows the data flow in the motion estimator /7/. The module performs a block matching function. The main purpose of the module is to compare HDTV frames at different times in order to detect the motion of pixel blocks within the frame segments. This information is used for data compression to reduce the amount of information to be transmitted.
The area of the two fields on which the comparison will take place is externally adjustable through a programming interface. The output signals are a (translational) vector, along which the sum of the difference in the gray scale of all the pixels available in the search pattern (DFD) is minimum, and this minimum DFD-value.
The MEP chips serve as a controller for the module. Through the circuitry they can operate in two different modes called AM and MP. The PAMEs are programmable processor arrays, that work in a pipeline way and perform the comparison between the two frames. MEP-AM chips manage the arrived data and take care that the PAMEs get the right part of the frame. The memory modules that the MEPs communicate with, consist of two (128k x 8 Bit) SRAMs and one FIFO with an organization of 1296 x 8Bit.
MEP-MP chip receives the DFD values that were calculated from PAMEs, determines for each pattern the minimum DFD value and delivers as a result this minimum DFD together with the related vector to the receiver outside the module. The data rate of all the paths in figure 1 except that of programming interface is 20 Mbyte/s.

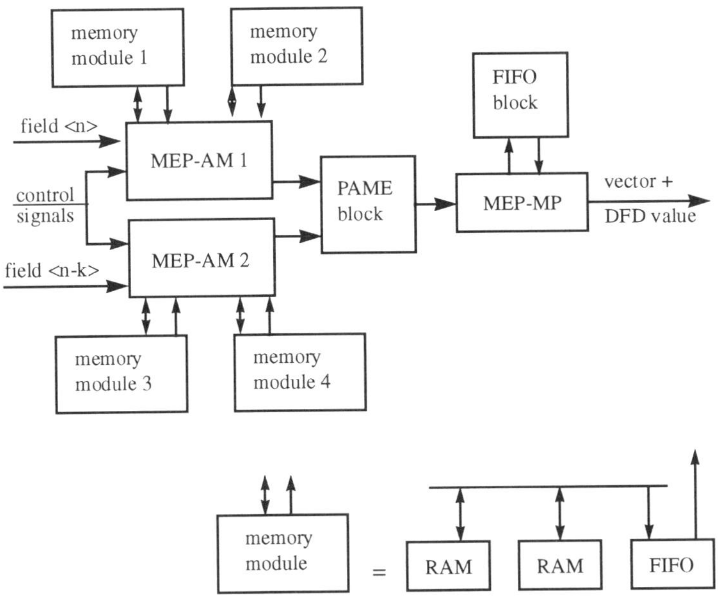


Figure 1: Principle of the motion estimator module.

### 3. Characteristics of the realized module

Table 1 shows the characteristics of the chips used in this module.

Table 1: Integrated circuits placed on the module

| | FIFO | SRAM | PAME | MEP | Driver (FCT244) |
|---|---|---|---|---|---|
| Technology<br>Manufacturer | CMOS<br>ITT | CMOS<br>Micron | CMOS<br>ITT | H4C<br>Motorola | CMOS<br>IDT |
| Number | 9 | 8 | 8 | 3 | 4 |
| Geometry [mm] | 4,93 x<br>4,45 x<br>0,28 | 6,12 x<br>13,82 x<br>0,5 | 9,6 x<br>12 x<br>0,4 | 9,93 x<br>9,93 x<br>0,45 | 1,44 x<br>1,95 x<br>0,5 |
| Number of signal / (power + ground) pins | 22 / 12 | 29 / 7 | 93 / 36 | MP: 217<br>AM1: 135<br>AM2: 139<br>/ 74 | 18 / 2 |
| Pad pitch (approx.) [µm] | 150 | 200 | 200 | 120 | 150 |

Table 1: Integrated circuits placed on the module

| | | | | | |
|---|---|---|---|---|---|
| Max. power dissipation [mW] | 250 | 700 | 1500 | 2500 | 75 |
| Max. allowed junction temp. [°C] | 110 | 105 | 110 | 105 | 105 |
| Number of simultaneous switching output buffers per chip (per GND pin) | 8 (8) | 8 (4) | 41 (8) | 100 (5) | 8 (4) |

Among these chips MEP is from different aspects one of the most critical one. It has the most number of pins, the shortest pin pitch and the highest power dissipation.
FIFO and PAME can have 8 simultaneous switching output buffer which makes them the most critical chips regarding the ground bounce problem. All these chips are commercial products so the junction temperature of them should not exceed 105-110°C.
Table 2 shows specifications of the module.

Table 2: Module specifications

| | | |
|---|---|---|
| 1.1 | Clock frequency | 20 MHz |
| 1.2 | Min. signal rise time | 1 ns |
| 1.3 | Number of nets | approx. 500 |
| 2.1 | Number of chips | 32 |
| 2.2 | Total number of chip contacts | approx. 2500 |
| 2.3 | Total number of external contacts | approx. 300 |
| 3.1 | Substrate technology | 4 layer Dycostrate™ |
| 3.2 | Substrate area | 110 x 125 $mm^2$ |
| 3.3 | Minimum line width (surface layer / signal layer) | 80 / 100 µm |
| 3.4 | Minimum line spacing (surface layer / signal layer) | 90 / 150 µm |
| 3.5 | Via diameter (hole / pad) | 90 / 250 µm |
| 4.1 | Chip contact method | aluminum wire bonding |
| 4.2 | Chip mounting | silver filled epoxy adhesive |
| 4.3 | Encapsulation | aluminum cover (nonhermetic) |
| 5.1 | Power dissipation | 28 W max. |
| 5.2 | Max. environment temperature | 45°C |
| 5.3 | Max. junction temp. of the Chips | 105°C |

Table 2: Module specifications

| | | |
|---|---|---|
| 5.4 | Heat sink | aluminum with ribs for forced convection cooling |
| 6.1 | Module mounting | flat, face-down |
| 6.2 | Module connector | 2 Cinch, 249 pins each, 1.27 mm, pressure contacts |

One of the most important data for the careful layout of the module is the minimum rise time of the output signals of the chips. Because of a rise time as low as 1 ns careful electrical design of the module is necessary.
The existence of 500 nets on the module together with high pin count and low pin pitch chips imply a foreseeable complex layout for the module.
The chips were dynamically tested on their wafer prior to mounting on the module, but they are not burn-in tested. Because there are 32 unpackaged chips on the module, the issue of repair is actual here.
There are approx. 2500 bond pads on a module with an extent of 110 x 125 $mm^2$. This makes the matter of testing of the module pretty difficult. A complex test adapter was developed to test the module with respect to connectivity of interconnects.
The module has approx. 300 external contacts consisting of about 150 signal pins, about 100 test pins and 50 power and ground pins. The test pins were necessary to trace the eventual errors in the prototypes both from the circuit design and from the assembly aspect. To cope with this number of necessary module pins two Cinch connectors were chosen each with 249 pins and with pressure contacts.

## 4. Package Design

A cross section of the module is shown in figure 2.

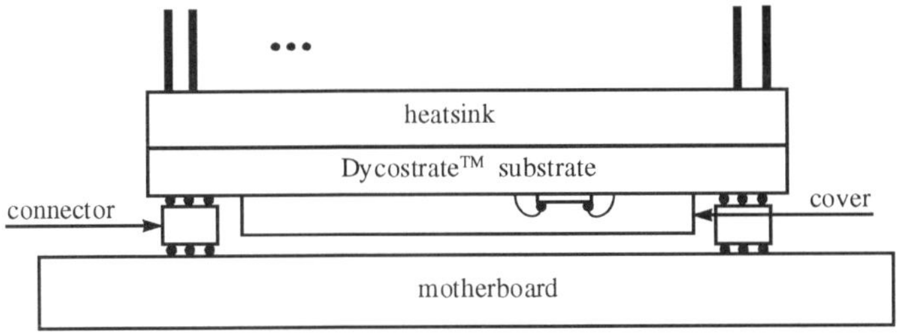


Figure 2: A cross section of the module.

The module itself consists of 0.2 mm laminated substrate on an aluminum heat sink. The bonded chips are protected nonhermetically with an aluminum cover which is screwed to the heat sink. The module is mounted face down on the motherboard through 2 Cinch connectors with an height of 7 mm. The connectors are screwed on the motherboard as well as on the module. The cover is designed with an height of 5 mm to allow the placement of some passive components below it on the motherboard.

The use of heat sink is also from another aspect necessary that is to make the substrate mechanically stable and bondable. Usually copper/invar/copper or copper/molybdän/copper is chosen as a stabilizer on the back side of this kind of substrate with a thermal expansion coefficient similar to that of silicon. But the maximum thickness of them is practically limited to about 2 mm and they increase considerably the weight of the module. The simultaneous use of these alloys together with the aluminum heat sink is not possible. Because the difference between their thermal expansion coefficient to that of aluminum causes the bending of module during the lamination process. The commercial environment of the operation of the module allows us to avoid the use of these alloys and accept the coexistence of aluminum heat sink with the silicon chips regardless of the difference in their thermal expansion coefficients.

## 5. Process flow for Dycostrate™ manufacturing

Figure 3 shows the process flow for manufacturing of two routing layers.

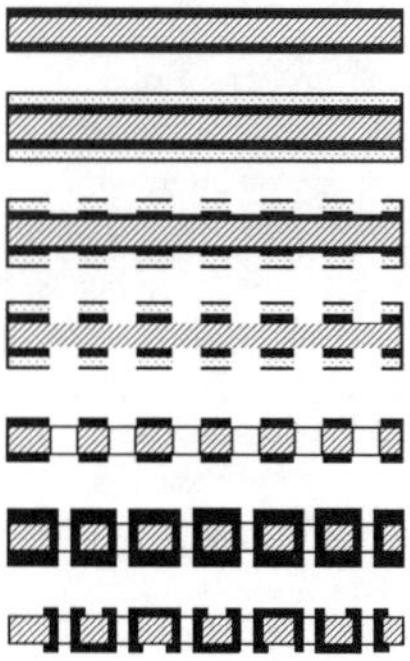


Figure 3: Process flow for Dycostrate™ manufacturing of two layers.

The starting point of manufacturing of our module was a double sided copper cladded polyimide with a nominal thickness of about 50 μm. Initial thickness of cladded copper is 3-5 μm. At the via sites copper is opened with a usual resist process which is then exposed to $O_2$-plasma etching in a barrel etcher. A following electrodeposition of copper in a galvanic bath deposits metal on the wall of the via structures and results in a final copper line thickness of about 20 μm. Patterning of metal on both sides results in the signal lines in x- and y-direction.

Manufacturing of a substrate with four metal layers is similar. On the already processed two layers two additional layers are laminated from both sides, which are one sided copper cladded and on the other side carry an adhesive layer (in this case epoxy adhesive). To create the vias the same procedure will be followed as described earlier. After finishing the processing of the vias, the pad side will be patterned and electroless deposition of nickel (about 4-5 μm) and gold (about 0.2 μm) will finish the processing of the end layer necessary for wire bond contacts.

## 6. Electrical simulation

Electrical simulations were performed to investigate whether the used design rules in

table 2 suffice to guarantee a signal noise below 500 mV for the critical lines as is required from the noise margin between the driver and receiver chips.
Two main causes of noise are signal coupling and ground bounce, which in the worst case can add up. For the simulation of signal coupling the lumped element approach was used.
First we consider the coupling effect. The line configuration shown in Figure 4a was used in the simulation. A transmission line model for the lines was used, because the rise time of the signal can become comparable to the time-of-flight delay /8/. Assuming that the middle line is a critical one, through the simulations we want to determine the voltage peek $V_L$ in the critical line as the driver of the two adjacent lines makes a $0 \rightarrow 1$ transition. We chose this transition because in our case it results in a steeper signal on the lines than the $1 \rightarrow 0$ one. The model of the inverter output buffer shown in figure 4a was available as SPICE parameter and it represents one of the critical chips on the module. Two dimensional field simulations were performed to calculate the parameters of the signal lines and the coupling coefficients from the adjacent lines with a program called LINE2d developed at TU-Berlin /9/. The static circuit model for one lumped element corresponding to a line length of 0.91cm is shown in figure 4b.

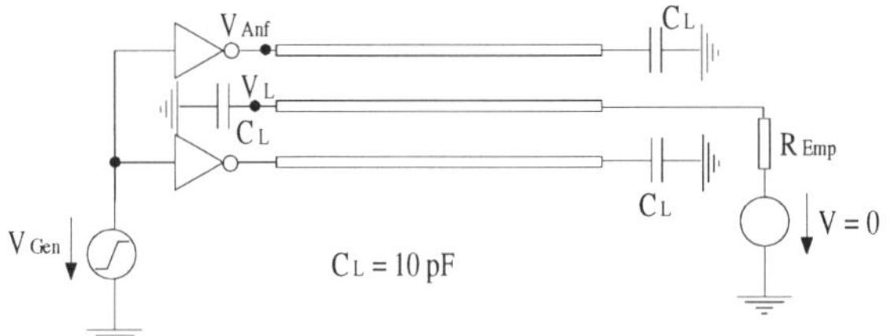


Figure 4a: Line configuration for the calculation of signal coupling noise.

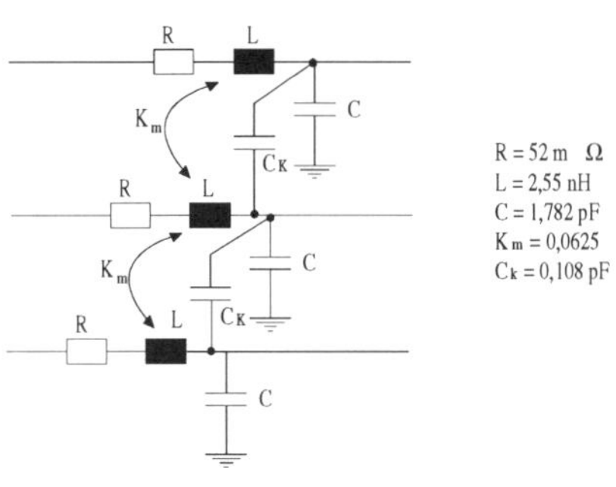


Figure 4b: Static circuit model for one lumped element (0.91 cm).

Because of the high thickness of the copper lines there exists a strong skin effect. Two dimensional simulations show that from 1 MHz to 100 MHz the resistance of the line increases more than twice. The worse case of signal coupling can be seen in the case that skin effect is not considered so that for these investigations just the DC resistance was used. The coupled noise for a line length of 6.37 cm consisting of 7 lumped Elements is

shown in figure 5a.
The voltage peek and the pulse width of the coupled noise as a function of coupled line length is shown in figure 5b.
It has a typical characteristic of a backward coupling effect. The forward coupling is not available because the dielectric of the strip line is almost homogeneous /10/. From this figure it is deducible that the contribution of coupling effect on the signal noise does not exceed 300 mV for the lines on the module.

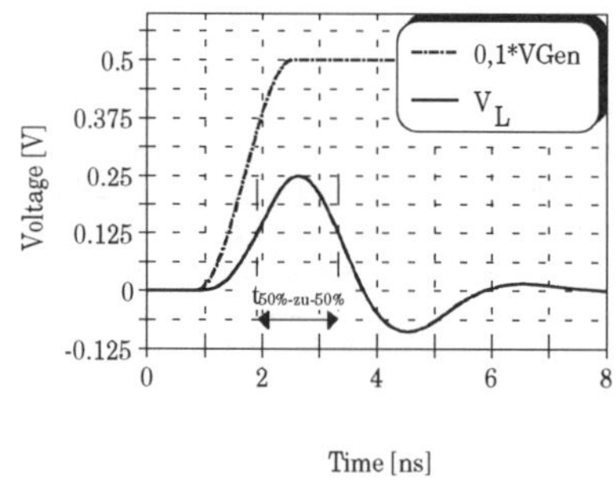


Figure 5a: Waveform of signal coupling noise for a line length of 6.37 cm (7 Elements).

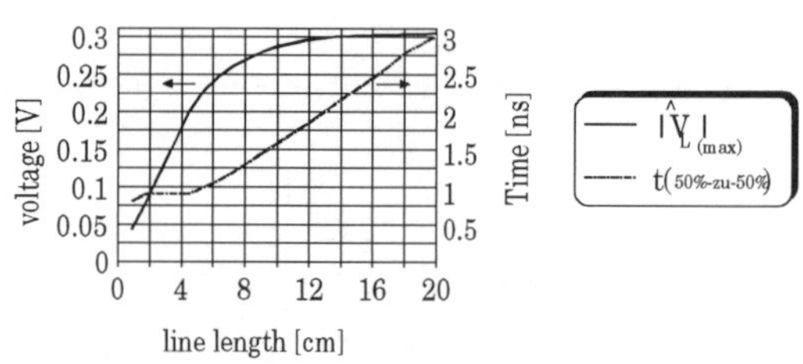


Figure 5b: Voltage peek and pulse width of noise as a function of line length.

For the estimation of an upper limit for ground bounce noise we consider the worst case of eight parallel switching output buffer which can be the case for some chips as was shown in table 1. Figure 6 shows the model and the simulation results.
The bond wires are modelled with an inductance of 1nH. The inductance of ground plane is estimated not to exceed 0.2 nH and that of power in the pad layer to be four times worse than ground path.
Between node 10 and 20 we modelled the decoupling capacitors used for these Chips. With 50 pF load capacitance of driver chip we took into account the maximum capacitance of interconnect connected to the driver chip plus the input capacitance of the receiver chip.
Figure 6 shows that the contribution of the ground bounce does not exceed 200 mV so that the sum of the contribution of coupling noise and that of ground bounce is kept with the design rule of table 2 below 500 mV for all critical lines of the layout. The measurements performed on the module show also no sign of violence of this noise margin.

model:

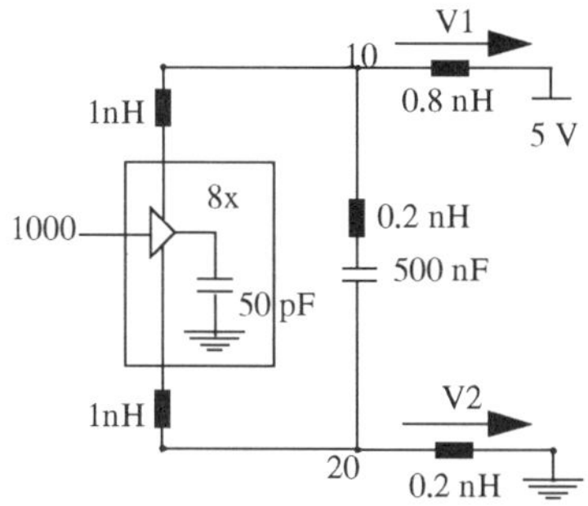


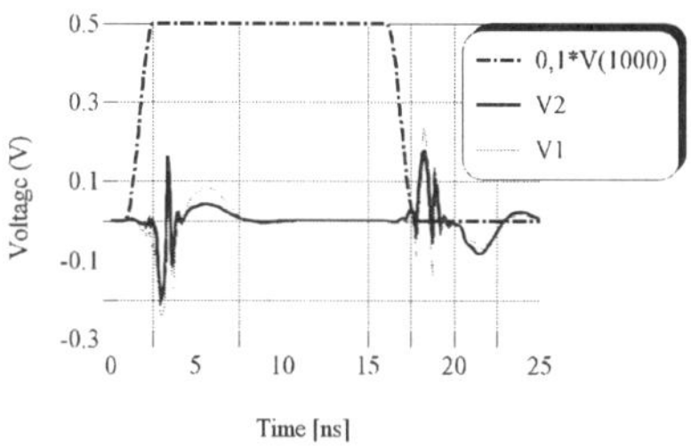


waveform

Figure 6: Simulation of ground bounce for the module.

## 7. Layout

The placement of the components can be seen from the layout of the pad layer shown in figure 7. The PAME block is situated on top of the module. MEP-MP with its five FIFOs is located on the bottom left side and the MEP-AMs with their memory modules fill the rest of the area. On the left and right side of the module the foot print of the connectors can be seen. The three holes shown in the figure outside the layout area are used to align the substrate relative to the heat sink.

The layout of the module was accomplished with two routing layers, one pad and one ground layer. On the pad layer all the areas under the chips are metallized and have fixed potential. The unused area of this layer is metallized and connected to power pins. This layer is not used for the routing of the module. As can be seen from the figure the pad layer is pretty well covered with metal connected to fixed potential. As a consequence the signal lines in the layer 2 and 3 have strip line structure and because of the symmetry of the substrate they have almost equal characteristic impedance of 30 Ω.

The design rules for the layout were already shown in table 2. The contact pads of the MEP chips are the only traces with a width of 70 µm and a spacing of 80 µm. To provide enough pads for this chip two rows of pads for each side of the chip were necessary with a spacing of 0.5 mm between them. The pads itself are also 0.5 mm long to provide enough reserved area for the contact of the substituted chips in the case that repair of the chips is necessary. The Layout was done with the board station software of MENTOR GRAPHICS. The routing was performed almost entirely with its autorouter. The routing was very dense and good distributed throughout the module area.

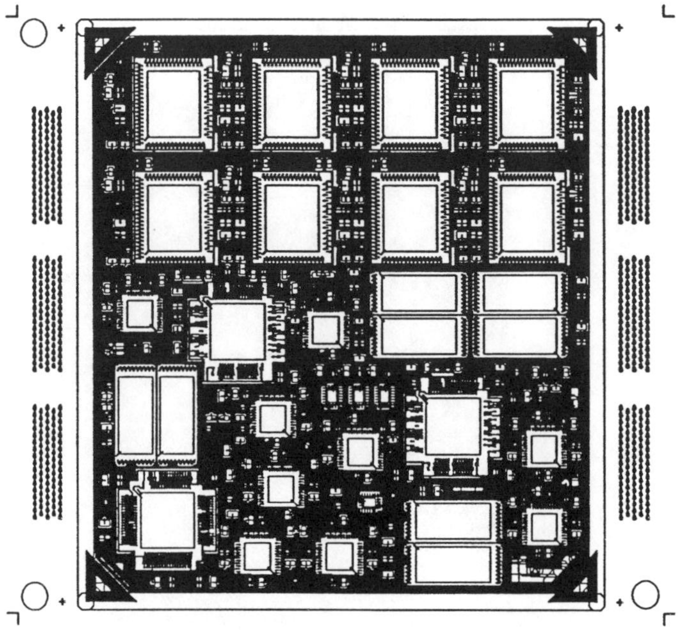

Figure 7: Layout of the pad layer of the module.

**8. Thermal simulation**

Thermal simulations for the module were performed to determine the necessary convection conditions and dimensions of the heat sink to cool the module.
The required conditions to hold are
1) to keep the temperature difference between the module and the environment below 60 K and
2) if forced convection is necessary to design a heat sink that cooling of the module with an air velocity less than 3 m/s is possible.
The thickness and the thermal conductivity of the materials used in the thermal simulations are shown in table 3.
The heat sink has a base thickness of 5 mm, a rib height of 15 mm, a channel width of 5 mm and a rib thickness of 1 mm. The model used for the thermal simulations is shown in figure 8.
The spreading effect of the metallization in the multilayer was only considered in the ground plane. Because of the face down mounting of the module on the motherboard only heat removal from the back side of the module was taken into account. The simulations were performed with the program FFTEMP developed at TU-Berlin which can handle the three dimensional structures in the assembly /6/. This program is based on Fast Fourier Transformation.
The three dimensional temperature distribution on the module simulated with this pro-

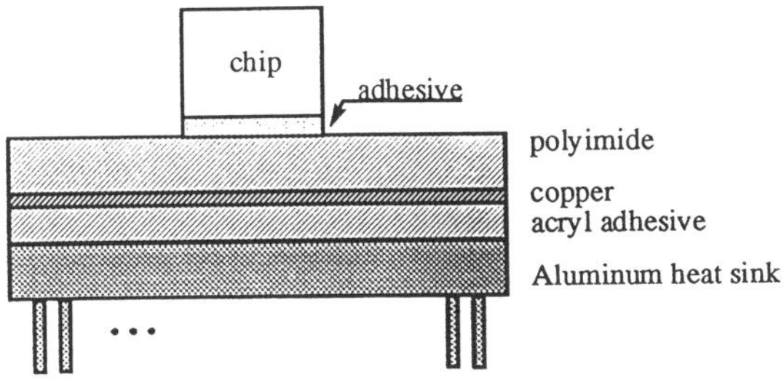


Figure 8: A thermal model of the module.

gram for a forced convection with an air velocity of 2.6 m/s is shown in figure 9. It can be clearly seen that the temperature peek is created by MEPs. A temperature difference of 25 K is available on the surface of the module which is mainly because of the low thermal conductivity of thick polyimide layers.

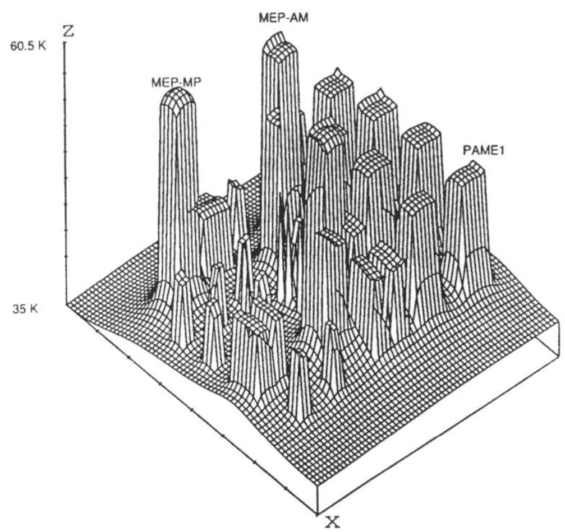


Figure 9: A three dimensional thermal simulation of the module.

**Table 3: Geometry and material data used in the thermal simulations**

| No. | Material | Thermal conductivity [W/mK] | Thickness [mm] |
|---|---|---|---|
| 1 | Silicon chip | 150 | 0.28 - 0.5 (see table 1) |
| 2 | Silver filled adhesive | 2 | 0.1 |
| 3 | Polyimide | 0.1 | 0.15 |
| 4 | Copper (ground layer) | 390 | 0.02 |

**Table 3: Geometry and material data used in the thermal simulations**

| No. | Material | Thermal conductivity [W/mK] | Thickness [mm] |
|---|---|---|---|
| 5 | Acryl adhesive | 0.1 | 0.05 |
| 6 | Aluminum base layer | 210 | 5 |

## 9. A cross section of the realized module

A cross section of the realized module was made to investigate geometry and profiles of the materials. Figure 10a shows top view of a track that ends in a via in the signal layer.

Figure 10a: Top view of a track and a via (magnification 205x).

This is the worst misalignment of via with respect to its pad that could be found on the module. It can be seen that a rest ring is available all around the via. No increase in line resistance was observed because of via misalignment. Regarding the extent of the module the degree of achieved alignment between the layers is good.
A cross section of a via connecting the pad layer with the first signal layer is shown in figure 10b. We see the isotropic etch behaviour of plasma barrel etcher in the triangle etch profile that it has left on polyimide. This profile could well be (over)compensated by the galvanically deposited copper layer. The step coverage of copper on polyimide is good and no void could be detected in this and other via cross sections. The via opening after metallization was about 75 μm in the middle and about 60 μm on the edges. From this figure it is obvious that even a rest ring of zero at a specific point on the pad perimeter could be tolerated. Figure 10c shows a cross section of a signal line on pad layer. The thickness of the copper layer was measured 20 - 22 μm and that of Nickel 4-5 μm. The

flush gold cannot be seen in the figure.

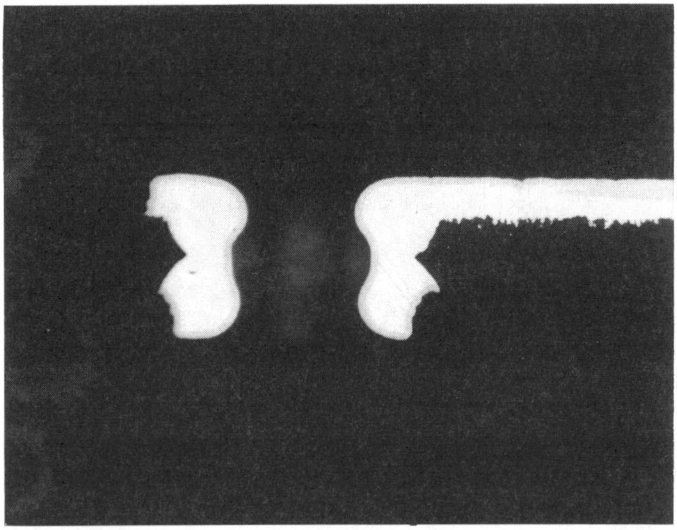

Figure 10b: Cross section of a via (magnification 260 x).

Figure 10c: Cross section of a track (magnification 670 x).

The roughness of the copper layer is very high and can be estimated to be up to 6 µm. We expect that although the roughness of copper has advantage for its adhesion strength on polyimide, for high frequency applications it can result in enhanced losses and perhaps not well defined impedance of lines. But in this technology polyimide with other depth of roughness of copper surface are also available and can be used if needed. In figure 10c copper has a steep angle of about 70° at the edges which is a sign of good controlled etch process. For the polyimide layer between both signal layers we have noticed in the photos of cross sections (not shown here) that sandwich layers of polyimide / (polyimide) adhesive / polyimide with a thickness of 16 / 8 / 16 µm exist which results in a total polyimide thickness of about 40 µm for this layer. The mentioned adhesive layer has the same dielectric constant as polyimide. The advantages of polyimide adhesives are stated elsewhere /11/.

## 10. Conclusion:

A multichip module as a motion estimator for HDTV applications is designed and realized, which is one of the most complex MCM of its own. A photograph of it is shown in figure 11.

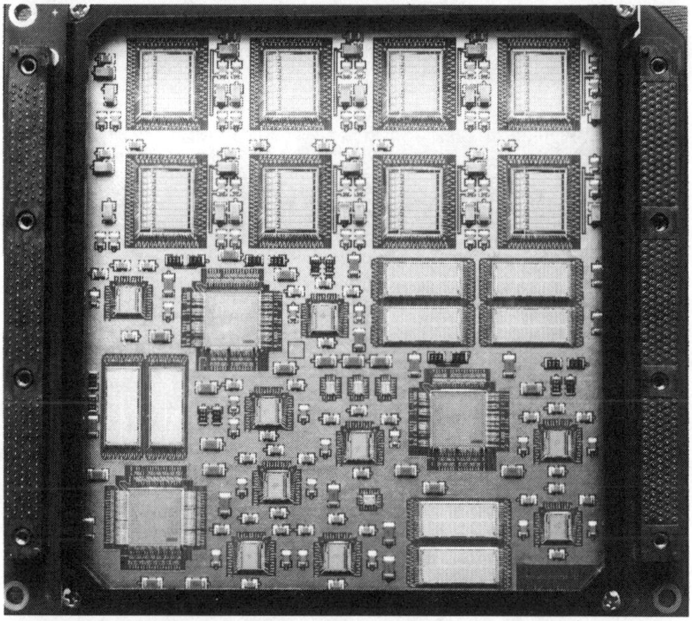

Figure 11: A photograph of the realized module.

The module is based on Dycostrate™ technology, which is a MCM-L one, and the chips are wire bonded on the module. 32 unpackaged chips together with their passive components and the connectors are placed in a module area of 110 x 125 mm². There are chips with approx. 300 pins and a pin pitch of 120 µm. The dielectric of the substrate is polyimide with via holes less than 100 µm in diameter.

Electrical simulations show that the design rules based on the minimum technological values still guarantee a noise peek of less than 500 mV as needed from the noise margin

between the driver and receiver chips.
Based on thermal simulations a heat sink with ribs was designed, which meet the thermal requirements of the module.
Thank optimum electrical and thermal design performed, the first prototypes of module proved to be functional.

## 11. references


/1/ Tummala, R.R. and Rymaszewski, E. J. (1989) *Microelectronics Packaging Handbook*, Van Nostrand Reinhold, New York.

/2/ Messner, G., Turlik I., Balde, J. W. and Garrou, P. E. (1992) *Thin Film Multichip Modules*, International Society for Hybrid Microelectronics, Reston, Virginia.

/3/ Harper, C. A. (1991) Electronic Packaging and Interconnection Handbook, McGraw-Hill, Inc., New York ...

/4/ Lam, C.-W., Ali, S. M., Nuytkens, P. (1993) Three Dimensional Modeling of Multichip Module Interconnects, *IEEE Transactions on Components, Hybrids, and Manufacturing Technology* **16**, 699-704.

/5/ Gong, S.-F., Strandberg, J., Thiede, H., Hentzell, H., Hesselbom, H. and Karner, W. (1993) Investigation of High-Speed Pulse Transmission in MCM-D, *IEEE Transactions on Components, Hybrids, and Manufacturing Technology* **16**, 735-742.

/6/ Reil, Martin (1994) *Simulations- und Berechnungsverfahren zum rechnerunterstützten thermischen Entwurf von Multichip-Modulen*, PhD Thesis at the Technical University Berlin, Verlag Shaker Aachen.

/7/ Manck, O. (1992) *Untersuchung und Entwicklung der Architektur und Schaltung für ein Modul sowie des zugehörigen Motherboards zum Aufbau einer Blockmatching-Unit für einen TV-Bewegungsvektorschätzer*, final report, Institut für Mikroelektronik, Technical University Berlin.

/8/ Bakoglu, H. B. (1990) *Circuits, Interconnections, and Packaging for VLSI*, Addison Wesley, Massachusetts. ...

/9/ Line2d user´s Guide (1992), Version 2.1, Technical University Berlin.

/10/ Seraphim Lasky and Li (19..) *Principles of Electronic Packaging* McGraw-Hill, Inc., New York ...

/11/ Mittal, K. L. (1984) *Polyimides Synthesis, Characterization, and Applications*, Volume 2, Plenum Press, New York and London.


## 12. Acknowledgements


We are truely in debt of our project partners: Mr. J. Reimers from "Deutsche Bundespost Telekom", Mr. C. Stoffers from "Technical University Berlin" and Mr. Talmi from "Heinrich Hertz Institut für Nachrichtentechnik in Berlin" for their cooperation. Without their contributions this work could not be accomplished.
This work was financed by "Deutsche Bundespost Telekom".

# MULTI-CHIP MODULE APPLICATIONS IN SATELLITE COMMUNICATIONS

**Dr Nihal Sinnadurai**
*Professor of Electronics Technology,*
*Middlesex University Microelectronics Centre, England*

## 1. The Challenge and The Opportunity

The use of Electronics Technologies in Space pose interesting but manageable challenges and significant opportunities for the use of Multi-Chip Modules (MCMs) and low cost packaging technologies - whose properties offer significant advantages in space and locations involving exposure to radiation. Communication via satellites located in space require performance at high frequencies in the microwave range (delivered by silicon or GaAs MMICs- monolithic microwave integrated circuits - and controlled by MCMs), hardware technologies which are compact and lightweight and will withstand the acceleration forces during launch and manoeuvre into orbit, semiconductor device and packaging and interconnection technologies which will withstand prolonged exposure to radiation, and an extremely high reliability because of the high telecommunications traffic carried and the enormous cost of repair if a failure should occur.

The performance requirements for the communications systems range from low gigaherz proposed for terrestrial mobile communications using low polar orbit satellites to many gigaherz for many channels of international trunk traffic using geostationary satellites. The reliability requirement is to achieve the equivalent of zero failures in 25 years. This would be statistically described as a particular low probability of failure bounded by confidence limits (governed by the component count and system Fault Tree Analysis), but the practical approach today is to achieve the right build of components and system, establishing confidence at each level.

## 2. The Global Communications Market

Communications is a rapidly expanding market and the total equipment market will grow from $55 billion in 1993 to $86 billion by 2000, subdivided as follows:



*W. K. Jones et al. (eds.), MCM C/Mixed Technologies and Thick Film Sensors,* 169–176.

| ***COUNTRY*** | ***1993*** | ***2000*** |
|---|---|---|
| | *$billion* | *$billion* |
| *USA* | *20* | *31* |
| *EUROPE* | *16* | *25* |
| *JAPAN* | *6* | *10* |
| *REST OF WORLD* | *13* | *20* |

Satellite Communication is a small fraction of the total market, but a number of opportunities are being explored to exploit the explosive growth in mobile communications, particularly with the potential arising from global personal mobility. Standards are becoming established in different geographic locations, and multi-standard lightweight personal communicators are being developed.

## 3. Opportunities for MCMs

The immediately evident opportunity for MCMs is to provide the compactness and light weight required of the payload of communications satellites.

Typical payload benefits obtained from the move to MCMs are:

| ***PARAMETER*** | ***SAVING*** |
|---|---|
| *VOLUME* | *6 x REDUCTION* |
| *WEIGHT* | *10 x REDUCTION* |
| *COMPONENTS* | *2 x REDUCTION* |
| *POWER* | *40% REDUCTION* |

Further opportunities arise from the potential for 3 dimensional interconnection and packaging being developed in Europe and in the USA. The European Trimod research project within the European Community ESPRIT research programme is indeed well advanced with collaboration led by Thomson-CSF between 6 partners. The concept makes use of IC dice bonded to Tape Automated Bonded (TAB) frames, protected by "Glob Top" (an inadequate terminology), stacked into a multilayer stack glued together with epoxy resin to form a 3D block which is then trimmed to expose the edges of the TAB fingers and then electroless plated and etched to form the interconnection pattern and package connection pads (**Figure 1**) (Courtesy of Thomson-CSF)

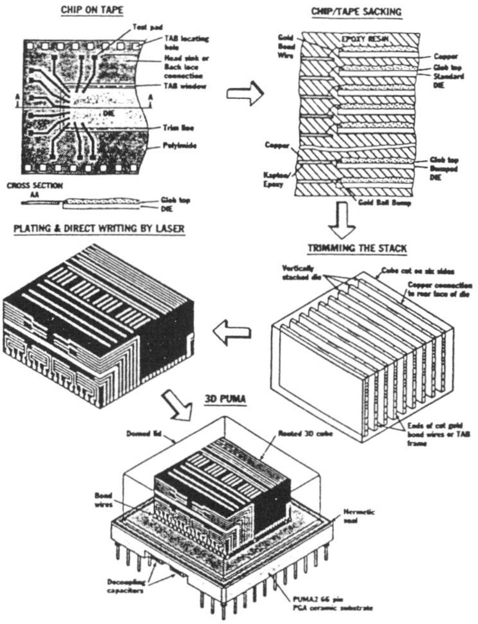


*Figure 1. 3D INTERCONNECTION*

The project intends to deliver 3D memory and 3D transputer blocks. A major payload benefit is that 3D assembly achieves more than a 100% 2D substrate area efficiency (**Figure 2**).(Courtesy of Thomson-CSF).

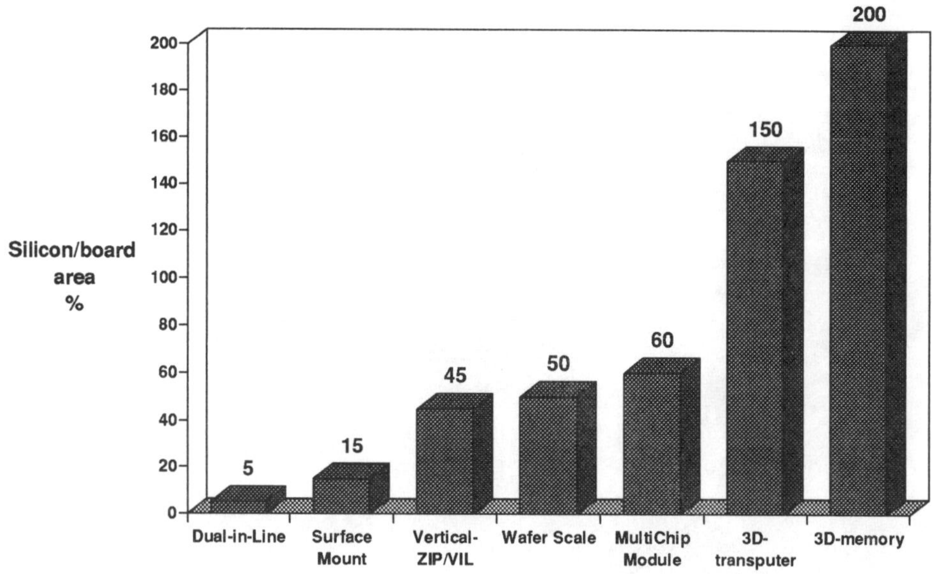


***Figure 2. Efficiency of Use of Substrate Area (= Active Silicon/Circuit Substrate Area)***

The reliability of the technology has yet to be evaluated. Further advantages are the resilience of the assembly to the acceleration forces, both because of the inherent robustness of TAB connections and the support of the connections by the Glob Top and the adhesive.

The Reliability potential arising from the use of plastic encapsulation is very high indeed, and can exceed that of hermetic packaging in tropical climates. Of course space is quite a different environment, with the hostility arising from launch and manoeuvre stresses and from radiation. Nevertheless, plastic encapsulation by the use of junction coatings (compliant thin coatings directly applied to the surfaces of the dice) offer very high protection of semiconductor devices. Glob Top is an unfortunate terminology, because it has been associated with low technology low reliability applications.

## 4. Reliability Opportunities

Reliability engineering of plastic encapsulation was pioneered during 1975-1981 at British Telecommunications Laboratories (BT Labs), England, with the active participation of major plastics materials suppliers. The evaluations were facilitated by the invention of the HAST reliability assessment technique by BT Labs in 1968. The high reliability capabilities were published in the early eighties, and plastic encapsulation was adopted for terrestrial telecommunications use at that time. Recent work discussed at an ESTEC Components Conference, presented at international reliability conferences and at an ISHM USA symposium, has illustrated the unreliability of hermeticity tested hermetic packages when used in tropical climates, where the problems are now avoided by using plastic encapsulations from selected suppliers. Therefore the myth of inherent unreliability of plastic encapsulation has been well and truly exploded. The way is now open for more widespread use of plastic encapsulated microelectronics in high reliability applications.

The potential for space applications arises because reliable dice can be protected by an intimate application of a junction coating. There are many excellent silicones which form intimate bonds with the surfaces of semiconductor dice by chemisorption. This requires the surfaces of the dice to be chemically clean, which can be achieved by either wet processes or ideally by a final plasma clean. Application in vacuum facilitates intimate contact. Coatings must be thoroughly cured, which means beyond the manufacturers' recommended curing times (which are defined for a high percentage of polymerisation). Effective curing times are, e.g. 150$^{o}$C for 30 mins + 3 hours at 180$^{o}$C. Such a coating achieves an intimate and robust seal which will safeguard the existing reliability of the dice against external climatic attack. For space applications, the encapsulation adds mechanical support of wire or TAB bonds, and adds the major advantage of attenuating particles which can give rise to soft errors. $\alpha$ particle activity of silicones is very low, of the order of 0.002 per cm$^2$ per hr for 1-8 MeV energies. On the other hand, silicone effectively attenuates $\alpha$ particles, for instance, resulting in zero errors in $10^6$ device hours of operation, compared with 200 soft errors if the silicone protection was not present. As $\alpha$ particles can be sourced also from alumina substrates and lead solder terminations, the silicone adds versatility to the range of materials that may be used in the assembly of the MCMs.

## 5. MCM Materials

The options for MCM materials for use in space are still wide open. The payload advantage currently remains with silicon substrates using thin film multilayers with either $SiO_2$ or polyimide dielectrics. Silicon substrates also provide advantages of heat conduction, which are certainly necessary for the 25W+ consumed by some of the signal processing systems, but not for some of the very high speed applications. While cost considerations may not be dominant, nevertheless, cost benefits do arise from alternative materials, which have to be weighed against packing density and even reliability (yet ill determined). Recent cost comparisons (cost comparisons are volatile) emerge as follows:

| *MCM TYPE* | *RELATIVE COST* |
|---|---|
| *MCM-L* | *1* |
| *MCM-C* | *2* |
| *MCM-D (thin film on $Al_2O_3$)* | *3* |
| *MCM-D (thin film on Si)* | *8* |

An analysis of MCM density, performance, development cost and investment costs gives a different perspective:

| *MCM Technology* | *Density* | *Very High Speed* | *Relative Development Cost* | *Relative Capital Investment* |
|---|---|---|---|---|
| *Thin film on Si* | *Very High* | *No* | *1* | *1* |
| *Thin film on polyimide* | *High* | *Yes* | *0.70* | *0.71* |
| *Low Temperature Co-fired $Al_2O_3$* | *High* | *Yes* | *0.73* | *0.45* |
| *Thick Film on $Al_2O_3$* | *Good* | *No* | *0.38* | *0.28* |

The Radiation Hardness of a number of MCM polymer materials have been tested to differing degrees of severity. For instance, polyimide has been found to retain essential properties with dosages in excess of 100 MRads of $\gamma$ radiation, while BCB materials have been tested successfully with dosages in excess of 100 MRads of $\gamma$ radiation. Therefore there are few fears for the inherent useability of polymers in satellite communications applications.

## 6. Performance

Space applications demand performance, and hence bare chip circuits become essential to avoid the parasitics contributed by the packages. A thorough analysis of package and bare chip parasitics has yielded a comparison of propagation delays (**Figure 3**):

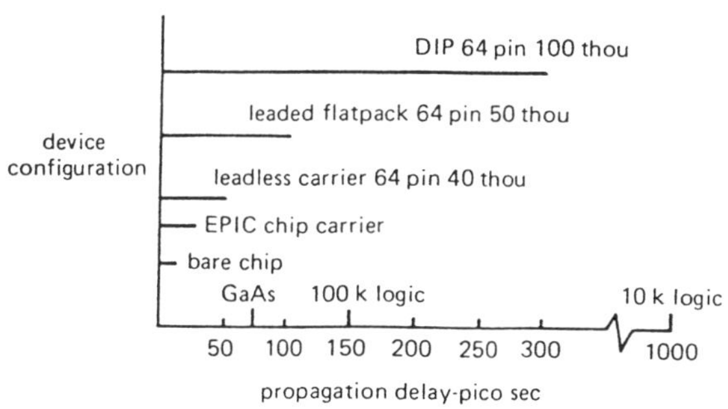


***Figure 3. Propagation Delay of Bare and Packaged Semiconductor Dice***

While the bare chips do indeed provide the best option, nevertheless they do have to be interconnected on substrates, and performance and power will be affected by the choice of substrate and the packing density. While thin film on silicon provides the highest density, the choice of dielectric will dominate the performance - the aim being to achieve HPI® (High Performance Interconnection) . Currently the very low dielectric constant polyimides ($\varepsilon \cong 2$) provide the best prospects for combining density and performance. Low $\varepsilon$ materials also improve crosstalk and cross-coupled noise. MCMs using VHSICs will require 15-20% of the bonding pads for power connections, and electrical noise can become a major hazard due to the increased currents (50-100mA) at the output buffers. As the rise times decrease below 1 ns, the solution to improve noise immunity is to reduce the equivalent inductances of the substrate and place a decoupling capacitor close to each IC between the power and ground planes. Signal layers should be sandwiched between ground planes to reduce crosstalk that nay be caused by capacitive coupling. High density also benefits power dissipation by diminishing the capacitive load on the devices, which can provide up to a 40% decrease in dissipation.

## 7. Testability

Testability of the individual die and the assembled dice in the overall MCM is essential for successful implementation of MCM technology. Therefore all the recently established successful approaches to design for testability must be exploited. Otherwise the quality and reliability of the MCM will be jeopardised. The design policy must ensure that all ICs conform to the IEEE Standard P1149.1 for embedded Boundary Scan (BScan) in all ICs and the full MCM circuit. BScan enables "virtual probing" of every IC and all interconnects for circuit diagnostics and to verify "good dice". The addition of a little more internal "tester" logic provides built-in self-test (BIST), which simplifies testing during wafer probing and delivers more authentic information on "Known Good Dice". Thus the initial quality of the MCM and the jeopardy to later performance is minimised.

## 8. System Reliability

The overall reliability of the system is dependent on the reliability of the MCMs, which are in turn subsystems with dependence on the individual component reliabilities, the hierarchy of failure dependence (Fault Tree Analysis) and any additional safeguards taken to minimise system failure due to separate parts. Techniques often used are replication with two out of three voting to safeguard restart of the system in the event of a failure. Often overlooked in such considerations are common-mode dependencies, e.g. common power supply degradation, or common software. Fortunately system designers are now wise to such pitfalls, thus there is a good track record of high reliability in high reliability communications systems including satellite systems.

# MATERIALS/DESIGN CONSIDERATIONS FOR MCMs

HARRY K. CHARLES, JR.
*The Johns Hopkins University*
*Applied Physics Laboratory*
*Laurel, Maryland 20723*

## Abstract

Multichip module (MCM) technology is a key driver for today's advanced electronic packaging. High density and speed requirements are placing increasing demands on materials parameters and packaging structures. Both physical and electronic properties of materials suitable for use in MCM construction are described. Module design considerations are presented along with material, structure, and performance trade-offs. Particular attention is focused on those elements of electrical, thermal, and mechanical performance necessary to ensure a high reliability MCM while ensuring both manufacturability and reasonable economic relevance.

## 1. Introduction

Multichip modules are rapidly becoming the mainstay of advanced high density packaging [1,2]. Table 1 captures current and future I/O and package pitch requirements for advanced multichip packaging structures. MCMs, such as the ones shown in schematic cross section in Fig. 1, combine several high-performance silicon or GaAs integrated circuits with a custom-designed multilayer substrate structure. These complex "hybrid" substrates can be fabricated on various base layers (e.g., ceramics, semiconductors, fiber-reinforced organic resins, and, sometimes, metals) using several different circuit technologies including thick and thin films, cofired [3], and printed wiring board type lamination. These substrate and conductor/dielectric technology differences give rise to three distinct MCM types (C for ceramic, D for deposited dielectrics and conductors, and L for laminated technology). Salient characteristics of the three module types have been presented previously [4].

A MCM typically forms a major system element such as a dynamic memory bank or a central processing unit [3]. Because of the high degree of functionality and logic partitioning within the MCM, the number of external I/O is usually higher than that predicted by Rent's Rule [5] based on the chip technology and the gate count. Although the I/O is higher, it usually can be accommodated by the increased



*W. K. Jones et al. (eds.), MCM C/Mixed Technologies and Thick Film Sensors,* 177–193.

dimensions of the MCM as compared to a single chip package. These larger package dimensions also provide increased area for thermal management (thermal vias) and even area type electrical interconnects. Within the module, heat and electrical load distributions can be improved by using several smaller chips rather than a single high-power IC. Because of the ability to tightly pack the smaller chips within the module, the use of multiple chips does not introduce a significant delay or other performance penalty. In fact, the improved thermal performance and the ability to mix chip and semiconductor technologies (e.g., CMOS, bipolar, silicon, and GaAs) within a single module can actually increase or optimize system performance. The larger module area also allows the use of special heat sink options that may not be easily employed on smaller units [3]. However, MCMs may also create handling and sealing problems, especially in the formation of large area hermetic seals. Finally, the use of MCMs may significantly reduce system cost. Estimates of cost for various MCM technologies [1,6,7] versus other electronic system implementations have been made previously and will be discussed briefly below.

TABLE 1. Current and future (year 2000) chip and package dimensions

| Parameters | Current | Future |
|---|---|---|
| Chip ($\mu$m) | | |
| Oxide Thickness | $1.25 \times 10^{-2}$ | $8.0 \times 10^{-3}$ |
| Metal Pitch (Line Gap) | 2.0-2.5 | 0.95-1.0 |
| Chip Size | $8 \times 10^{3}$ | $12\text{-}13 \times 10^{3}$ |
| Chip Connection Pitch | $1 \times 10^{2}$ | 20-25 |
| Package (mm) | | |
| Current Grid Array Pitch | 2.5 | --- |
| Grid Array Requirement | 0.8 | 0.15 |
| MCM Grid Array Pitch Requirement | 1.25 | 0.5 |
| Current QFP Pitch | 0.5 | --- |
| QFP Pitch Requirement | 0.3 | 0.08 |
| MCM QFP Pitch Requirement | 0.5 | 0.1 |
| MCM Interconnection Pitch Requirement @ Silicon | 0.05 | 0.01 |

## 2. Substrates

The major differences between MCM-C, D, and L are the host substrate materials and the dielectric/metallization schemes. MCM substrates must be strong without bowing for multilayer circuit fabrication and possess a reasonably high thermal conductivity (to allow uniform heat distribution and removal) and a coefficient of thermal expansion (CTE) compatible with that of the semiconductor material. Although alumina ceramic ($Al_2O_3$) has been the workhorse of the hybrid industry for many years, MCMs are also being fabricated with both silicon and aluminum nitride (AlN) substrates. Silicon's CTE is totally compatible with that of silicon-based very large scale integrated circuit chips. Silicon also provides a smoother surface than ceramic ($Al_2O_3$, AlN, etc.), which allows finer line definition, and has a thermal conductivity many times that of $Al_2O_3$ (145 W/m•K vs. 25 W/m•K, respectively). In addition to conductors, both active and passive devices can be fabricated in the silicon substrate [8]. AlN has a high thermal conductivity (>200 W/m•K) and a CTE compatible to both Si and GaAs. It is more difficult to metallize and is not as smooth as other MCM substrate materials, even with extra surface polishing. Another thin film multilayer substrate is sapphire (single crystal $Al_2O_3$), which has a high thermal conductivity (70 W/m•K) and an extremely smooth surface, allowing greater circuit density. Diamond (both films and wafers) [9] may prove to be the ultimate in MCM substrates, with a low dielectric constant for excellent electrical performance and an extremely high thermal conductivity ($\geq$1000 W/m•K) for thermal management.

Cofired ceramic substrates have been used in combination with organic thin film technology to provide high-performance MCMs [3]. A typical arrangement is shown in Fig. 1. The power and ground planes are located in the high dielectric constant ($\epsilon_r > 5$) ceramic (providing excellent power supply decoupling) while all the signal layers are in the lower dielectric constant organic material. In other applications where density and speeds are not as stringent, all cofired MCM substrate structures can be used for both the signal layers and the power/ground planes. Thick film technology, on both $Al_2O_3$ and AlN, has also been used in MCM substrate fabrication. Thick film offers design flexibility and economy over cofired technology, especially in low production volumes.

Organic board materials (epoxy-glass and polyimide-glass) have also been used for MCM substrates. Polyimide-glass provides excellent stability and low dielectric constant ($\epsilon_r < 5$), thus ensuring efficient signal propagation. Substrate thermal conductivity, high temperature stability, and compatibility with the semiconductors' CTE remain major concerns. Because of electrical design rules for controlled impedance and photolithographic limitations on organic boards, circuit density on fiber-reinforced organic materials remains somewhat limited. While the basic costs of organic board material is low, because of the low photolithographic resolution, the cost per substrate interconnect in an MCM-L may be high.

Metal substrates such as copper, copper-clad Invar, and copper-clad molybdenum are also suitable for both heat sinks and substrate applications. In a substrate role, the clad metals offer controlled and tailored CTE, but they must be coated with nickel, titanium, chromium, or some other corrosion-resistant barrier layer that adheres well to the dielectric material such as polyimide. Metal matrix composites [10] also offer significant promise as substrate materials.

## 3. Dielectrics

For thin film multichip modules (MCM-Ds), polyimide is the mainstay dielectric material because of its low dielectric constant ($\epsilon_r \sim 3.5$) good solvent resistance, high glass transition temperature ($T_g$ > 280°C), and reasonable planarization capability (allows patterning at the micrometer level). Polyimide comes in a standard and a photo-imagable form [11], which facilitates patterning and via formation. A recent report has shown polyimide to be an excellent dielectric for fabrication of reliable MCMs, with documented lifetimes greater than 2000 hours at 85°C and 85% relative humidity [12].

Other organic materials for MCMs that show significant promise are: 1) fluorinated polyimides [4], which have moisture absorption less than 0.5% (compared to standard polyimide in the 1-3% range); 2) acetylene-terminated polyimides [13], which show excellent planarization and can be formed in thicker layers in a single application; 3) solventless epoxy resins (ultraviolet-curable cycloaliphatic epoxies), which have excellent planarization characteristics but low thermal stability [14]; and 4) an organic resin derived from benzocyclobutene (BCB) that offers a lower cure temperature than polyimide (i.e., 250°C compared to ≥350°C for polyimide), a lower dielectric constant ($\epsilon_r \simeq 2.60$), and a compatibility with standard thin films [15]. BCB is also available in a photo-imagable form [16].

Low dielectric constant ceramics are of significant interest in the fabrication of MCMs (MCM-Cs). With such materials, ceramic printed wiring boards (MCM substrates) can have the electrical characteristics of their organic counterparts while maintaining the desirable thermal, mechanical, and electrical insulating characteristics of ceramic. For the successful application of low dielectric constant ceramics ($\epsilon_r \leq 5.0$) to multichip modules, two primary conditions must be met: 1) the production of large area multilayered ceramic structures (> 65 cm$^2$) with CTEs compatible with that of silicon; and 2) the development of low resistance metallization schemes capable of supporting chip solder attach (inverted reflow), wire bonding, or tape automated bonding. To achieve low dielectric constant in a ceramic structure, two basic technologies are used: thick film and cofired. In thick film, a mixed composition, low dielectric constant ink is printed and fired on a standard alumina substrate; for cofired, either a mixed composition or a controlled porosity tape cast structure is required. In most multichip ceramic module applications, the metallization scheme has been gold (or palladium gold complexes). A combination of two technologies - a multilayer thick film signal structure on a cofirable power and ground substrate assembly - offers advantages in providing both superior electrical performance and extremely tight impedance control (±10%) over a large area ceramic board.

In printed wiring board systems (MCM-Ls), the technology drivers for MCMs include effective utilization of space and weight (due to larger design rule spacings and added heat sinks or thermal risers); low coefficients of thermal expansion (due to organics' normally high CTE); and good thermal stability (due to some organics' normally low $T_g$).

Inorganic dielectrics should not be overlooked in the fabrication of MCMs. The MCM schematic cross section in Fig. 1 uses silicon nitride as a dielectric layer for separating the signal planes on a semiconductor substrate type MCM. These materials are deposited at low temperatures (200-350°C) by plasma-assisted chemical vapor deposition techniques. In large MCMs, it is important to minimize the stress in these deposited films. Low stresses improve the film layers' performance during temperature cycling. The addition of $P_2O_5$ (phosphorus pentoxide) during deposition has been shown to reduce the stresses in $SiO_2$ films.

## 4. Metallizations

Metallizations for MCMs fall into three primary categories: 1) pure elemental or compound materials deposited over the full (or selected) substrate areas by various vacuum deposition and/or plating techniques and then patterned by photolithography (i.e., thin films, MCM-D); 2) mixtures of various noble and refractory metals (such as gold, silver, tungsten and molybdenum with fillers, binders, and solvents) forced through pre-patterned screens or stencils (and, in some cases, forced through a capillary) onto substrates and then dried and fired to affix the film layer (i.e., thick films or cofired, MCM-C); and 3) copper metal sheets or layers that are laminated (glued) to various reinforced organic layers (i.e., printed wiring board technology, MCM-L).

Newer organic insulators such as polyimide are being used to fabricate thin film multilayer conductor structures for MCMs. Copper is the most widely used conductor material for these applications because of its high electrical conductivity and good solderability (for both flip-chip and solderable tape). It can also be electroplated to thicknesses greater than 5 $\mu$m, thus making extremely low resistance and net-shaped (controlled impedance) interconnects even for metallization lines as narrow as 1 $\mu$m. Copper can be gold flashed for surface layer applications and those involving gold or aluminum wirebonding. Some MCMs have used all-gold surface layers to eliminate passivation requirements. Aluminum has been used in MCM structures with polyimide, but is more resistive and there have been some concerns with reliability in nonhermetic applications [12].

## 5. Die Attach

The electrical connection to most integrated circuits is made to bonding pads on the surface or face of the chip. Except for flip-chip (inverted reflow), chips are mounted face up on the substrate (or package bottom). This mounting provides mechanical support (and sometimes a back surface electrical and/or thermal interconnect). For the back surface attach (or die attach) is accomplished by three primary techniques: 1) eutectic die attach; 2) metal alloy attach; and 3) organic adhesive attach. By eutectic attach, one typically means the gold-silicon eutectic (94 wt. % Au, 6 wt. % Si), which melts at 370°C. Various techniques for forming the gold-silicon eutectic have been described [17] but, regardless of the technique, it is essential that sufficient gold be present to ensure that the alloy forms properly.

Other metal alloys with either unique melting points or practical solidus-liquidus ranges have been used for die attach (e.g., gold-tin, tin-lead). Because they do not rely on the interdiffusion of gold with silicon (and the formation of the gold-silicon eutectic), either a thin sheet of the alloy or pre-tinning (in the case of tin-lead solders) must be used between the metallized back of the chip and the metallized region of the substrate or package bottom. Alloy die bonds provide good mechanical strength, electrical conductivity, and low thermal impedance. Some alloys or solders require flux to aid in the reflow and prevent excessive oxidation. If flux is used, it must be removed, thereby introducing extra processing, cleaning, and inspection steps.

Organic adhesives such as epoxy or polyimide can be used for die attach. Organic adhesives are typically poor thermal conductors and natural electrical insulators. To enhance performance, the organic resins are loaded with conductive particles. Gold and silver are the materials commonly used to provide electrical (and thermal) conductivity. Ceramic particles such as $Al_2O_3$, BeO, SiC, and AlN are used to provide enhanced thermal conductivity while still maintaining electrical isolation properties. A perceived disadvantage of the organic adhesive bond is the potential for outgassing or the subsequent release of absorbed gas or moisture during periods of thermal stress. These outgas products (known and unknown contaminants) have the potential for adversely affecting device reliability and, hence, product lifetime. The poor thermal stability of organic adhesives (especially above the glass transition temperature), coupled with the outgassing of trapped solvents and reaction by-products, have produced significant concerns for large area die attach reliability, especially in MCMs.

Silver-filled, low melting point ($\sim$400°C) glass materials have emerged as a bonding alternative for die attachment. The use of 400°C temperatures and oxidizing ambients (necessary for achieving proper adhesion reactions) has caused major concerns in processing. These materials also contain solvents and binders, thus producing conditions similar to the outgassing problems associated with organic adhesives.

## 6. Electrical Interconnect

Electrical interconnect is accomplished by several techniques including wirebonding, tape automated bonding, flip chip bonding (inverted reflow), wafer-scale integration (WSI), and laser pantography. The last two techniques are relatively experimental in nature, requiring costly amounts of extra processing and/or die shaping, and are not expected to make an impact in the MCM arena for some time. Additional details on both WSI and laser pantography have been presented previously [18]. Another technique is using multilayer thin film metallizations with planarized chip substrate combinations as shown in Fig. 2. Performance of this interconnect scheme should be comparable to the numbers presented for laser pantography in Table 2.

The three remaining interconnect methods (wirebonding, tape automated bonding, and inverted reflow) have all been used in the production of MCMs (see

TABLE 2. Current size and performance criteria for various chip interconnect methods

| Method | Diameter, μm | Length, mm | Pitch, μm | Typical I/O Number | Inductance, nH | Mutual Inductance, pH |
|---|---|---|---|---|---|---|
| Wirebond | 25.4 | 1 | 200 | 256[a)] | 1-2 | 100 |
| TAB | 50[b)] | 1 | 100 | 400 | 1 | 5 |
| Flip-Chip | 100 | 0.1 | 400 | 625[c)] | 0.05-0.1 | 1 |
| Laser Pantography | 5.5 | 1[e)] | 25 | 1600 | 0.25 | 1 |

[a)]Ultrasonic wedge bonded chips with I/O's up to 320 have been reported.
[b)]Effective ribbon diameter = (thickness + width)/2. (50 μm thick x 50 μm wide; yields a 50 μm effective diameter)
[c)]Plus area interconnects up to 16,000.
[d)]A 10 μm wide x 1 μm thick laser written line yields a 5.5 μm effective diameter.
[e)]Laser pantography lines can be written to any practical length because they are entirely supported by the substrate.

Fig. 2). Wirebonding is the most common method in use but, perhaps, the most limited (see Table 2) for future development of high-density, high-performance MCMs. There are three wirebonding methods: thermocompression, thermosonic, and ultrasonic. Gold wire can be bonded by all three techniques, while aluminum wire is usually bonded ultrasonically. For MCMs, thermosonic and ultrasonic bonding [19] are most common. Because of the close pad spacings and high I/O densities of MCMs, automatic wirebonding is necessary. Even with automatic bonding, however, the relatively long, round wire interconnect necessary in most modules will ultimately limit the performance of packaged integrated circuits and modules because of its high inductance. The size of the ball (and capillary) in thermocompression and thermosonic methods also limits the density of interconnect [19]. Wedge size can also be a hindrance, but, in general, closer densities or pitches can be achieved with automatic wedgebonding [19]. Size parameters and electrical performance of the major interconnect schemes are given in Table 2.

Tape automated bonding (TAB) is a planar lead interconnection process that uses prefabricated metallic interconnection patterns (either single- or multi-level) on an organic carrier film (typically polyimide). The film is usually one circuit interconnection pattern (or lead frame) wide and hundreds of patterns long, making it suitable for winding on reels or spools for use in automated assembly equipment [18]. Very few MCMs have been fabricated to date with tape automated bonding because of its inherent disadvantages of cost and long production lead times. A specific interconnect pattern is required for each different chip; thus, new masking and tooling are needed for each chip (or lead connection pattern) to be mounted. Accordingly, this method has been considered practical only for high-volume production. Because of its electrical performance advantages (see Table 2) and its high-density 50-μm-wide leads on 100-μm centers, it is still being considered as a potentially important interconnection method for current and future high-performance chips. In MCM applications, TAB requires greater chip-to-chip spacing than either wirebonding or flip chipping [21].

The third major chip interconnection scheme is direct interconnection by means of plated solder interconnect metallization, typically in the form of spheroidal bumps. In this method, solder metallization is attached to the chip bonding pads and/or the substrate lead pattern. The chip is then inverted over the substrate pattern and the bond is formed by solder reflow. The solder metallization melts and gravity, coupled with surface tension, causes a perfectly shaped and aligned solder joint to be formed. Inverted chip solder reflow provides very short, low-resistance interconnections that minimize inductance and capacitance - especially important for a high-frequency operation such as that encountered in MCMs. Inverted chip solder reflow is also amenable to full area attachment (i.e., bonding pads over the full active device area, not only on the perimeter) since no force is used in the bonding process. It has three disadvantages: the inverted geometry prevents inspection of the interconnection, heat sinking is poor (if no thermal post is brought into contact with the back of the chip) because the only thermal escape path is through the solder joints, and the devices are difficult to replace if they fail.

## 7. Package Sealing

Hermetic MCMs can be produced by placing the MCM substrate in a pre-shaped metal or ceramic packaging. These packages contain the pins, leads, or solder feet necessary to connect the MCM to the next level packaging [2]. The package can then be sealed by welding or soldering a lid to a specially prepared seal area on the package. Metal packages are typically made of Kovar (iron-nickel-cobalt alloy), whose CTE matches the glass used for the hermetic lead feed throughs. Because of its large iron content, Kovar must be protected from the atmosphere by nickel and gold platings. Lid sealing (metal) is accomplished by welding the plated Kovar lid to the sidewalls of the Kovar package. Solder can also be used - both gold-tin (80 wt % Au, 20 wt % Sn) and tin-lead solders have been used. Seal rings are usually brazed to the metallizated ceramic to allow lid attach. The substitution of low melting point seals for either lid welding or solder has the potential to significantly reduce packaging cost.

Instead of putting the MCM in a separate package, the MCM substrate itself can form part of the package structure. A seal ring would be brazed to the outer exterior of the substrate. This brazing, because of the substrate size and CTE mismatches [20], must be done with care to effect a good hermetic seal. Once the seal ring or package sidewall is in place, a lid can be attached by either welding or soldering. Metal lids are typically thicker on MCMs than on single chip packages due to the need for mechanical integrity over a much larger area.

The third sealing technique is to use organic overcoats and/or encapsulation. Epoxy, polyimide, Parylene, polyurethanes, silicones, acrylics, and glass-base systems have been used for circuit protection. Ultraviolet-curable acrylics can be easily applied and cured in a few seconds. Epoxies require temperatures of at least 70°C and about 2 hours to cure. Parylene is vacuum deposited at room temperature. Polyimide is suitable for performance at high temperatures (>150°C) but requires curing at temperatures between 300 and 400°C for times in excess of 30 minutes.

## 8. MCM Design

Design begins by selecting the circuit to be converted to an MCM. The integrated circuit type(s) must be identified (silicon vs. GaAs, high power [BJT] vs. low power [CMOS]). These parameters will affect electrical impedance and thermal considerations as well as help select the type of signal interconnect, die attach, and, perhaps, the substrate technology itself. Based on the number of integrated circuits in the module, its application, and functionality of each chip, the number of module inputs and outputs is determined. If available, the actual input/output (I/O) counts from the circuit diagram or schematic are used; otherwise, an estimate must be made based on module functionality using a technique such as Rent's Rule [5]. Examples of I/O partitioning using the Rent's Rule approach have been presented recently [7,21].

Once the internal and external I/O numbers are established, minimum module size can be determined from chip dimensions and the space around the chip required for the interconnect. Table 3 presents interconnect spacings dimensions for various interconnect technologies. Recent studies [7,21] have shown that only two interconnect methods - wirebonding and flip chip (inverted reflow) - consistently allow the close chip spacing required for today's high performance MCMs.

TABLE 3. Chip dimensions and spacing parameters for various chip interconnect methods

| Technology | $\ell_G$, mm | $\ell_c$, mm[a] | $\ell_p$, cm | $\ell_I$, cm[b] |
|---|---|---|---|---|
| Flip chip | 0.38-0.64 | 7.62-12.70 | N/A | 0.80-1.33 |
| Wirebond | 1.78-2.03 | 7.62-12.70 | N/A | 0.94-1.47 |
| Flip TAB | 2.54-3.05 | 7.62-12.70 | N/A | 1.02-1.57 |
| Regular TAB | 3.81-10.16 | 7.62-12.70 | N/A | 1.14-2.29 |
| Chip carriers | 0.51-1.91 | 7.62-12.70 | 2.03-3.05 | 2.08-3.24 |
| Q flat packs | 5.08-11.43 | 7.62-12.70 | 2.54-3.81 | 3.05-4.95 |

[a]Dimensions for high I/O complex chips (nominally ≥ 200 I/O).
[b]$\ell_I = \ell_c + \ell_G$ for all nonpackaged parts; for packaged parts $\ell_I = \ell_p + \ell_G$, where $\ell_p$ = length of package edge.

In order to select design rules and substrate technology for the MCM, the substrate track interconnect density (cm/cm$^2$) must be determined from the routed trace length and the substrate area. Trace length calculations require knowledge of the I/O partitioning, the chip size and spacing (due to interconnect method), and the relative priorities of interconnect between the various chips in the chip set. Fan-in and fan-out statistics for each class of I/O (internal, external, data busses, power and ground, etc.) are also important. Systematic combinations of the above information have shown that large MCMs routinely have interconnect densities between 400 and 600 cm/cm$^2$ while comparable surface mount technology boards range from 40 to 60 cm/cm$^2$ [21]. From the interconnect density, the number of board layers can be determined based on the track pitch of the substrate technology and the allowed interlayer via penalty factor. Via penalty burdens of greater than 10% of a layer's line density are not uncommon and may approach 15% or more as highly complex chips are packaged with a low-density substrate technology. Column via constructs tends to have higher via penalty factors than boards or substrate technologies using staggered vias. Minimizing the number of layers (and vias) should be a goal in all MCM implementations to reduce cost and increase both performance and reliability.

## 9. MCM Performance

At this point, performance and its relation to selected MCM technology must be considered before a final trade-off in size, structure, and cost can be made. Track density (track width and thickness) along with the dielectric constant will determine the impedance limit of the technology using models for both stripline and microstrip. The dielectric constant in high speed (short rise time applications) will determine the critical spacing for adjacent chip components. Fortunately, the spacing parameter produced by both wirebonding and flip chipping are under the critical spacing lengths for all but the most stringent conditions. A lower dielectric constant produces smaller time delays and longer critical lengths, thus reducing pressure on tight chip placement. The influence of structure and dielectric constant on trace impedance has been described previously [4].

At very high speeds (10-ps rise times) or frequencies, the interconnect technology will also play a role in determining the module performance. Flip chip-type interconnects currently provide the best electrical performance and the smallest spacing penalty of all the major interconnect schemes, yet even in volume that technology is relatively costly. Average line lengths and the dielectric constant of the selected technology can be used to determine interchip time delays and overall module speed. Typical time delays plotted against dielectric constant for different line arrangements on a large MCM are shown in Fig. 3. Because of circuit speed and band width considerations (i.e., 35 GHz for 10 ps rise times), all conductors must be of the transmission line type. To achieve nominally high impedance ($\sim$50 $\Omega$) thick dielectric layers (>25 $\mu$m) must be used. Power and ground planes must be distributed with high decoupling capability.

Ground plans should be gridded to improve interlayer adhesion in these large area modules. Grid size is dependent upon the processing requirements and the required electrical impedance. In all cases, the grids should be placed at a 45° angle to the orthogonal conductor pattern to minimize changes in electrical behavior between conductors in on-grid and off-grid positions. The signal line resistance must be low, which produces requirements for wide and thick lines, high conductivity metals, and large plated-up vias for reduced contact resistance. Wide signal lines also reduce conductor self-heating, which can aid thermal performance but may have an adverse impact on density.

Thermal performance is another major consideration. High power dissipation will force certain substrates (carriers) such as aluminum nitride, metal, and, perhaps, diamond to be selected. Inherently poor thermally conducting substrates and dielectrics such as organic materials will require the use of plated thermal vias (risers) or cutouts in the dielectric for direct chip mounting to a carrier material with higher thermal conductivity. In either case, interconnect real estate will be lost, causing an effective reduction in routing or track density. For direct heat sink mounting, this routing penalty can be as large as the chip area. For thermal vias, the area consumed can be as little as 5% of the chip area, and the assembly will still produce effective thermal performance.

## 10. MCM Cost

Table 4 presents the cost factors for MCM substrate structures, interconnect, and assembly operations. The table contains two cost numbers for each category: a low-volume number and a high-volume number. The low-volume number assumes a limited production run where processes are new and most individuals are just starting to learn the processes. This would be typical of most of today's limited MCM development runs. The high-volume number assumes a mature product where everyone has gained sufficient experience to produce product at minimum cost. Technologies based on semiconductor processing also assume yield and wafer size enhancements for high volumes. The cost of chips and the payment for "known good die" have been addressed by several authors [22]. In cost analysis based on the numbers presented in Table 4, the cost of chips for all MCM technologies would be the same, although a packaging premium would be added to the die (chip) cost for surface mount technology implementation.

Cost comparisons [21] across technology types indicate that in low volume ($\sim$1000 to 10,000) MCM-C technology has a cost edge over the others (D and L) by as much as a factor of two, depending on selected interconnect method. SMT implementations are also significantly cheaper than MCM technology, but suffer severe weight, size, and performance (electrical and thermal) penalties.

In high volume, MCM-D technology appears to have the potential to outperform the ceramic-based technology in the cost arena. In fact, in high volume, the MCM-D costs are comparable to those of the surface-mount implementations, while the MCM-D significantly outperforms the surface mount implementation in speed and all physical features.

TABLE 4. Cost parameters for MCM substrates, interconnect, and packaging operations

| Technology | Low Volume | High Volume |
|---|---|---|
| Board/Substrate ($/layer/cm$^2$) | | |
| MCM-L | 8 | 3-5 |
| MCM-C | 16 | 6-8 |
| MCM-D | 80 | 30-50 |
| STD-ML PWB | 0.50-1.0 | 0.10-0.15 |
| Interconnect ($/interconnect) | | |
| Wirebond | 0.05-0.02 | 0.01-0.002 |
| TAB | 0.16 | 0.055 |
| Flip chip | 0.10 | 0.030 |
| Solder joint (manual) | 0.50 | 0.50 |
| Solder joint (machine) | 0.05 | 0.02 |
| Package ($/chip) | | |
| Die attach | 1.15 | 0.30 |
| Over coating | 0.10 | 0.01 |
| Plastic single-chip package | 2.20 | 0.50 |
| Hermetic single-chip package | 25.00 | 8.00 |

In smaller modules (fewer chips and reduced trace density), MCM-L technology could come into play. If MCM-L technology could support the application in about twice the numbers of layers as the MCM-D technology with comparable size, MCM-L technology would become a cost-effective option. Selecting a technology on the basis of cost alone appears to be a sensitive function based on module size, track density, and, of course, the normal electrical and thermal performance requirements.

## 11. Summary

MCMs are becoming the mainstay of modern high-performance packaging. Different structures are required to address the diverse application areas of today's electronic technology. High-performance materials and stringent process control are required to successfully execute a viable MCM technology. Materials and structures that currently play a major role in MCM technology are aluminum nitride and silicon for substrates; polyimide for dielectrics; copper metallization systems; bumped solder joints and tape automated bonding interconnects; and Parylene and urethane overcoats. These materials and structures, which are appropriately modelled and analyzed to ensure optimum performance and reliability, are the basis of advanced packaging technology today. In the future, new materials such as diamond may be extremely important in thermal management. Laser formed (laser reflow) or deposited interconnects may provide the ultimate flexibility while maintaining high density, electrical performance, and reliability.

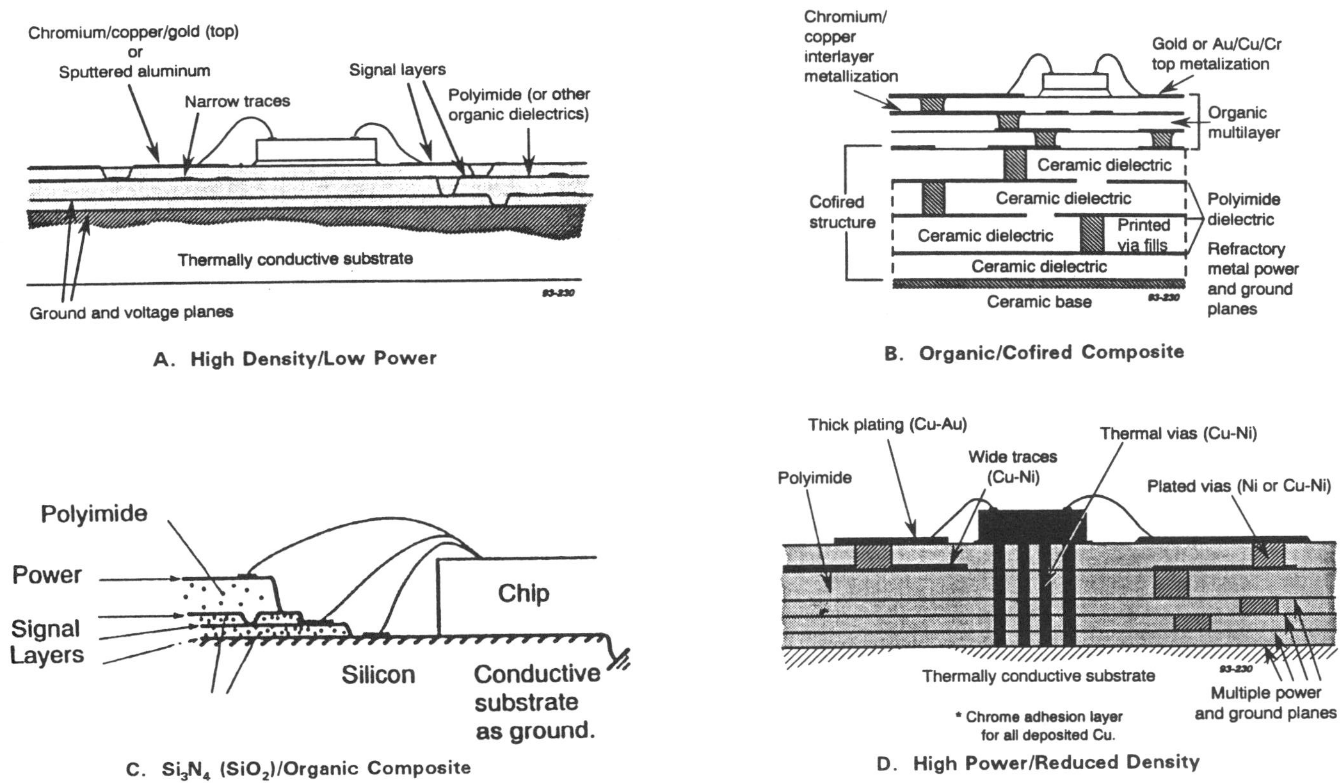


Figure 1. Multichip module structures for high performance applications.

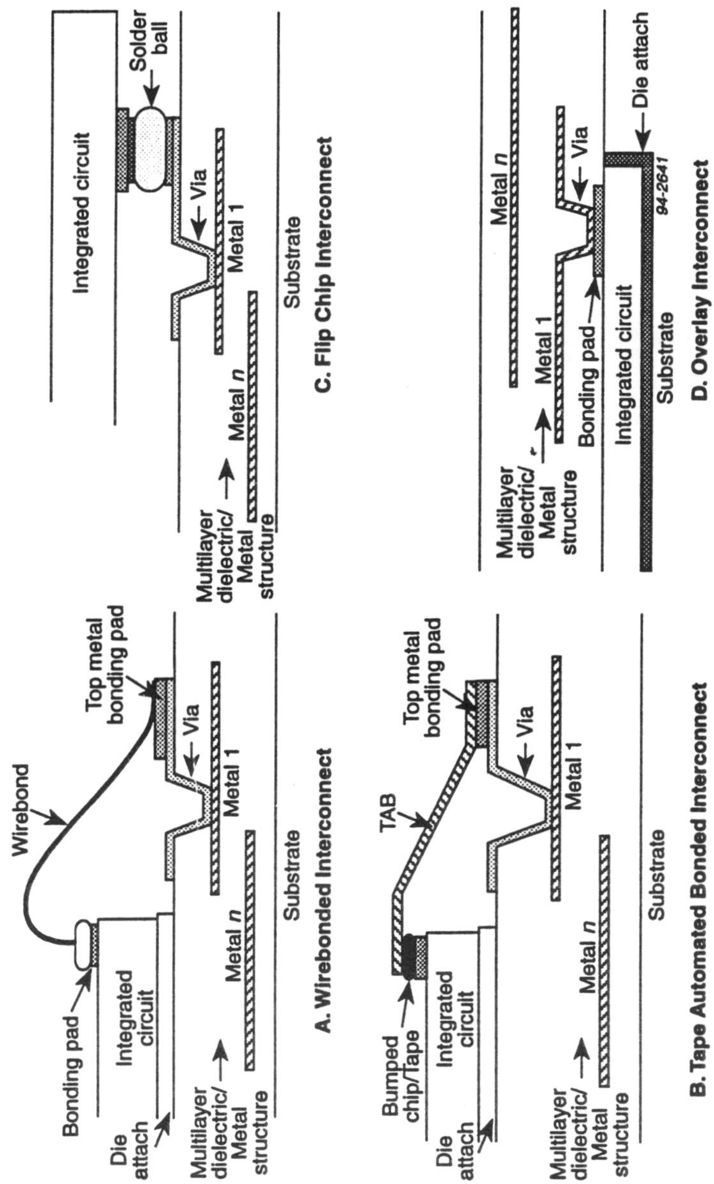


Figure 2. Multichip module interconnect methods.

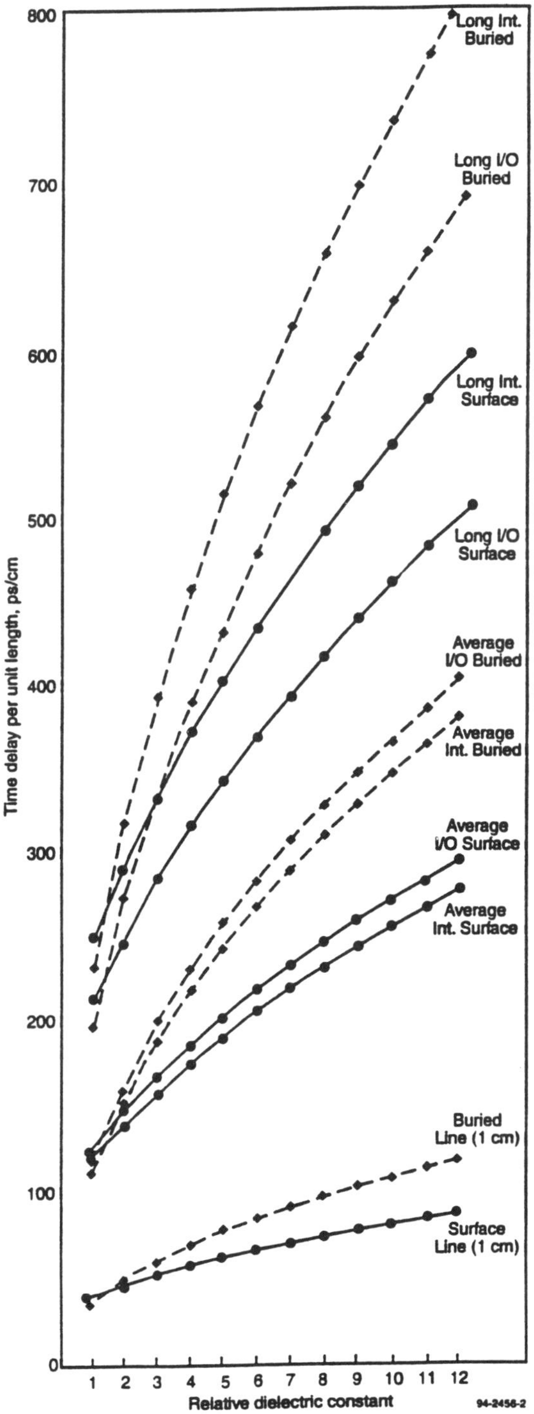


Figure 3. Time delay versus dielectric constant for various internal and external I/O lines in 6 x 6 multichip module. The bottom two curves present basic time delay per cm for a surface line (microstrip) and a buried line (stripline). The upper curves, when multiplied by the spacing parameter, $\ell_1$, yield the time delay for the average and longest I/O and internal trace for the technology selected.

## 12. Acknowledgement

I thank Mrs. S. Lynn Hoff for word processing and manuscript preparation.

## 13. References

1. Messner, G. (1988) Price/Density Tradeoffs of Multi-Chip Modules, *Proc. 1988 Int. Microelectronics Symposium*, Seattle, WA, 28-36.

2. Charles, Jr., H. K. (1991) Applications of Adhesives and Sealants in Electronic Packaging, *Proc. 1991 Int. Microelectronics Symposium*, Orlando, FL, 139-147.

3. Blodgett, A. J. and Barbour, D. R. (1982) Thermal Conduction Module: A High Performance Multilayer Ceramic Package, *IBM J. Res. Dev.*, **26**(1), 30-36.

4. Charles, Jr., H. K. (1992) Electronic Materials and Structures for Multichip Modules, *ASME Journal of Electronic Packaging*, **26**, 226-233.

5. Schmidt, D. C. (1982) Circuit Pack Parameter Estimation Using Rent's Rule, *IEEE Trans. Computers*, **CAD-1**(4), 186-192.

6. Neugenbauer, C. A. (1985) Comparison of VLSI Packaging Approaches to Wafer Scale Integration, *Proc. IEEE 1985 Custom Integrated Circuits Conference*, 32-37.

7. Charles, Jr., H. K. (1993) Design Rules for Advanced Packaging, *Proc. Int. Microelectronic Symposium*, Dallas, TX, 301-306.

8. Bodendorf, D. J., Olson, K. T., Trunko, J. F., and Winnad, J. R. (1972) Active Silicon Chip Carrier, *IBM Tech. Disclosure Bulletin*, **15**(2), 656.

9. Yoder, M. N. (1987) Applications of Diamond Technology, *Proc. Second Annual Technology Initiative Seminar*, Durham, NC, 3 pages.

10. Aghajanian, K. K. (1991) Processing and Properties of Silicon Carbide Reinforced Aluminum Metal Matrix Composites for Electronic Applications, *Proc. 1991 Int. Microelectronics Symposium*, Orlando, FL, 358.

11. Clatterbaugh, G. V. and Charles, Jr., H. K. (1988) The Application of Photosensitive Polyimide Dielectrics in Thin Film Multilayer Hybrid Circuit Structures, *Proc. 1988 Int. Microelectronics Symposium*, Seattle, WA, 320.

12. Roy, K. K. (1991) Multichip Module Substrates: Reliability Issues, *Abstract Book ISHM 1991 Joint Technology Conference*, Dallas, TX, 6 pages.

13. Rossi, R. D., Jobe, P. G., McMahon, J., and Puglisi, C. (1991) Acetylene Terminated Polyimides as the Interlayer Dielectric for High Density Multichip Modules, *Abstract Book ISHM 1991 Joint Technology Conference*, Dallas, TX, 3 pages.

14. LaVergne, D. B. and Hofer, D. C. (1985) Modeling Planarization with Polymers, *Proc. SPIE*, **539**, 115.

15. Johnson, R. W., Phillips, T. L., Hahn, S. F., Bordeaux, D. C., and Townsend, P. (1988) Multilayer Thin Film Hybrid on Silicon, *Proc. 1988 International Microelectronics Symposium*, Seattle, WA, 365-373.

16. Takahashi, T., Rutter, Jr., E. W., Moyer, E. S., Harris, R. F., Frye, D. C., St. Jeor, V. L., and Oaks, F. L. (1992) A Photodefinable Benzocyclobutene Resin for Thin Film Microelectronic Applications, *Proc. Int. Microelectronics Conference*, Yokohama, Japan, 64-70.

17. Charles, Jr., H. K. (1990) Electronic Packaging Applications for Adhesives and Sealants, *Engineered Materials Handbook: Adhesives and Sealants*, ASM International, Metals Park, OH, 3, 579-603.

18. Charles, Jr., H. K. (1989) VHSIC Packaging for Performance, *APL Technical Review*, 1(2).

19. Harman, G. G. (1992) Wire Bonding: Towards 6σ Yield and Fine Pitch, *Proc. 42nd Electronic Components and Technology Conference*, San Diego, CA, 903-910.

20. Zhuge, R., Jones, W. K., and Maganti, S. S., (1992) Stress Analysis of Multi Layer Multichip Module (MCM-D) Structures Using the Series Solution of Elasticity Theory," *Proc. Int. Microelectronics Conference*, Yokohama, Japan, 441-447.

21. Charles, Jr., H. K. (1994) MCM Design and Cost Parameters, *Proc. Int. Microelectronics Conference*, Omiya, Japan, 271-274.

22. Bauer, C. E., Posse, K. W., and Wilson, K. T. (1994) Good Enough Die? A Cost Model for KGD Test Decisions, *Proc. Int. Microelectronics Conference*, Omiya, Japan, 266-270.

# MCMs: MATERIAL CHOICES FOR ELECTRONICS AND OPTOELECTRONICS

**N. CHANDLER, N.E. SELLARS, R.K. BARTON, S.J. FOSTER, P.D. SLEEP, G.N. BLACKIE and I.R. CROSTON**
*GEC-Marconi Research Centre*
*Great Baddow, Essex, UK, CM2 8HN*
*Phone +44 (0) 245 473331 Fax +44 (0) 245 475244*

**KEYWORDS/ABSTRACT:** MCMs, MultiChip Modules, materials, electronic interconnect, optoelectronics,

As systems requirements and device developments combine to demand more from packaging in terms of density and performance, so are developments in materials and processes fostered. There are already several techniques used to make MultiChip Modules; in each there are evolutionary enhancements and revolutionary changes. This paper describes the MCM technologies presently available and being further developed at the GEC-Marconi Research Centre: using laminated organic dielectrics (MCM-Ls); ceramic dielectrics (MCM-Cs); organic dielectrics deposited onto passive or active silicon substrates (MCM-Ds); inorganic optical interconnect deposited onto silicon to make optoelectronic MCMs.

## 1 Introduction

Future products can look to MultiChip Modules (MCMs) to provide many benefits, the most frequently cited being miniaturisation. Table 1 gives some comparative data. MCMs also offer improved system performance (speed and functionality), reduced power consumption, improved reliability and lower costs. However, the different types of MCM, with their various materials properties and other attributes, allow different trade-offs to be made. Indeed, even within one type of MCM there can be significant differences depending on the choices of material, design, assembly method, etc.

MCM-L substrates use lamination techniques to combine organic dielectrics and bulk metal layers, essentially the same techniques as for advanced flexi-rigid printed circuit boards (PCBs) but with considerably finer features [1]. The very tiny holes, an order of magnitude smaller than in PCBs, are cut by non-mechanical means. The dielectric layers are considerably thinner than in conventional PCBs but several times thicker than in a typical MCM-D, which results in low capacitance. The thin fine-feature layers are supported by a base, usually a reinforced polymer (such as a PCB) or, alternatively, a metal or ceramic plate. As well as providing mechanical support, these bases can also be used to adjust the TCE, to extract heat or to provide any other features normally associated with a PCB, a metal box or a ceramic substrate (e.g. package base, resistors and capacitors). In



*W. K. Jones et al. (eds.), MCM C/Mixed Technologies and Thick Film Sensors,* 195–207.

the future, resistors will also be available within the fine-feature MCM-L layers. High permittivity layers may also be available for integral capacitors. The copper conductors in MCM-Ls are relatively thick (compared to MCM-Cs and MCM-Ds). They therefore have much lower resistance. They can readily accept not only unpackaged ICs assembled by wire-bond, Tape Automated Bonding (TAB) or solder-bump flip-chip but also packaged components.

TABLE 1. MCM density comparisons

| Parameter | MCM-L | MCM-C | | MCM-D |
|---|---|---|---|---|
| | | Fine Line Thick Film | Co-fired | |
| Minimum Practical Track Pitch (μm) | 80 | 150 | 200 | 40 |
| Typical Via-Hole Diameter (μm) | 50 | 100 | 150 | 30 |
| Maximum Interconnection Density/Layer ($cm/cm^2$) | 80 | 40 | 30 | 160 |
| Maximum No. of Signal Layers | >12 | 7 | 60 | 4 (fine feature layers) |
| Maximum Interconnection Density ($cm/cm^2$) | >960 | 240 | 1800 | 640 |
| Maximum Area Ratio Silicon: Substrate (%) | ~50 single side<br>~100 double side<br>>100 (3D) | <80 (single side)<br>>100 (double side)<br>>100 (3D) | | <80 (passive)<br>120 (active)<br>>100 (3D) |
| Maximum Substrate Size* ($cm^2$) | >400 | 100 | >200 | 100 |

* The finest features cannot necessarily be used for the largest substrate sizes of any particular technology.

MCM-Cs have ceramic substrates made by either a thick film or a co-firing technique (at either "low" or "high" temperatures). The thick film and low temperature techniques have the advantages of being able to include the familiar wide range of printed resistor values and of having tracks more conductive than the high temperature co-fired variant. Recent developments in materials and process technology have greatly increased the interconnection density per layer and will no doubt continue to improve. Substrates having higher thermal conductivity (such as aluminium nitride) will greatly improve the thermal performance of MCM-Cs. New high permittivity materials are being developed, to make capacitors of reasonably large values within the substrate; in digital applications these will principally be for decoupling. Lower permittivity materials (typically glass-ceramics) are now available for MCM-Cs. The alternative approach is to add one or more layers of organic dielectric, as in MCM-Ds. Such substrates are sometimes called MCM-C/Ds. A ceramic substrate can also serve as the base of a non-metallic hermetic package, with cost and performance advantages.

MCM-D substrates use deposition techniques for dielectric and metal layers in a manner similar to silicon device fabrication. The dielectric layers can be either inorganic, such as silicon dioxide, or organic, such as a polymer. At present inorganic dielectrics can be deposited only in thin layers, therefore to reduce capacitance or provide a useful characteristic line impedance the tracks must be narrow. This combination of thin and narrow dimensions gives rise to relatively high resistance. Another variable is the means of patterning the dielectric (mainly to create the via-holes between layers), either by removal (e.g. by plasma or laser) or by photo-imaging. The latter promises to be by far the cheaper method but at present forces compromises in other properties of the dielectrics.

The thin-film technology also enables resistors (of moderate value) to be included, at very small size, as well as very fine spiral inductors. Developments are also in progress on high permittivity material deposition processes to provide reasonably high values of capacitance. An MCM-D can also use a ceramic substrate, to take advantage of higher values for resistors and capacitors or to use as the package base, with components on one or both sides.

The use of a semiconductor substrate, however, gives the MCM-D the unique advantage of being able to incorporate circuitry into the underlying substrate. This can be a "standard" wafer, such as memory, or a custom wafer including, for example, logic circuitry, protection devices, power devices, etc. Another use for the active substrate is to incorporate test circuitry, either to pre-test the interconnection on the substrate before committing to the assembly of the added-on ICs, or to test the circuit after assembly or for in-service testing/reconfiguration. In the future, the substrate could perhaps be made from alternative semiconductors such as gallium arsenide, for high speed, or silicon carbide or diamond, for very high temperature operation.

MCMs can include optical or optoelectronic components and/or optical interconnection media. A silicon substrate is an ideal base for a silicon dioxide layer into which optical waveguides are incorporated and micromachined features can be used for aligning optical fibres and optoelectronic components. Its thermal conductivity is also useful for removing heat from the high power-density optical transmitters. The use of optical interconnect, either within a module or between modules, can greatly increase the data rate and reduce crosstalk, interference and the effects of line capacitance.

## 2 Key Material Properties

The important material properties can conveniently be divided into electrical (dielectrics and conductors), thermo-mechanical, optical and fabrication-related.

### 2.1 ELECTRICAL PROPERTIES

The dielectric constant is often considered to be the most important parameter because of its effect on propagation delay. At low signal rates this is not likely to present a problem. As rates increase it becomes necessary to consider the line capacitance and resistance so as to avoid line charging problems. Clearly, a low dielectric constant, $\varepsilon_r$, will reduce capacitance (for given values of track width, *w*, and dielectric thickness, *h*). As rates rise towards 100 MHz and controlled impedance tracks become necessary, a lower value of $\varepsilon_r$ allows the characteristic impedance to be achieved with a wider track, so minimising line resistance. The lower dielectric constant also helps reduce crosstalk [2]. Track thickness also reduces line resistance but as the signal risetimes approach 0.5 nanoseconds, the skin effect begins to limit the benefit of extra thickness, so extra width may be needed, again pointing to lower $\varepsilon_r$. The resistivity of the metallisation affects not only the dc/low frequency resistance but also the frequency at which the skin effect becomes significant (i.e. the skin depth depends on resistivity).

As the frequency continues to rise, it becomes important to provide well-matched terminations and a tight tolerance on the characteristic impedance. A low value of $\varepsilon_r$ will help here too, because the greater the track width the easier it is to achieve a tight tolerance in manufacture. It is also important, of course, to achieve an accurate value of dielectric thickness. Another important attribute is for $\varepsilon_r$ to vary as little as possible with frequency, otherwise the signal will become distorted due to dispersion.

Dielectric losses are unlikely to be a problem at low frequencies, in fact with most dielectrics likely to be used in MCMs the dielectric losses will be negligible up to the GHz region, although any absorbed water will add to the losses.

For high speed/microwave analogue signals, the lengths of the tracks take on a special significance. In this so-called "distributed circuitry", the electrical length of the track is all-important and it is usually helpful to use a medium or high value of $\varepsilon_r$ to achieve the desired electrical length in the smallest distance.

### 2.2 THERMO-MECHANICAL PROPERTIES

The principal thermo-mechanical properties in relation to MCMs are the Temperature Coefficient of Expansion (TCE), tensile modulus, dimensional stability and thermal conductivity.

For constant performance and reliability, the ideal would be for all the materials (in the MCM substrate and its components, its package and the next interconnection medium) to have the same TCE and to be at a uniform temperature. Different TCEs and thermal gradients cause stresses and strains which can lead to warping (which can create assembly problems in severe cases), delamination or fracture within the substrate and fatigue failure of materials such as solders. The levels of stress and strain depend on the TCE values, tensile moduli and thicknesses of the various materials. Polymers have a temperature below which their thermo-mechanical properties are well-behaved and above which there are large changes, such as greatly increased TCE or reduced modulus. Ideally this temperature, the glass transition temperature, $Tg$, should be significantly higher than any temperature which the material will subsequently experience, either during the later manufacturing stages or in service.

Good thermal conductivities ensure low thermal gradients and hence low thermal stresses and strains, also low thermal impedances and hence lower device temperatures and increased reliability. Most dielectrics have poor thermal conductivity. However, by providing a good thermal path through the low conductivity dielectric, by inserting another material (such as copper filled vias), the problem is solved without having to change the dielectric.

In some cases weight is a major factor, so the combination of size and density can be a crucial factor and can affect the choice of materials for the substrate and for the MCM package.

MCMs must also withstand environmental stresses (temperature cycling, thermal shock, vibration, shock, etc.), the severity of which depends on the intended application.

## 2.3 OPTICAL PROPERTIES

The material used for the optical waveguides should have a reasonably low loss (0.1 dB/cm or less) at the wavelengths to be used. For wavelength division multiplexing, the refractive index should be essentially constant over a range of wavelengths, to avoid variations in velocity and hence phase delay problems. The process for defining the position, size and geometry of the waveguides (a doping process in the case of silica) must be able to create the very precise dimensions needed for integrated functions such as couplers and to interface with the assembled optoelectronic components (emitters and detectors). The optical waveguide material and the underlying substrate must enable the very precise alignment of the components to be achieved and maintained.

## 2.4 FABRICATION-RELATED PROPERTIES

This category includes the materials properties such as water absorption, shrinkage/dimensional stability and the ability to withstand subsequent process steps in manufacture and assembly.

Water absorption can affect the electrical performance (through its high dielectric constant), cause corrosion and create delamination. Susceptible materials may need to be sealed or hermetically packaged, which may add to the cost. Shrinkage during processing can cause stress; in MCM-Ls during the lamination process, in MCM-Cs during the co-firing process and in MCM-Ds when a layer of dielectric continues to contract as it cools down after it has adhered to the underlying (non-shrinking or previously cured) material. If the shrinkage is not reproducible and uniform, the designer may be unable to use the smallest possible features. The materials should be able to be processed with fine features, e.g. cutting miniature holes in dielectric and making fine tracks, gaps and pads in conductor layers. As the desired feature sizes become smaller, the substrate surface must be sufficiently flat and smooth to permit the definition and manufacture of the narrow tracks and gaps and the assembly of flip-TAB and flip-chip ICs. On MCM-Ls, for example, this drives a move away from weave-reinforced laminates to reinforcement by short fibres or particles or to the omission of reinforcement. The use of a laser or plasma to cut tiny holes also favours laminates with small reinforcers or none at all. The materials must also withstand the subsequent processing of succeeding layers and operations such as the assembly of ICs and other items, cleaning, packaging, next-level assembly etc. Among the "other items" we can include "integrated" resistors and capacitors, i.e. materials incorporated as part of the substrate fabrication process.

To make multi-layer substrates, the laminating methods (MCM-Ls and co-fired MCM-Cs) must ensure that each conductor layer is encapsulated by the overlying dielectric. With the "fluid" deposition methods (thick film dielectrics and MCM-Ds) the dielectric must coat the underlying patterned conductors and present a reasonably planar surface on which to deposit the next conductor layer - this is known as "planarisation".

Values for the most important materials properties for the various types of MCM are summarised in Tables 2 and 3.

## 3 Discussion and Examples

In addition to the material properties described above, there is another major criterion: cost. At present, MCM-Ls tend to be the cheapest and MCM-Ds the most expensive. However, these are complex items and the cost depends on many interacting factors including production quantities. Over a period of time all the MCM technologies will become more cost-effective and it is likely that the price differentials between MCM types will diminish significantly.

### 3.1 MCM-Ls

The simplest MCM-Ls can readily be made using conventional weave-reinforced materials and this represents a very cost-effective approach, available from a number of suppliers. As the required density increases, smaller via-hole diameters and narrower tracks and gaps require thinner dielectric layers with smooth surfaces and a material which can have very small holes cut by laser or plasma etching. The most common approach at present is to use unreinforced polyimide, the multilayer structure being bonded together with adhesives. To facilitate the manufacture of the finest features, the individual laminates (copper and polyimide) should be free of adhesive. This is the technique currently used at GEC-Marconi Research Centre, for example in the microprocessor MCM-L shown in Figure 1. The

TABLE 3. Materials Properties for MCM-Cs and Ds

| Material | | Dielectric Constant | TCE (x, y) (ppm/K) | Tensile Modulus (GPa) | Tg (°C) | Thermal Conductivity (W/mK) | Density (gm/cm$^3$ | Water Absorption (%wt) | Hole cutting methods (see note 1) |
|---|---|---|---|---|---|---|---|---|---|
| Alumina (HTCC) | | 9 to 10 | 6 to 7 | 370 | - | 32 | 3.9 | - | L, M |
| Glass-ceramics (LTCC) | | 4 to 8 | 3 to 8 | | - | 3 to 7 | ~3 | - | L, M |
| Aluminium nitride | | 8.8 | 4.1 | 310 | - | 160 to 220 | 3.3 | - | M |
| Silicon carbide | | 40 | 3.8 | 380 | - | 70 to 260 | 3.2 | - | M |
| Diamond | | 5.8 | 1 to 4 | | - | >2000 | | - | M |
| Polyimide | - "standard" | 3.3 to 3.5 | 40 to 60 | 2.4 to 3.4 | ~300 | 0.2 | | 1.1 to 5 | L, P |
| | - low stress | 2.9 to 3.3 | 3 to 20 | 5 to 12 | >400 | 0.2 | | 0.4 to 1.2 | L, P |
| | - fluorinated | 2.8 to 2.9 | 35 | 2.8 | <300 | 0.2 | | 0.8 | L, P |
| | - photosensitive | 2.8 to 3.5 | 13 to 45 | 3 to 6 | 280 to 400 | 0.2 | | 0.5 to 3 | C, L |
| PPQ | | 2.7 to 2.9 | 40 to 55 | 2 | >350 | | | 0.8 | L, P |
| BCB | | 2.7 | 65 | 2.3 | >350 | 0.2 | | 0.23 | L, P |
| Ptfe | | 2.1 | 20 | 1.5 | 240 | 0.1 | | 0.01 | L, P |
| Silicon dioxide | | 3.9 | 0.5 | | - | ~5 | 2.2 | - | C, P |
| Silicon | | 11.8 | 2.5 | | - | 108 to 168 (see note 2) | 2.5 | - | C, L, M |
| GaAs | | 12.9 | 6 | 112 | - | 38 | 5.3 | - | M |
| Kovar | | - | 5.9 | 7 | - | 17 | 8.3 | - | C, M |
| Copper-tungsten | | - | 6 to 7 | 133 | - | 205 to 250 | 15 to 17 | - | M |

Notes:
1. Hole cutting methods: C = chemical, L = laser, M = mechanical, P = plasma/RIE.
2. Thermal conductivity of silicon decreases as temperature rises, these two figures are for 100°C and 0°C respectively.

TABLE 2. Materials Properties for MCM-Ls

| Material | Dielectric Constant | TCE (x, y) (ppm/K) | Tensile Modulus (GPa) | Dimensional Stability (%) | Tg (°C) | Thermal Conductivity (W/mK) | Density ($gm/cm^3$) | Water Absorption (%wt) | Hole cutting methods (see note) |
|---|---|---|---|---|---|---|---|---|---|
| Epoxy/glass (FR4) | 4.8 | 18 | 17 | 0.04 | 130 | 0.3 | 1.7 | 0.1 to 1.3 | M |
| BT epoxy/glass | 4.1 | 18 | | 0.05 | 190 | 0.3 | 1.7 | 0.5 to 1.3 | M |
| Polyimide/glass | 4.2 | 18 | 19 | 0.04 | 260 | 0.3 | 1.8 | 0.2 to 1.3 | M |
| Polyimide (sheet) | 3.5 | 24 | 5.5 | 0.05 | | 0.3 | 1.4 | 0.1 to 0.5 | L, M, P |
| Cyanate ester/ptfe | 2.7 | 23 | | 0.02 | 225 | 0.3 | 1.5 | 1.0 | L, M |
| Cyanate ester/aramid | 3.3 | 7 | | 0.05 | 240 | 0.3 | 1.3 | 0.08 | L, M |
| Epoxy/aramid | 3.9 | 8 | 14 | 0.06 | 180 | 0.3 | 1.3 | 0.85 | L, M |
| Ptfe/ceramic | 2.6 to 10 | 17 | 0.7 to 2 | 0.03 | - | ~0.4 | 2.1 to 2.9 | <0.1 | L, M |
| Copper | - | 17 | 130 | - | - | 390 | 8.9 | - | C, L, M |
| Aluminium | - | 24 | 70 | - | - | 210 | 2.7 | - | C, L, M |
| Al/SiC | - | 6 to 17 | 100 to 265 | - | - | 160 to 200 | 2.7 | - | C, L, M |
| Copper/Invar/Copper | - | 4.6 to 5.2 | 135 | - | - | 107 to 170 (x, y) | 8.4 | - | C, M |

Note: Hole cutting methods: C = chemical, L = laser, M = mechanical, P = plasma/RIE.

fine-feature layers are supported by a base of conventional PCB materials, in this case epoxy/glass. Improved performance, environmental ruggedness, miniaturisation and cost can be achieved by bonding the fine-feature layers together with a fully-curing, homogeneous material, exactly as in PCB manufacture, instead of with adhesives. The laminates which use cyanate ester or multi-functional epoxy can be used this way and some of them provide a lower TCE, higher Tg and lower water absorption into the bargain. Add to that the ability to cut miniature vias by laser and it's easy to see why these materials are being evaluated by many, including GEC-Marconi.

**Figure 1: Microprocessor MCM-L containing a total of 22 unpackaged ICs, a clock IC in a ceramic package and surface mount capacitors (on both sides)**

In order to reduce the TCE of the MCM-L substrate to decrease the mismatch to the assembled silicon, thin (0.1 mm) metal planes are incorporated. Presently copper/Invar/copper, as in PCBs, these planes also extract heat, sometimes with the aid of thermal vias, and increase the mechanical strength and stiffness of the MCM-L, making it perhaps the most rugged of all MCM substrates, including under very high acceleration/shock. As an alternative, metal matrix composites, such as aluminium/silicon carbide, are becoming available and are being evaluated.

The same technology, albeit with slightly relaxed design rules, is being used for fine-feature PCB motherboards to carry MCMs and other packaged components, with greatly reduced layer count, weight and cost, compared to conventional PCBs.

### 3.2 MCM-Cs

Among the most interesting developments in materials for MCM-Cs are the numerous low-temperature co-firable dielectrics. Other future possibilities include co-fired aluminium nitride and materials with extremely high dielectric constant.

Since many of the other contributions to this Workshop deal exclusively with MCM-Cs, two examples will suffice to highlight how careful design and material selection can solve problems. For high power modules, a natural choice for the substrate is aluminium nitride, since it has low density, low dielectric constant and low TCE as well as very high thermal conductivity. This can be augmented by direct-bonded copper for the surface/die-attach layer (which is thick enough to have good electrical and thermal conductivity) and by using copper-tungsten for the package base.

Many materials commonly used in hybrids and MCMs are ferromagnetic and must be avoided in some applications. Figure 2 shows a circuit which is used in quantity in a medical body-scanner. The ceramic substrate/package base avoids having a metal package and the pins are made of an alloy which does not need a nickel solder barrier. Nickel barriers on the components are also avoided by using a conductive adhesive instead of solder attachment.

**Figure 2: A ceramic-based circuit used in a medical body-scanner, this assembly avoids all ferromagnetic materials.**

3.3 MCM-Ds

Silicon provides a very good basis for MCM-Ds because of its thermal expansion coefficient, its smooth surface (which facilitates ultra-fine feature lithography) and its good thermal conductivity. Silicon is the most common material chosen as the basis of MCM-Ds but ceramics are an acceptable alternative, particularly if they have smooth surfaces and some other particularly desirable attribute such as a high thermal conductivity or combinations of low and high dielectric constant (for interconnection layers and capacitor layers respectively). There is a growing army of polymer dielectrics, each seeking to offer a favourable balance of properties such as dielectric constant, water absorption, adhesion, TCE, stress, cost, etc. At present there are no clear overall winners since none of the materials can provide the desired value of all these properties simultaneously. Using a "standard" polyimide for the multilayer dielectric, our first silicon MCM-Ds have been memory modules. Figure 3 shows a prototype which is part of our programme to develop active substrate MCM-Ds in which a commercially available memory wafer is used as the substrate. The added-on ICs are memory and interface chips. The project includes material selection and process adjustment in order to build more complex processor modules, also using memory in the active substrate. The ability to de-select defective "cells" in the active substrate is a key issue, both during the manufacturing cycle and for reconfigurability in-service in high reliability applications.

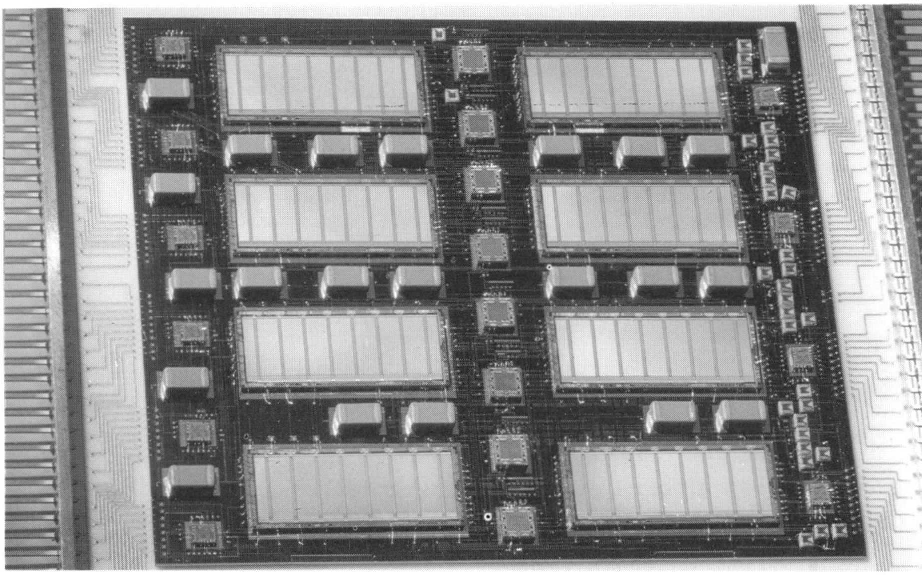

**Figure 3: A 16 Mbit memory module MCM-D incorporating an active silicon substrate**

3.4 OPTOELECTRONIC MCMs

Three distinct technology streams are converging towards one technique for the production of optoelectronic MCMs: high density electrical interconnection; micromachining in silicon, for precise alignment of optoelectronic components or fibres; silica waveguides, for optical interconnection and to create optical functions such as wavelength multiplexing and coupling. At GEC-Marconi Research Centre we have adopted silica because it satisfies the optical and fabrication-related requirements and is compatible with the use of silicon as the base of the MCM for either a purely optical MCM or an optoelectronic circuit [3]. The use of a common substrate onto which the optical, electronic and optoelectronic components are integrated, using micromachined structures for precise, passive, optical alignment, is advantageous. Since the optical components will occupy only a small part of the substrate area determined by the optical waveguide layout, the addition of electronic functions, such as amplification, logic circuitry or analogue to digital conversion, will save space elsewhere in the system and decrease the susceptibility to signal corruption of the electrical output. Figure 4 shows schematically a typical optoelectronic MCM, incorporating an integrated optical circuit, silicon and optoelectronic ICs, singlemode waveguide on the MCM and fibres, aligned to the waveguide. For connection to the outside world, the extremely high bandwidth which the fibres can accommodate can be used to avoid the difficulties associated with very large numbers of electrical connections to MCMs.

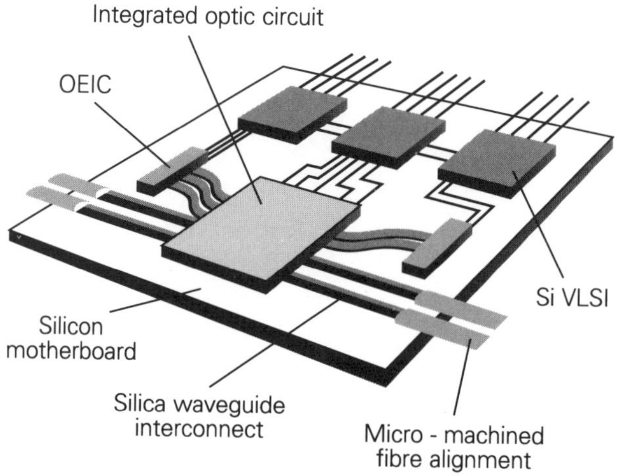


**Figure 4: A schematic of a typical optoelectronic MCM**

## 4 Conclusion

There is a wide and increasing range of materials and processes with which MCM substrates and packages can be made. All types of MCM technology can provide considerable benefits compared to previous technologies. The designers can select for the MCM substrate, assembly methods and packaging, the materials, processes and design features which are most suitable for a particular product or range of products. Inevitably, the choice will be constrained by factors such as cost, availability in the quantity and timescale required and environmental performance, but it is clear that the use of new materials and processes and of carefully selected existing techniques for advanced packaging will improve system performance.

## 5 Acknowledgements

Parts of the work described here were supported by GEC-Marconi Product Units, by the ESPRIT projects APACHIP and CHIPPAC (supported by the European Commission).

## 6 References

1. N. Chandler and S.G. Tyler (1994). "Ultra-fine feature printed circuits and multichip modules", GEC Journal of Research, 11(2) 90-98.
2. L.G. Salmon (1993). "Evaluation of thin film MCM materials for high speed applications", IEEE Trans. CHMT, 16(4), 388-391.
3. G.N. Blackie and I.R. Croston (1993). "Silicon optohybrids for advanced optoelectronic multichip modules", GEC Journal of Research, 10(2) 106-110.

# LOW PERMITTIVITY POROUS SILICA THIN FILMS FOR MCM-C/D APPLICATIONS

ROSARIO A. GERHARDT*, JULIE R. KOKAN and PAUL A. KOHL
*Georgia Institute of Technology*
*School of Materials Science and Engineering*
*Atlanta, GA 30332-0245*
*Tel: (404) 894-6886*
*Fax: (404) 853-1940*
*e-mail: rosario.gerhardt@mse.gatech.edu*

## ABSTRACT

Preliminary results indicate that nearly crack-free 2.5-12 $\mu$m thick low permittivity porous silica thin films can be made by dipping aluminum coated slides into gel solutions prepared by a colloidal processing method. We have been successful in making reproducible films with dielectric constants less than 2 and dielectric losses less than 5%(for kHz-MHz frequencies). The films permittivity and loss increase if exposed to high relative humidities but may be lowered again by an appropriate heat treatment. Suitable hydrophobic coatings are currently being sought. Adaptation to spin coating for further implementation of these films into MCM-C/D applications is being carried out.

## 1. Introduction

Silica has been used as an insulator in microcircuits for many years but for high speed applications, it is generally replaced by polymeric materials such as polyimide or teflon because they have lower dielectric constants(~3 vs. 3.8). The most promising polymeric material is benzocyclobutene because of its low dielectric constant of 2.8 [1]. Nevertheless, an even lower dielectric permittivity material is desirable due to the dependence of signal propagation delay on the square root of the dielectric constant of the medium in which the signal travels. The most ideal dielectric would therefore be vacuum or void space. Since the conducting lines need to be mechanically supported and electrically screened, the next best solution is a porous dielectric.

The major objective of on-going work at Georgia Tech in this area has been to develop a method for the formation of reproducible and homogeneous low permittivity porous silica thin films for application as interlevel dielectrics in integrated circuits and multichip modules. We have chosen colloidal sol-gel processing because porous silica



*W. K. Jones et al. (eds.), MCM C/Mixed Technologies and Thick Film Sensors*, 209–219.

can be prepared by a variety of sol-gel methods and the chemistry is fairly well understood [2]. Earlier attempts to make porous silica thin films for use in integrated circuits have included rf sputtering [3], colloidal processing [4] and mixing polymers with colloidal silica [5]. In all cases, difficulties in drying of the films without cracking or not achieving a high enough porosity content to sufficiently lower the dielectric constant discouraged them from pursuing using these films further.

A method which results in a homogeneous and reproducible distribution of pores in thin film form is therefore necessary to minimize defect formation and short circuiting. Most conventional thin film deposition methods such as sputtering, e-beam evaporation and even sol gel are designed to result in dense films. For these reasons, a colloidal processing method that we are familiar with was chosen [6]. This method results in as-dried bulk gels that are 15% dense with a porous microstructure which is homogeneous on a scale smaller than 1 $\mu$m [7-10]. The microstructure and dielectric properties of this material have been studied at length in the bulk form [7-15] and it has been shown that bulk porous silica may be prepared this way with a dielectric constant as low as 1.6 [11]. Making this material in thin film form in a reproducible way would be a revolutionary breakthrough for high density electronic systems, because silica's thermal conductivity is about ten times that of most polymers and its cost is far less. Early attempts at making porous silica thin films by this method have been reported elsewhere [16].

## 2. Experimental

The thin films were prepared by mixing varying ratios of potassium silicate and colloidal silica and hydrolyzing the mixtures with diluted formamide [7]. After forty-five minutes of gelation, aluminum coated slides were dipped in the gel solution for ten seconds twice. They were then left to dry in a controlled humidity environment for twenty-four hours. After gelation, the films were leached in order to remove the potassium and sodium introduced during sample preparation. It is important to eliminate the residual alkali ions not exuded during gelation because they lead to the formation of crystallites and cause a high dielectric constant in the films. An earlier account of this work [16] described the effect of various leachants on the microstructure and dielectric properties of the films. Because leaching in deionized water resulted in the best dielectric properties(3.0 at 1 kHz), dielectric data reported here will be for films that were leached in deionized water for varying amounts of time ranging from 5-40 minutes.

After leaching and a period of drying in air, films were examined in an optical microscope to determine whether any cracks had formed. For coherent uncracked films, a top electrode was applied using silver paint. After the silver paint was dry, the capacitance and dissipation factor of each sample were measured and the dielectric constant calculated. Because of the high porosity of these films, humidity affects their dielectric properties so measurements were carried out after drying films and also as a function of time in a 60%RH environment. The microstructures were monitored at all these steps and photomicrographs were obtained.

## 3. Results

Reproducible films of approximately 2.5-12 $\mu$m thickness have been successfully manufactured. These films are originally crack free and homogeneous (see Figure 1). Adherence to aluminum is reliable as determined by a scotch tape test. If films are not leached shortly after gelation is complete, they will form potassium-rich crystallites as shown in Figure 2. The dielectric properties of these films are unsuitable for the current application but may have application as microelectronic humidity sensors [12,14,17,18]. Even though the crystallites are water soluble and may be removed, it is necessary to prevent their formation.

Bulk porous silica prepared by this method has been successfully leached by immersing it in several baths of $NH_4NO_3$ and $HNO_3$ [7]. The alkali content remaining after this treatment was less than 50 ppm and the dielectric properties were optimal [11]. However, this treatment is too harsh for the films and large cracks that can lead to short circuiting tend to form. A representative film leached in $NH_4NO_3$ is depicted in Figure 3. Several other leachants were attempted[17] but the best results were obtained by leaching with deionized water.

Several films were prepared simultaneously and were then leached in deionized water for periods ranging from 5-40 minutes. Their microstructures were recorded before and after leaching. Figures 4(a) and (b) show that the longer the films remain in the deionized water, the more cracking they developed. A final low temperature heat treatment done in order to minimize remnant water content does not affect the microstructures very much.

The dielectric constant vs. frequency response for fully dried samples leached for 5-40 minutes was measured. The data show a fairly small frequency dispersion with the lowest values being present at the frequencies of interest (> MHz). As shown in Fig. 5, the values obtained ranged as high as 4.6 at the lowest frequencies and as low as 1.6 at the highest frequencies. Similar trends can be seen for the dissipation factor depicted in Figure 6. It is worthwhile to remark that above 1 MHz the losses are less than about 5%. While longer leaching times appear to result in lower dielectric properties, effects of cracks forming and remnant alkali content affect the order in which they vary and further experiments are needed to sort this out. Nevertheless, the data is very encouraging and fairly reproducible. The best dielectric constant and accompanying dielectric loss obtained so far are shown in Fig. 7.

As has been observed before in the bulk [12], the films will quickly re-absorb water and increase their dielectric constant by as much as a factor of 10 (especially at lower frequencies) when placed in a humid atmosphere. The increase in dielectric constant at 1kHz due to exposure at 60%RH for the film reported in Fig. 7 is shown in Fig. 8. With proper re-drying, however, the dielectric properties can be restored to the initial dried values shown in Fig. 6 without apparent damage to their microstructure. Similar effects are observed for the dissipation factor.

## 4. Discussion

We have been successful in fabricating porous silica films of thicknesses ranging from 2.5-12 um. The dielectric properties of these films are excellent. If films are dried and kept at a low %RH, the dielectric constant is found to be 1.6 in the MHz range (and presumably lower at higher frequencies) making these films serious competitors in the low permittivity arena. A dielectric loss of 0.05 is not yet optimum but further improvements in the leaching procedures should result in lower losses.

Leaching with deionized water has been effective but at longer periods than 20 minutes, the formation of cracks seems to degrade the structure so much that it can cause short circuiting of the films. More work needs to be done to assess the best compromise between the length of time needed to remove as many alkali ions as possible but yet minimize cracking. Another area of concern is the ability of these materials to pick up moisture from the ambient. Silica surfaces are inherently hydrophobic, but the presence of silanols, OH radicals, or chemisorbed water causes them to become hydrophilic [19]. At low frequencies, the adsorbed water has a greater effect on the permittivity than it does at higher frequencies [12] but the water content needs to be kept low nevertheless. It has been found that physisorbed water may easily be removed by an appropriate heat treatment. Since for many applications these films would be used under hermetic conditions, water adsorption may not be as large an issue at it may first appear. For those applications where the films would not be in a hermetic package, it will of course be necessary to identify suitable coatings to prevent re-adsorption of water under humid conditions.

## 5. Future Plans

The films have so far been prepared by dipping aluminum coated slides in a gel solution, but efforts are underway to introduce spin coating since this method would be more compatible with the current processing of integrated circuits and multichip modules. Metallization without short circuiting has been accomplished using commercial silver paint but conversion to vapor deposition, sputtering or screen printing needs to be made. Characterization of the inherent porosity and alkali content of the films also need to be made, although the dielectric results indicate many similarities with the material in the bulk form suggesting that the microstructure must be similar as well. Exploration of the proper method for via formation in these materials will also be necessary in order to fully implement them as interlevel dielectric layers in MCM-C/D applications.

## 6. Conclusions

A promising new processing method for producing porous silica thin films for use as

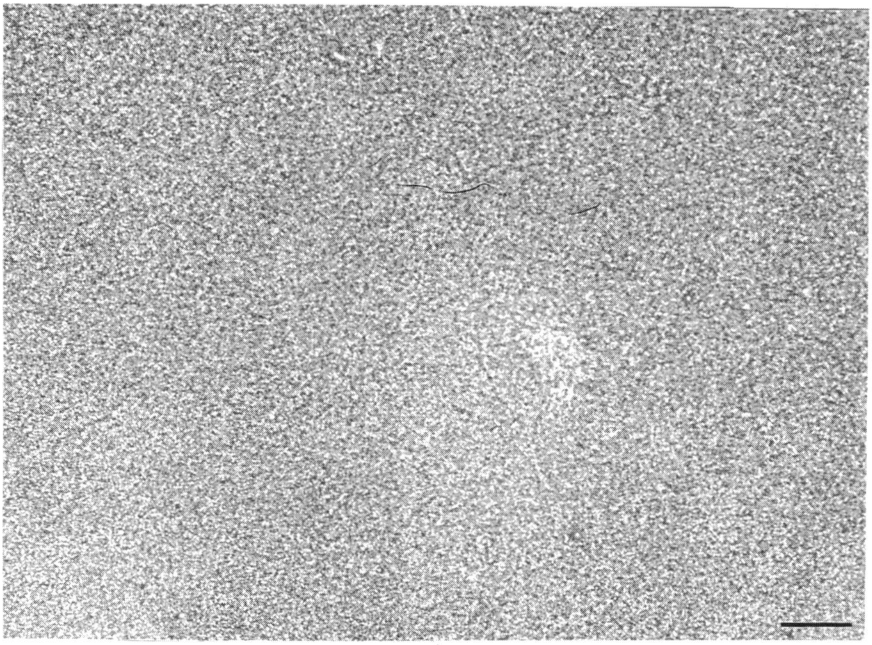

Figure 1: Unleached porous silica thin film showing no cracks or inhomogeneties. Bar represents 0.01cm

Figure 2: Formation of potassium-rich crystallites in unleached thin films. Bar represents 0.01 cm.

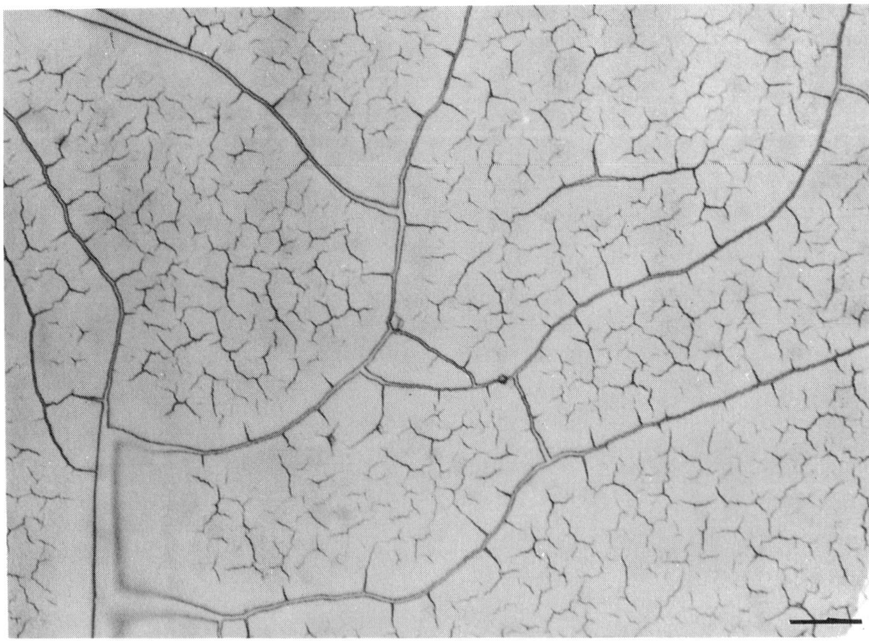

Figure 3: Cracks formed due to leaching films in $HNO_3$ and $NH_4NO_3$. Bar represents 0.01 cm.

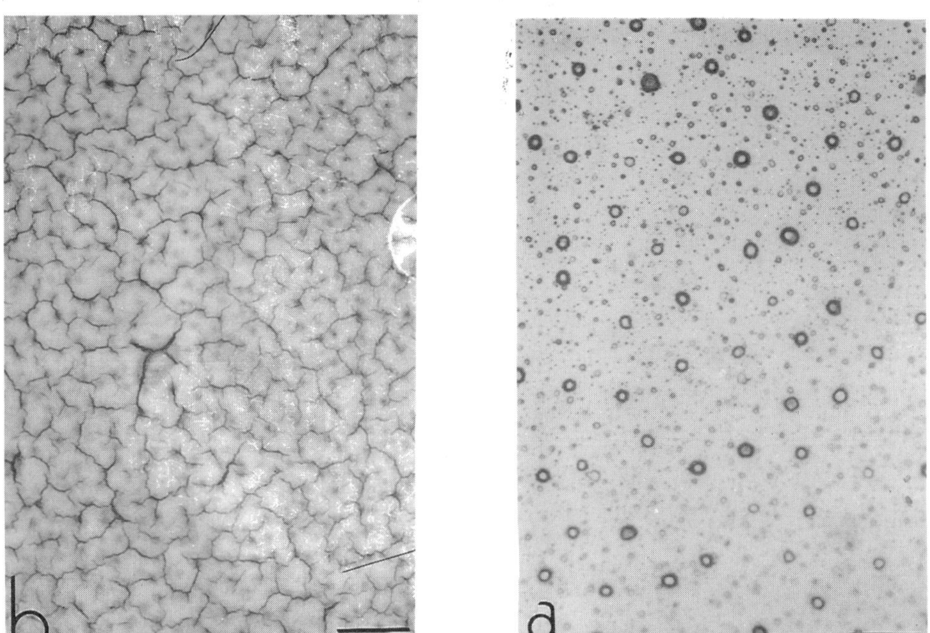


Figure 4: Microstructure of films leached in deionized water for (a) 5 minutes (b) 40 minutes. Bar represents 20 $\mu$m.

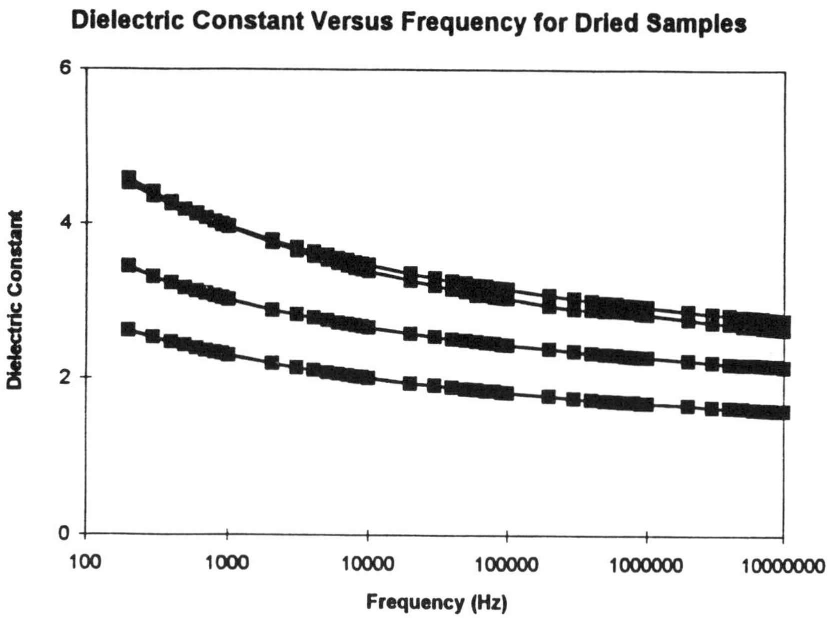


Figure 5: Dielectric constant vs. log frequency for several samples leached in deionized water.

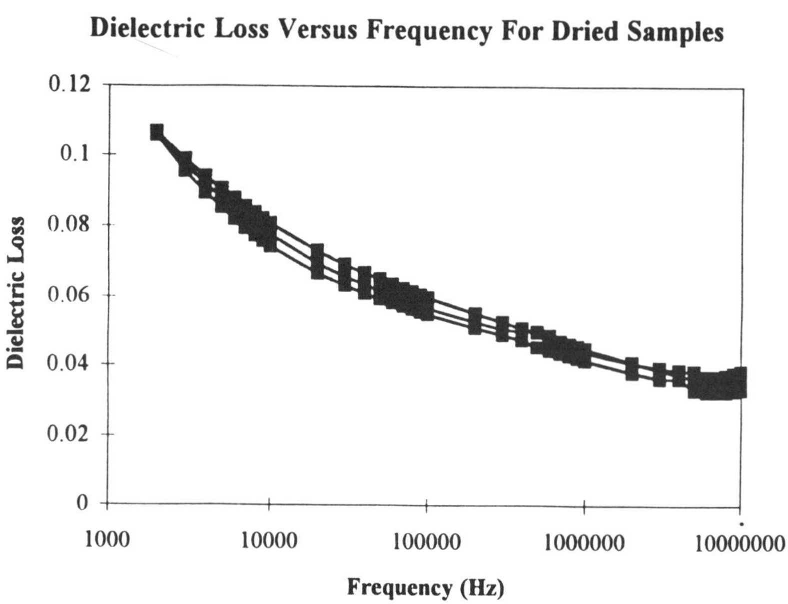


Figure 6: Dielectric loss vs. log frequency for several samples leached in deionized water.

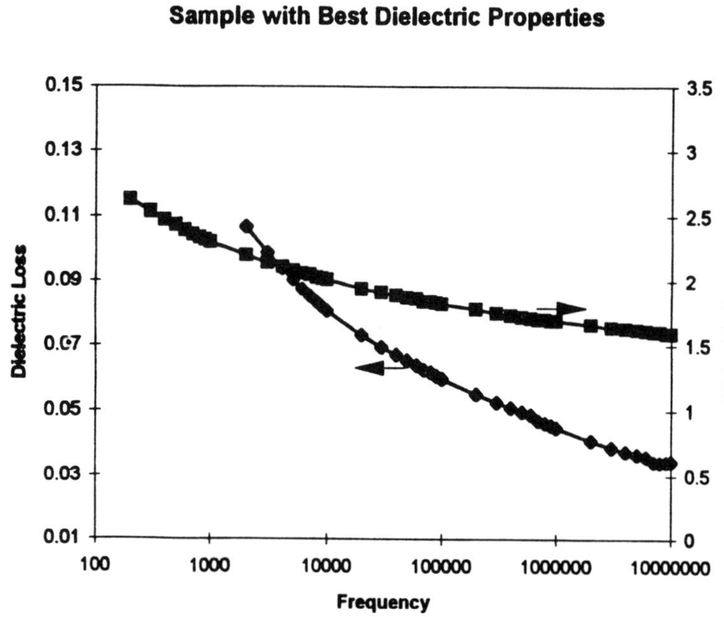


Figure 7: Dielectric constant and loss vs. log frequency for best sample obtained.

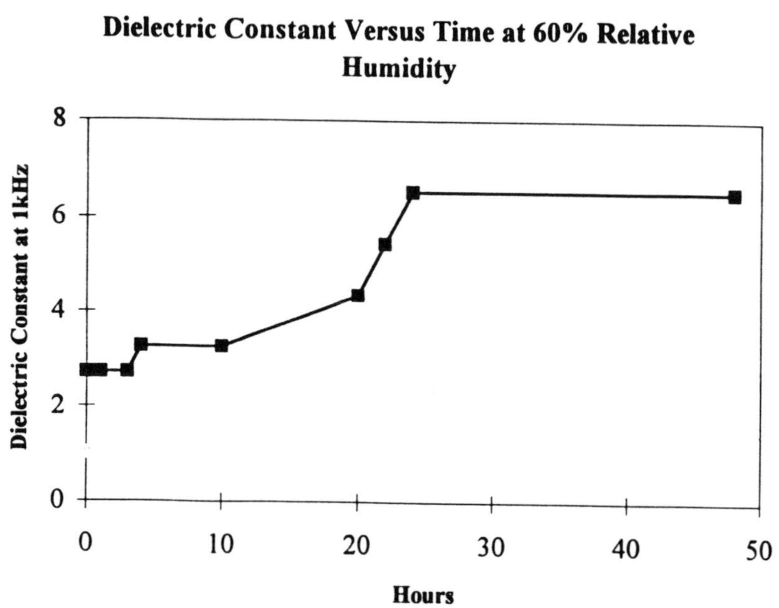


Figure 8: Increase in dielectric constant of sample in Fig. 7 as a result of exposure to 60% RH for over 50 hours.

interlevel dielectrics in integrated circuits and multilayers on MCM's has been found. The best dielectric properties of these films obtained so far indicate a dielectric constant of 1.6 and a dielectric loss of 0.05 at 1 MHz and presumably lower at higher frequencies. These properties make these films competitive with the best polymers currently being used. Additional work needs to be done to fully incorporate these films into ICs and MCMs so that they can be used as the dielectrics in those layers where the design requires minimum propagation delay.

## 7. Acknowledgements

The authors would like to acknowledge the Focused Research Program of GTRC for funding this research.

## 8. References

1. Balde, J.W., (1991) Overview of Multichip Technology, *Multichip Modules: System Advantages, Major Constructions, and Materials Technologies*, ed. R.W. Johnson, R.K.F. Teng, and J.W. Balde, IEEE Press, pp. 297-312, 1991.

2. Brinker, C.J. and Scherer, G.W., (1990) *Sol-Gel Science: The Physics and Chemistry of Sol-Gel Processing,* Academic Press..

3. Das, A., Messier, R., Gururaja, T.R. and Cross, L.E.(1986) Low Permittivity $SiO_2$/Void in Nanocomposite Films, *Electronic Packaging Materials Science II*, MRS Proc. **72**, pp. 27-33.

4. Yarbrough, W.A., Gururaja, T.R. and Cross, L.E. (1987) Materials for IC Packaging with Very Low Permittivity via Colloidal Sol-Gel Processing, *Ceramic Bulletin* **66** [4], pp. 692-698.

5. Mohideen, U., Gururaja, T.R., Cross, L.E. and Ray, R. (1988) Ultra-low Dielectric Constant Porous Silica Thick Films for High Speed IC Packaging, *IEEE Trans. Comp. Hyb. Manufact. Tech.* **11** [1], pp. 159-162.

6. Shoup, R.D. and Wein, W.J. (1977) Low Temperature Production of High Purity Fused Silica, U.S. Patent No. 4059658, November 22, 1977.

7. Cao, W., Gerhardt, R. and Wachtman,Jr., J.W.(1988) Preparation and Sintering of Colloidal Silica: Alkali Silicate Gels, *J. Amer. Ceram. Soc.* **71** [12],1108-1113.

8. Bhattacharja, S., D'Orazio, F., Tarczon, J.C., Halperin, W.P., and Gerhardt, R., (1989) Internal Structure of Porous Silica: A Model System for Characterization by Nuclear Magnetic Resonance, *J. Am. Ceram. Soc.* **72** [11] 2126-2130.

9. Kerch, H., Gerhardt, R., and Grazul, J., (1990) Quantitative Electron Microscopic Investigation of the pore structure in 10:90 Colloidal Silica-Potassium Silicate Sol-Gels, *J. Am. Ceram. Soc.* **73** [8] 2228-2237.

10. Kerch, H.M., Cosandey, F., and Gerhardt, R., (1993) Imaging of Fine Porosity in Silica Gels by a Defocus Contrast Technique, *J. Non-Cryst. Solids* **152** [1], 18-31.

11. Cao, W., Gerhardt, R., and Wachtman, Jr., J.B., (1989) Low Permittivity Porous Silica by a Colloidal Processing Method, Advances in Ceramics, Vol.26, *Ceramic Substrates and Packages for Electronic Applications*, ed. M. Yan, K. Niwa, H. O'Bryan, Jr. and W.S. Young, pp. 409-418.

12. Cao, W., Gerhardt, R., and Wachtman, Jr., J.B., (1990) Dielectric Relaxation of Water Adsorbed on Monolithic Porous Silica Gels, in *Ceram. Trans. vol. 8: Ceramic Dielectrics*, ed. H.C. Ling and M.F. Yan, 175-184.

13. Long, G.G., Krueger, S., Jemian, P.R., Black, D.R., Burdette, H.A., Cline, J.P., and Gerhardt, R.A., (1990) Small Angle Scattering Determination of the Microstructure and Morphology of Porous Silica Precursors, *J. Appl. Cryst.* **23**, 535-544.

14. Cao, W., Gerhardt, R., and Wachtman, Jr., J.B., (1990) Effect of Sodium Ions on the Dielectric Conductivity of Porous Silica in Humid Environments, in *Physical Phenomena in Granular Materials, Mat. Res. Soc. Symp. Proc.* **195**, 471-476.

15. Long, G.G., Krueger, S., Gerhardt, R.A., and Page, R.A., (1991) Small Angle Neutron Scattering Characterization of Processing/Microstructure Relationships in the Sintering of Crystalline and Glassy Ceramics, *J. Mat. Res.* **6** [12] 2706-2715.

16. Kokan, J.R., Tadayon, F., Gerhardt, R., and Kohl, P., Sol-Gel Processing of Porous Silica Thin Films for Interlevel Dielectrics in Integrated Circuits, (1994) *Proc. 44th Annual Electronic Components and Technology Conference, Washington, D.C., May 1994, IEEE Proc. 1994*, pp.704-711.

17. Zhang, G., M.S. Thesis (1993) Effects of Alkali Ions on the Dielectric Properties of Porous Silica Prepared By a Colloidal Processing Method, *The Georgia Institute of Technology*.

18. Gerhardt, R., Zhang, G. and Cao, W. (1992) Analysis of Impedance Spectra for Dried Porous Silica Gels Exposed to Various Humidities, *2nd International Conference on Electrochemical Impedance, University of California, Santa Barbara, CA*, July 1992, to be published.

19. Cao, W., Ph.D. Thesis (1990) Processing, Pore Microstructure and Dielectric Properties of Bulk Porous Silica, *Rutgers, The State University of New Jersey*, University Microfilms, Ann Arbor, MI.

# MCM PASSIVATION STUDIES FOR ENHANCED PRODUCIBILITY AND RELIABILITY

THOMAS J SANDERS, C. ROY MILLER, KURT E. GSTEIGER
GLENN T. HESS AND DURAISWAMI NEDUNCHELLYAN
*Florida Institute of Technology*
*Melbourne, Fl 32901*

## Abstract

This paper presents the research activities of the Florida Institute of Technology in the area of advanced electronic packaging . A number of research projects have been conducted all aimed at increasing producibility and reliability of multichip modules. The primary concentration of effort has been on improved passivation processes. Data is reported on silicone gels as well as silicon nitride for protective overcoat on integrated circuits.

## Introduction

The advanced packaging research program at the Florida Institute of Technology has focused on improving the producibility and reliability of multichip modules(MCM). These two attributes are often trade-off parameters in that to achieve increased producibility, process simplifications are made which result in decreased reliability. the thrust of this research has been to study these trade-offs and to develop advanced processes and design techniques which give the best of both worlds, high producibility and high reliability.

The need for electronic systems which have lower cost, increased packing density, and higher speed has been the primary driver[1] for semiconductor packaging. In the past few years, the dominant packages have evolved from dual-in-line packages, to chip carriers, to the present day multichip modules. Each change brings a reduction in size, an increase in speed, and a potential reduction in cost. Each change also increases the interconnect density and the complexity of the package, providing a variety of challenges for the packaging engineer. Providing a reliable MCM while remaining cost effective involves determining the proper combination of materials and technologies. Therefore, investigations of new materials for MCM manufacturing is a necessary research activity[2].



*W. K. Jones et al. (eds.), MCM C/Mixed Technologies and Thick Film Sensors,* 221–226.

A primary concern for reliability of multichip modules is proper passivation. MCM,s require special passivation materials in that most often processing temperatures during passivation must be kept very low. To solve this problem, the IEEE Computer Society Packaging Committee Special Task Force(IEEE Gel Task Force) [3] has investigated the effectiveness of silicone gels as a good moisture and ion barrier. They reported that silicone gel coatings over the integrated circuit had improved but not perfect reliability performance. In addition, it was found the specific processing parameters are very important in the gel's performance.

**Multichip Module Fabrication**

An MCM fabrication process has been developed at Florida Institute of Technology to achieve a number of performance goals. These goals include:
1. High reliability and capable of withstanding specific qualification tests,
2. Fast speed performance compatable with the university fabrication equipments,
3. Encapsulated chips with overall MCM thickness less than 0.05 inches.

To achieve these goals an alumina substrate was chosen which is about 10 mils thick as the base of the MCM. The substrate is metalized with sputtered titanium (400A), palladium (2000A), and gold (300A). For reliable interconnect, the gold was increased to 100 microinches by electroplating.

The layout required for patterning the metal is designed using Finesse software from Harris EDA running on a Sun sparkstation. The pattern which consists of interconnect lines and bondpads is imprinted into a mask and subsequently etched into the metallization on the substrate using a conventional photoresist and etch process. This is a low cost process capable of linewidths down to a few mils.

After the substrate preparation is complete, the active components are attached. In the test case reported here, these components are Sandia National Labs assembly test chips (ATC01)[4]. The die are normally attached using epoxy or special conductive adhesive. The die and metal lines are then connected using aluminum ultrasonic wirebonds.

A series of experiments have been conducted to investigate different passivation methods of the multichip modules. The techniques investigated are shown in Figures 1 thru 3. Figure 1 illustrates a passivation which consists of a layer of silicone gel to envirnomentally protect the gel followed by a ceramic lid for mechanical protection. This technique potential for good reliability at the lowest cost.

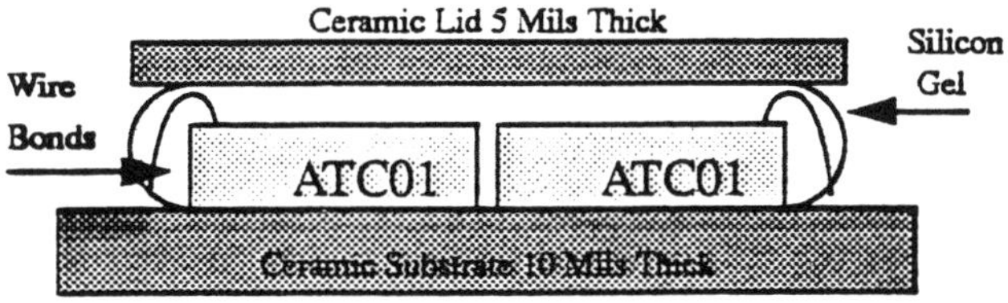


Figure 1. Ceramic lid on top of gel.

Figure 2 illustrates an improvement on the first technique from a mechanical standpoint. Here, dams are constructed around the chips and wirebonds for the purpose of containing the gel and forming a solid base for attaching the lid. In the prototype packages developed, the dams are made of thin ceramic strips attached with epoxy around the perimeter of the MCM.

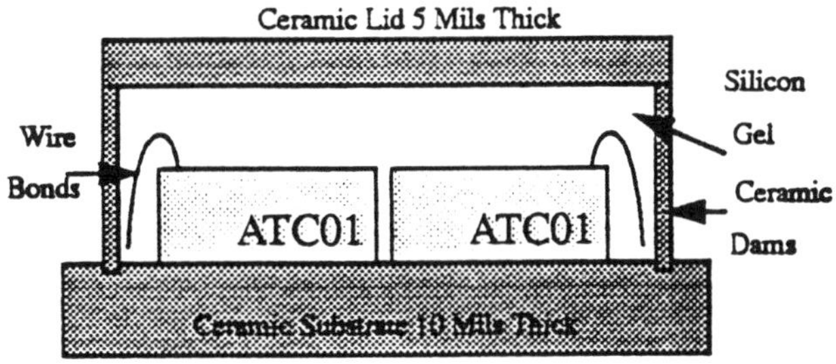


Figure 2.Ceramic lid on the dams.

Figure 3 shows a different approach which does not involve the silicone gel. In this process, an organic coating is applied and hardened to provide the mechnical protection. This is lower cost since a ceramic lid is not required, however, there is an expected loss of reliability without the silicone gel overcoat.

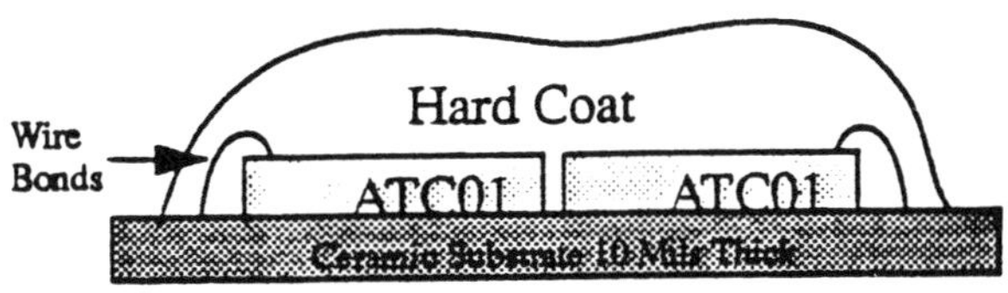


Figure 3. Hard coated MCM.

The reliability testing has shown mixed results. Electrical testing of the MCM was performed by measuring the resistance of several triple track structures on each chip. Each encapuslation technique provided the MCM with a high level of reliability for many of the tests, but no technique proved perfect for all tests. Complete details of the fabrication of the MCM and the reliability results are given in the referenced paper[5].

**Organic Passivation**

The use of silicone gels in the MCM fabrication process has led to questions concerning reliability on the interactions between the gel and other assembly materials. Specifically, a study has been conducted to investigate the interaction between the gel and the organic materials used for die attach[6].

The research at Florida Institute of Technology again used the ACT01 test chip with the triple track structures. Four different die attach adhesives have been employed in the study and in most cases put into a single chip ceramic dual-in-line package. After the die attach adhesive was cured per the manufacturers directions, aluminum wire bonds were made and conventional silicone gel was applied to the die. No lid was put on the units under test: however, lids were put on control units.

The units are then subjected to temperature cycling from -65C to 150C and the triple tracks tested for continunity. Units passing this test were then exposed to Highly Accelerated Stress Test (HAST) [7]. The HAST equipment at Florida Institute of Technology runs at 135C with 85% relative humidity. The test units experienced these conditions for a total duration of 1000 hours.

The units were removed from the HAST chamber at intervals and the aluminum triple track resistance was measured. The results of these measurements are illustrated in Figure 4. As seen in this graph, the resistance of the triple track structures increases in the initial 24 to 48 hour time period and then slowly decreases. The two curves shown are for structures which had different bias during the HAST exposure.

These results were consistant for all the die attach adhesives tested and all show good reliability characteristics. Further discussion of this work and of the testing capabilities at Florida Institute of Technology is given in reference 6.

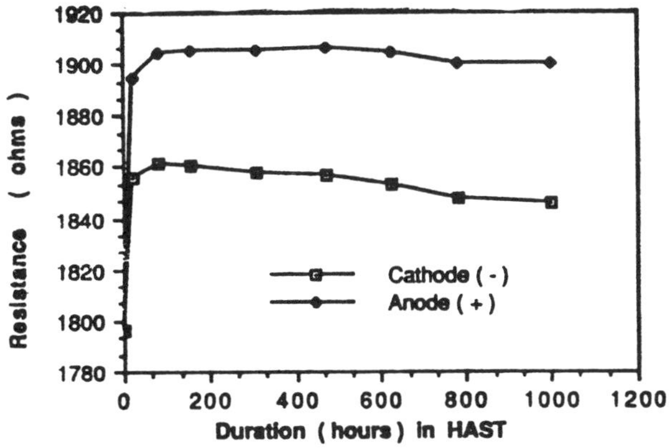


Figure 4. Triple track resistance as a function of HAST testing time.

**Inoganic Passivation**

In the search for an improved passivation material, the Florida Institute of Technology researchers decided to study silicon nitride. Because of the need to apply the passivation to multichip modules and possibly chip-on-board structures, the search has been restricted to silicon nitride deposited by chemical vapor deposition at very low temperatures.

The first silicon nitride studied was deposited at room temperature with a process which uses two plasmas with silane and nitrogen gas as the source materials. This proces holds much promise because a silicon nitride film with very low hydrogen content should be deposited. At Florida Institute of Technology the properties of films deposited with this technique were investigated using a variety of analytical and electrical techniques. The results have shown good potential for depositing a silicon nitride film which serves its primary purpose of blocking the diffusion of ionic contamination. In addition, the film conformally coats all surfaces and adheres well to the silicon chips and MCM components. However, the work to date have resulted in films with high oxygen content and questionable electrical properties. Further work needs to be performed to make this a manufacturable process.

In a similar direction, Florida Institute of Technology is investigating a new chemical vapor deposition technique which uses low frequency plasma. This process is of interest because the deposition equipment should be very low cost to purchase and to operate. This will allow a large number of small and medium size manufacturers of MCMs and PC boards to apply inorganic passivation such as silicon nitride and thus increase the reliability of their products. This research is ongoing and will be reported in the future.

**Conclusions.**

This paper has presented many of the research capabilities, projects and results in the area of advanced packaging at Florida Institute of Technology. Much of this effort is continuing and will result in improved processes and techniques for fabricating highly reliable multichip modules. In addition, this paper has presented much of the advanced fabrication and reliability testing available at Florida Institute of Technology. This work has been funded by Harris Corporation and the State of Florida.

**References**

1. Tummala, R. R., and Rymaszewski, E. J. Microelectronics Packaging Handbook, 1989.

2. Balde, J. W. "Multichip Packaging and the Need for New Materials", J. of Electronic Materials, vol., no., pp221-227, Feb 1998.

3. Balde, J. W., "The Effectiveness of Silicone Gels for Corrosion Prevention of Silicon Circuits: The Final Report of the IEEE Computer Society Packaging Committee Special Task Force", IEEE Transactions on CHMT, June, 1991.

4. Sweet, J. N., et. al., "Short and Long Loop Manufacturing Feedback Using a Multisensor Assembly Test Chip", IEEE Proceedings of the Ninth Int. Electronics Manu. Tech. Sym., 1990.

5. Hess, Glenn T., et. al., "Fabrication of Thin-Reliable Multichip Modules", Proceedings of the IEEE UGIM Symposium, May, 1992.

6. Nedunchellyan, D., et. al., "The Impact of Various Die-Attach Adhesives on the Aluminum Protection Characteristics of Silicone Gels", Proceedings of the Florida Microelectronics Conference, Tampa, FL, 1991.

7. Gavin, F., "Accelerated Stress Testing Reduces Test Time and Cost", Electronics Test, Dec., 1985.

# BURIED THICK FILM CAPACITORS BUILT UP WITH HIGH-K DIELECTRICS FOR MCM-APPLICATIONS

W. SMETANA
*Institut für Werkstoffe der Elektrotechnik, Technische Universität Wien, Gußhausstraße 27, 1040 Wien, Austria*

## 1. Introduction

Thick film multi-chip modules are characterized by a high interconnection density on the basis of multilayer techniques. Beside conductors and insulating layers, passive components like resistors can be produced in thick film technique while capacitors usually are applied as SMDs on the top of the multilayer. Nevertheless, thick film technique offers also the possibility to realize capacitors. But there is only a low capacitance capability associated with thick film systems. The interest in thick film capacitors increases as low firing high-K thick film dielectrics on the base of $BaTiO_3$ or relaxor ceramic are obtainable[1, 2]. The electrical properties of high-K thick film dielectrics are sensitive to added fluxes and conductor binder systems. Chemical reactions between the electrode binder and the dielectric may result in deleterious low K phases which lower the effective capacitance. Since high-K ceramic systems are porous and therefore sensitive to humidity, they have to be protected by an appropriate passivation. Often a two layered glass passivation has to be applied. This means an additional cost factor which must be taken into consideration. Beyond that a larger area has to be provided for a thick film capacitor in a plate configuration than for the attachement of a chip capacitor of the same capacitance value. Primarly this fact impedes a wide spread practical application of integrated thick film capacitors.

As the trend to further miniaturization continues, it is necessary to minimize the space on the surface of microcircuits occupied by passive components such as chip capacitors. The embeddement of thick film capacitors inside the multilayer structure can be considered as an approach to increase the packaging density of thick film multilayer circuits. Besides, no additional passivation of the high-K dielectric is necessary. In turn, electrical properties of the high-K dielectric are sensitive to the processing conditions, especially to the firing in the presence of conductors and multilayer dielectrics.

Since the high-K capacitors are more suitable for applications in which the tolerance is not very tight, one can bury them under other structures. In this case one must use a cross over or a multilayer dielectric to separate the capacitor from the circuit on top. It is well documented that the permittivity increases as the silver content in the electrode increases [1]. This must be related to the lower binder frit in Ag-conductors in comparison to conventional PdAg-conductors. To prevent bridging in multilayer dielectric due to migration of silver often the application of a buffer-dielectric is recommended.



*W. K. Jones et al. (eds.), MCM C/Mixed Technologies and Thick Film Sensors,* 227–237.

This paper will discuss the properties of different high-K dielectrics selected for buried capacitor applications. The embedded dielectric is exposed to multiple firing. This procedure enhances certainly the interaction between the high-K dielectric, the multilayer dielectric, and the electrode. A change of capacitance must be expected due to the sensitivity of high K-dielectrics to species diffusing from electrode and multilayer dielectric.

## 2. Experimental

### 2.1. TEST STRUCTURE

Elements which are not part of the dielectric composition may diffuse into a buried capacitor in two ways: One is peripherally from around the electrode pads, the other is through or from the electrode. The sum of these effects affects the dielectric properties. In the current study an attempt has been undertaken to specify the influence of parameters which contribute to a change of performance characteristics of high-K thick film dielectrics. Test samples are capacitors with an interdigitated electrode configuration which are overprinted with dielectric layers or combinations of layers under investigation. If the dielectric is applied in an appropriate thickness the stray field of the capacitor pattern should develop only inside the dielectric (Figure 1a). The total capacitance of the interdigital pattern results from the stray fields inside the alumina and the applied dielectric layer. It may be calculated using following equation:

$$C_{total} = (\varepsilon_{substrate} + \varepsilon_{layer}) \frac{K(k)}{K(k')} (N-1) l$$

where $\varepsilon_{substrate}$ and $\varepsilon_{layer}$ are the permittivities of the substrate and of the dielectric layer covering the interdigitated pattern, respectively [3]. K(k) an K(k') are complete elliptic integrals and are obtained from the geometry. N (10) is the number of sections comprising the periodic structure and l (8mm) is the length of the electrode fingers.

If the gap between the finger electrodes is larger than the thickness of the overprinted dielectric a certain fraction of the stray field will protrude beyond the surface of the dielectric (Figure 1b) which results in a lower total capacitance. If an additional conductive layer is applied, the total capacitance of this configuration is formed by the two serially connected capacitances which arise between each row of the interdigitated pattern and the overprinted conductor layer (Figure 1c).

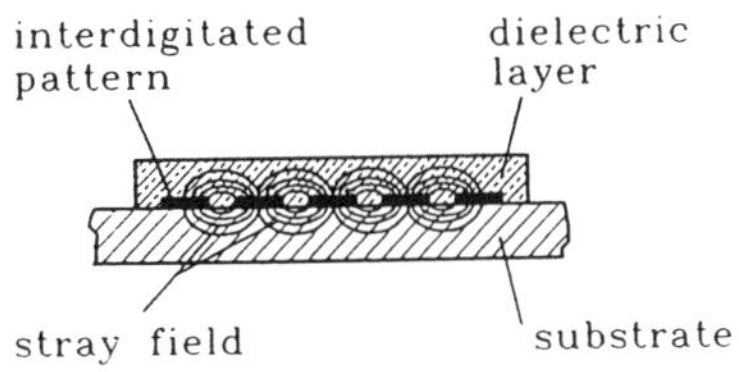


Figure 1a

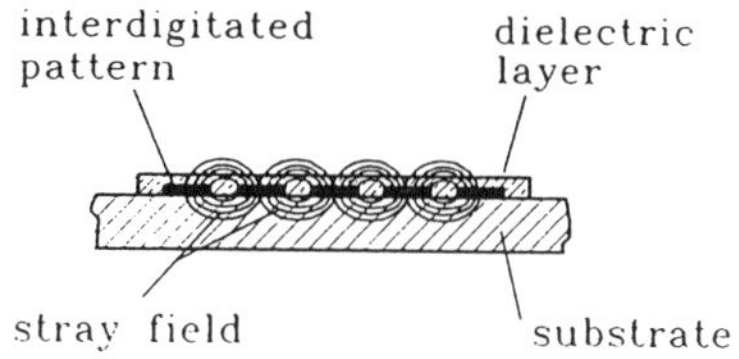


Figure 1b

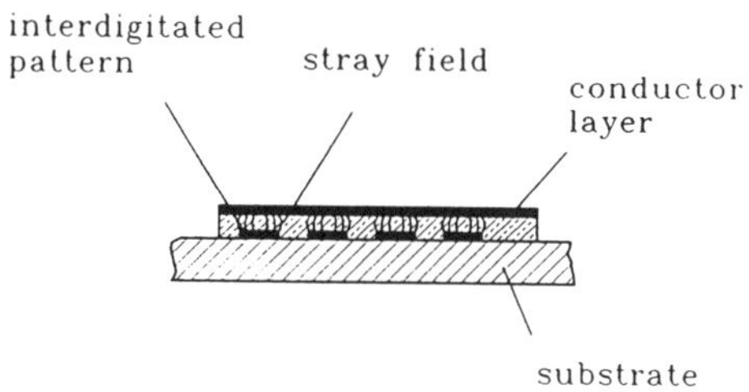


Figure 1c

Figure 1a-c. Distribution of stray fields of an interdigitated capacitor structure.

Standard thick film techniques were employed to prepare capacitor test samples on 96% alumina substrates. Table 1 lists the materials used in this study.

Table 1. Thick film materials

| Multilayer system: | Conductor (metal type) | High-K dielectric (K at 1kHz) | Multilayer Dielectric | Buffer |
|---|---|---|---|---|
| DuPont (1) | QM 14 (Ag) | 5540 (6000) | QM 42 | 5520 |
| DuPont (2) | QS 170 (AgPt) | 5540 (6000) | QM42 | 5520 |
| EMCA | 3325* (AgPd) | 7236 (6000) | 7001 | 7601 |
| Experimental | C (AgPt) | K(10000) | D | - |

* a not recomended system!

The capacitor structure consists of the interdigital conductor pattern, followed by two, up to four layers of dielectric. Conductors were printed with a 325 mesh screen coated with a 20μm thick emulsion. Dielectrics were applied with a 200 mesh screen with an emulsion thickness of 30μm. After printing each paste was dried for 10 minutes at 150°C before firing. The firing was conducted in a belt furnace with cycle duration of 60 minutes and 10 minutes at peak temperature of 850 °C. All layers were separately fired. The target fired thickness of the high-K dielectric was 70 to 80 μm. The thickness of the multilayer dielectric was 40 or 80μm respectively.

Due to the geometrical proportions of the gap width between the electrodes (150 μm) and the thickness of the dielectric (80 μm) the total capacitance becomes sensitive to any additionally applied layer with differing permittivity (Figure 1b). To evaluate the influence of multilayer compositions on the performance of the high-K dielectric, batches of different samples with varying sequences of layer deposition have been prepared (Figure 2):

Arrangement A: electrode pattern overprinted by high-K dielectric (thickness: 70-80μm).

Arrangement B: structure identical with arrangement A, but overprinted -in case of need- with buffer-layer (thickness:20μm) and multilayer dielectric (thickness: 40μm or 80μm).

Arrangement C: structure identical with arrangement A, but overprinted with a conductor layer.

Arrangement D: structure identical with arrangement C, but covered with multilayer dielectric (thickness: 80μm).

The printed areas of all applied layers are identical.

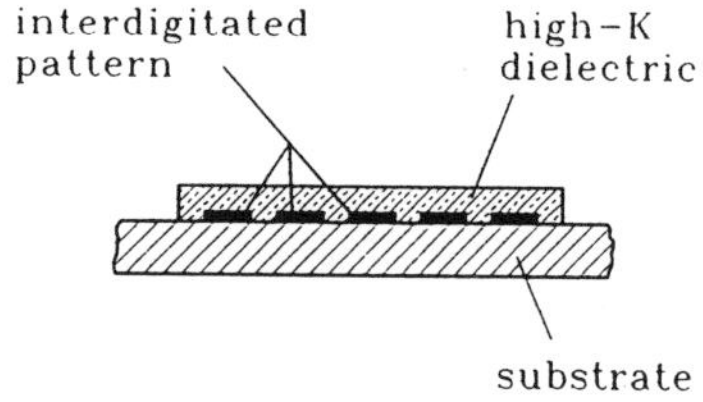


arrangement A

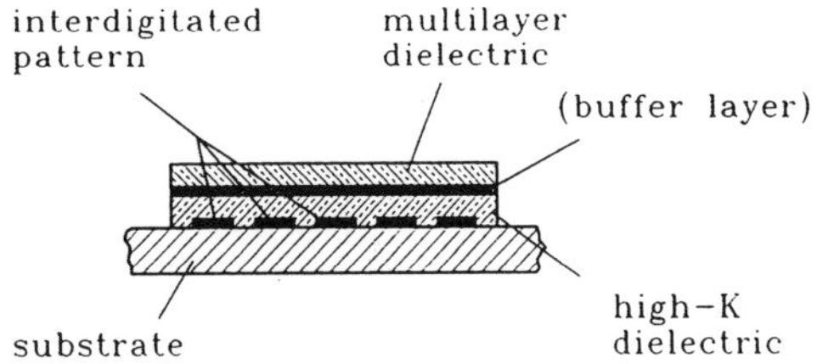


arrangement B

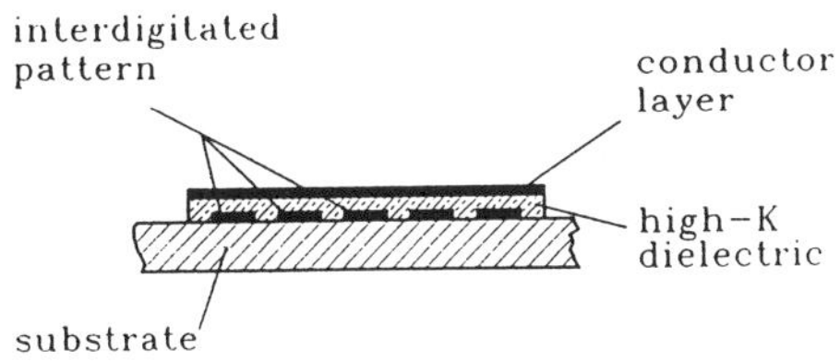


arrangement C

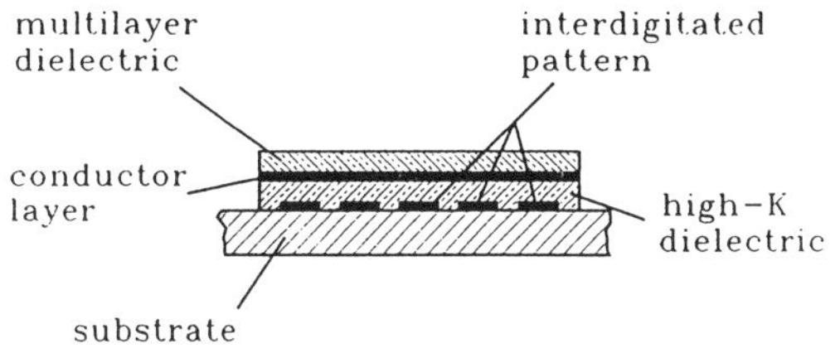


arrangement D

Figure 2. Arrangement in layers

*2.1.1. Results*

Capacitance and dissipation factor were measured with a HP4192A impedance analyser. Measurements were carried out using an automated equipment over the temperature range -55°C -125°C realized by a thermostream (Froilabo).

The performance of dielectrics is characterized by a change on capacitance. In order to compare the performance of different high-K dielectrics a normalized graphical representation for the capacitance shift at a frequency of 1kHz has been selected. Figure 3 shows the temperature characteristics for the high-K dielectrics under investigation (arrangement A, Figure 2).

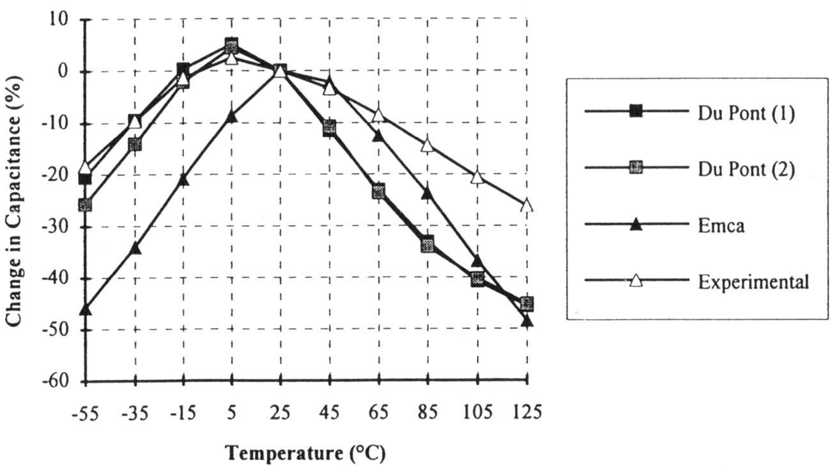


Figure 3. Temperature dependence of capacitance.

The maximum of the capacitance curve (Curie-temperature) shifts from 5°C to 35°C and the curve becomes broader as the high-K dielectric of the Du Pont(1) system is overprinted by the buffer layer as well as by the multilayer dielectric (thickness: 40 μm) according to arrangement B (Figure 2 and Figure 4). Due to the dilution of the ferroelectric phase the capacitance drops slightly as an additional layer of multilayer dielectric (total thickness of multilayer dielectric: 80μm) is applied. The shift on capacitance at 25°C for different compositions is shown in Figure 5.

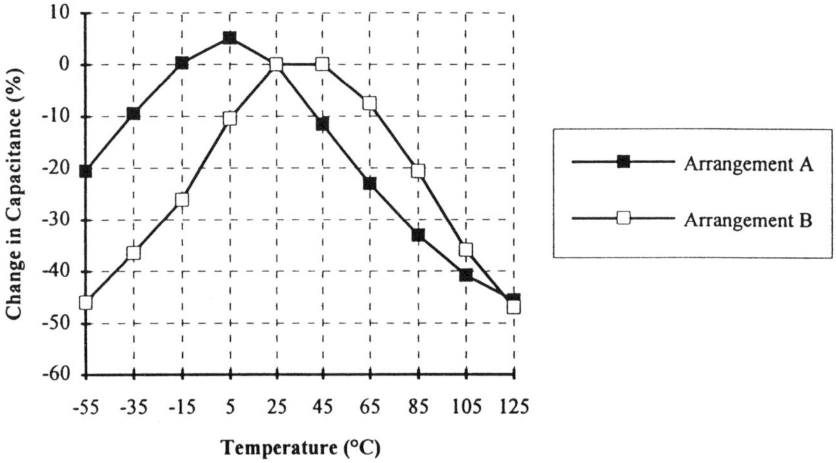


Figure 4. Influence of buffer and multilayer dielectric on the temperature characteristic of a high-K dielectric.

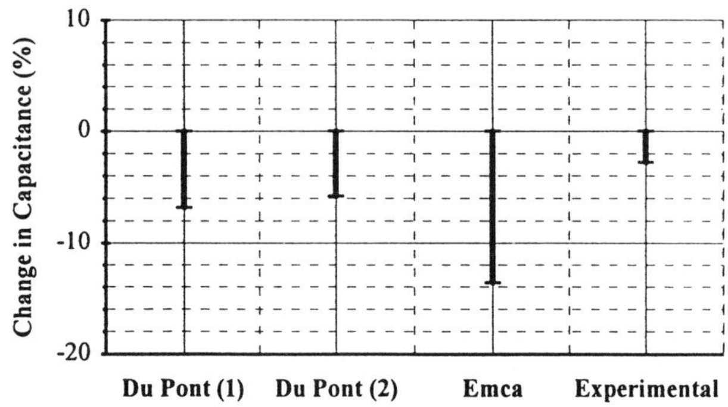


Figure 5. Capacitance shift due to the increase of multilayer dielectric thickness.

In order to evaluate the influence of the electrode on the performance of high-K dielectrics a conductor layer is applied (arrangement C, Figure 2). In accordance to the actual configuration of a buried capacitor, additional layers of multilayer dielectrics are printed on the conductor layer (arrangement D, Figure 2). This results in a shift of capacitance which varies for the applied paste systems (Figure 6).

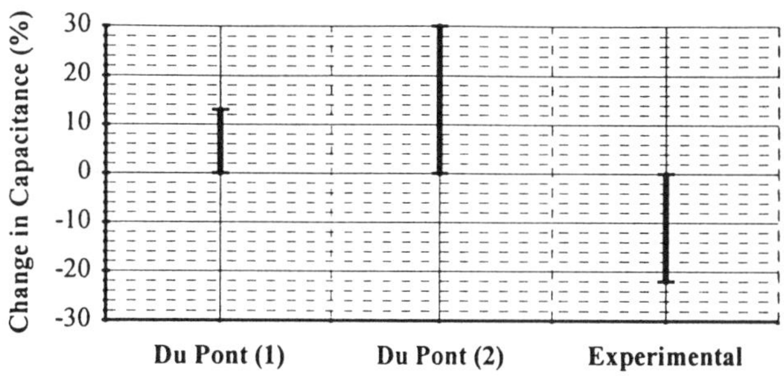


Figure 6. Capacitance change (at 25°C) due to deposition of multilayer dielectric on conductor layer.

In accordance to the processing conditions of thick film multilayers the test samples with different arrangement in layers (Figure 2) have been exposed to multiple firing cycles in order to evaluate the sensitivity of the high K-dielectric to refiring. Figure 7 depicts the changes in capacitance versus additional firings.

The experimental dielectrics as well as the Emca high-K dielectrics exhibit fairly large shifts especially upon the first five refiring cycles (arrangement A,). With exception of the latter composition the refire shifts in capacitance are negative. The capacitance response of arrangement D to extra firing must be related to two opposite effects: The refiring process contributes to an increase of capacitance (arrangement C), as well to a drop of capacitance on behalf of the sensitivity of the high-K dielectric to refiring (arrangement A). The net effect is the sum of these. The sum can be positive or negative depending on their effectiveness. Migration of electrode material during firing might be a plausible reason for the increase on capacitance of the Du-Pont-composition and the experimental paste system because a further densification of the dielectric could not be detected. Measuring the dielectric thickness before and after refiring cycles showed no change.

The temperature characteristic of arrangement B for the Du Pont(1) system shows evidently an interaction between the dielectric layers (Figure 8) due to extra firings which results in a further shift of the Curie temperature.

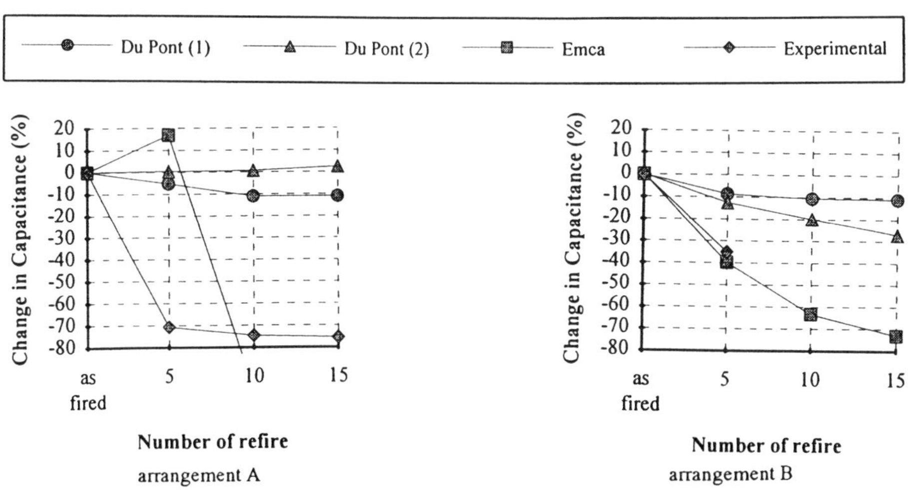

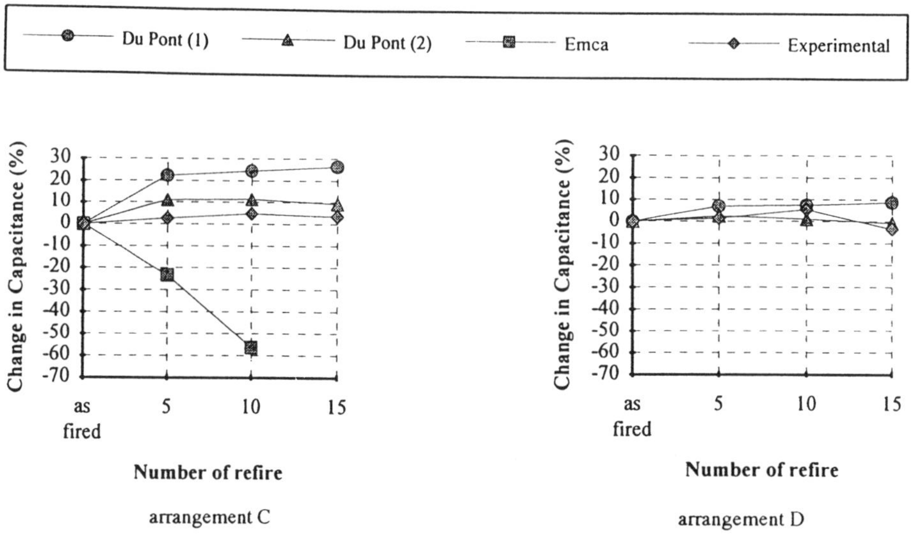


Figure 7. Change in capacitance in dependence of refiring cycles.

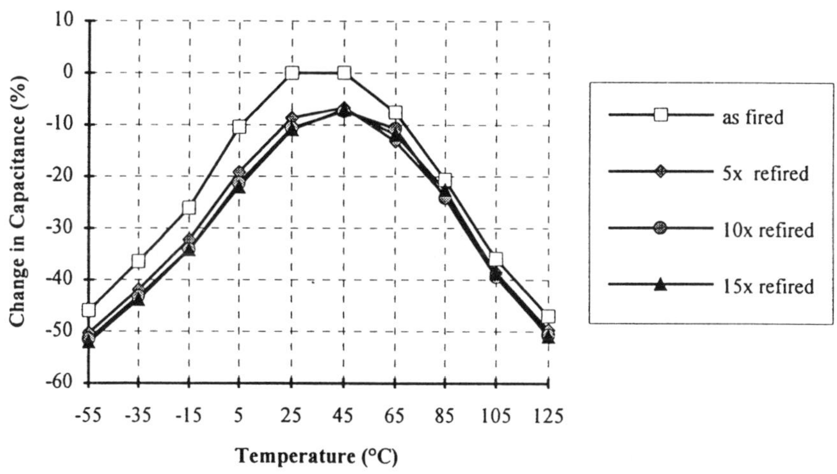


Figure 8. Shift of temperature characteristic due to extra firing.

## 2.2. BURIED CAPACITORS

In a next step of study the viability of realizing buried capacitors will be demonstrated by application of the Du-Pont(1)-system and the experimental-paste- system. To get familiar with the size dependent aspects of the capacitance shift, square parallel plate capacitor structures of 2mmx2mm, 4x4mm, 6mmx6mm and 12mmx12mm have been produced, followed by firing a buffer layer (only for the Du-Pont-system) and a multilayer dielectric on top. The thickness of the high-K dielectric is 48 μm for the Du-Pont-system and 55 μm for the experimental-system. Because a number of additional firings are necessary to build up a multilayer, the samples were exposed up to ten additional firings. Figure 9 and Figure 10 depict changes in the capacitance and dissipation factor versus additional firings in dependence of the size of the capacitor. The buffer layer has been recommended for the application of the Du-Pont-system. For comparison purpose batches of capacitors have been realized without a buffer layer (Figure 11). These samples show an improved stability to refire processes.

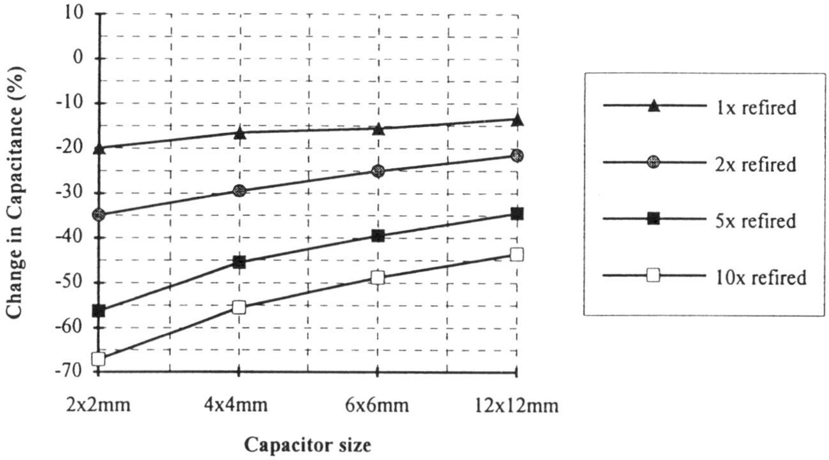


Figure 9a

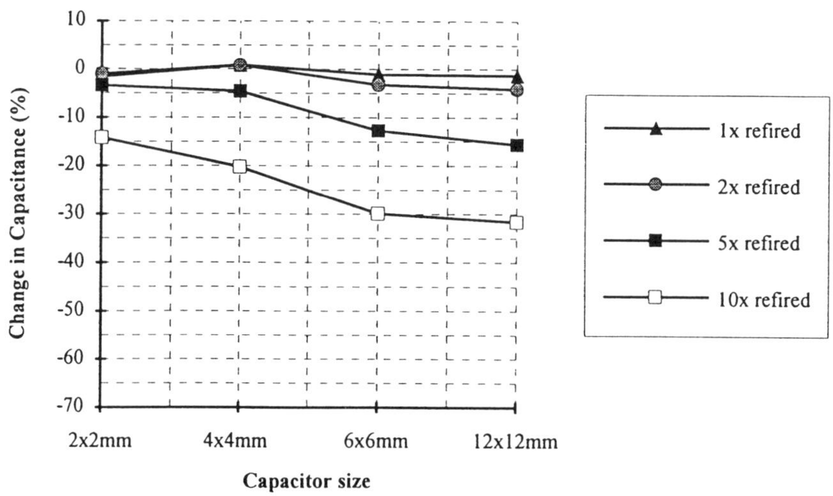


Figure 9b

Figure 9. Capacitance change due to additional firings (a.: Du Pont with buffer layer, b.: experimental).

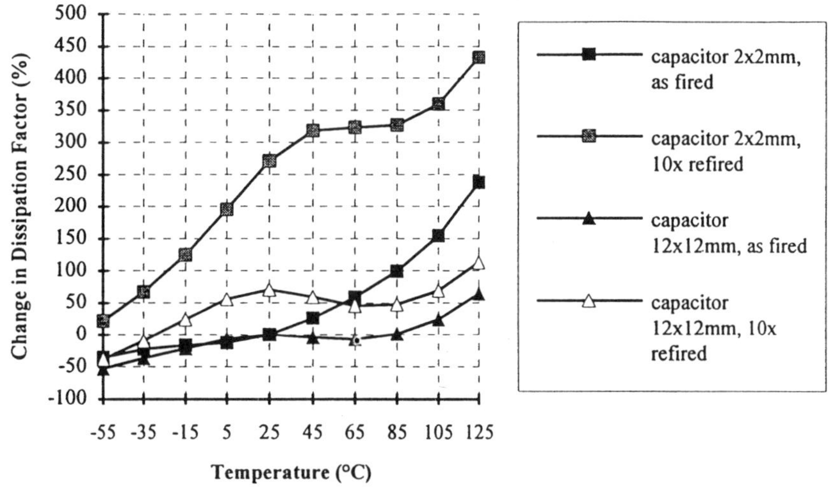


Figure 10. Changes in dissipation factor for refired capacitors (Du Pont).

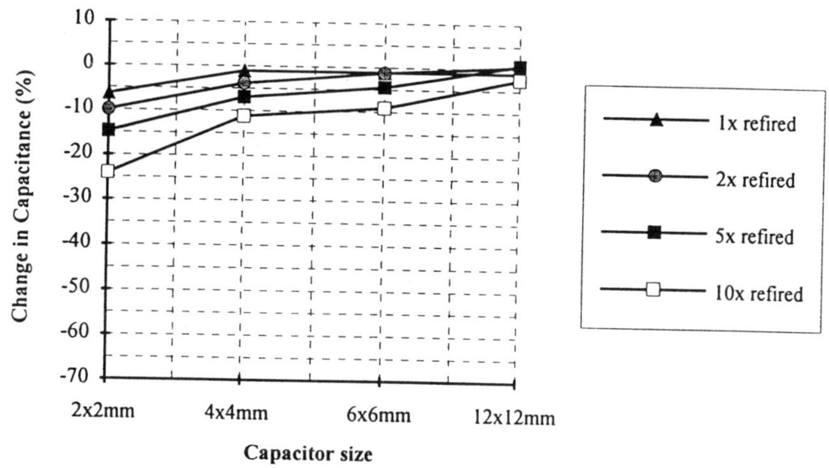


Figure 11. Capacitance change due to refiring cycles (Du Pont without buffer).

The buried capacitors realized with the Du Pont system exposed to refiring exhibit a shift of Curie temperature (Figure 12), which is characteristic for the interaction between high-K dielectric, the buffer layer, and the multilayer composition in accordance to the results shown in Figure 4. The capacitance shift becomes greater as the capacitor size is reduced. This must be related to a peripheral diffusion because the silver electrodes form a nearly impenetrable barrier for diffusion. Therefore the diffusion becomes more effective as the area to circumference ratio of the electrodes becomes smaller.

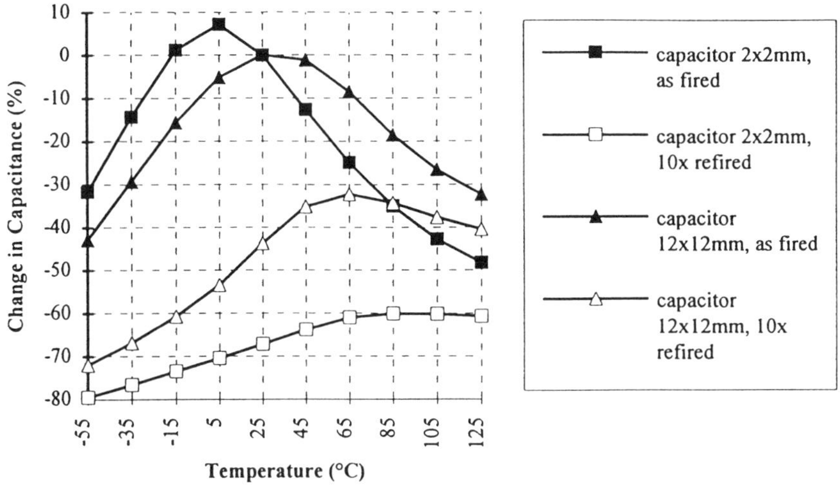


Figure 12. Shift in temperature characteristic due to refiring (Du Pont with buffer).

## 3. Summary

Due to the sensitivity of high-K dielectrics to multiple firings and to diffusion effects the fabrication of buried capacitors becomes a crucial process. But nevertheless the shift of capacitance for a particular geometry and for a distinct arrangement in layer due to multiple firings is predictable. This enables to design capacitors even in tight tolerances. Therefore buried capacitors must be considered to be an attractive method to increase the packaging density on thick film MCMs.

## 4. References


1. Bless, P.W., Wahlers R.L., Huang, C.Y.D. and Stein, S.J. (1992) New capacitor dielectrics covering K=2.000 to 12.000 for printing and firing applications below 1000°C, *Proc. ISHM,* 445-450.
2. Burn, I. and Drozdyk, L. (1992) Reliability of thick-film capacitors, *Proc. ISHM,* 439-444.
3. Lim, Y.C. and Moore, R.A. (1968) Properties of alternately charged coplanar parallel strips by conformal mappings, *IEEE Trans. on Electron Devices* **15**, 173-180.

# ADVANCES IN MATERIALS FOR SENSORS

SYLVIA MERGUI
*Electrical and Computer Engineering Department,*
*Florida International University, Miami, Florida 33199.*

**Abstract**

This paper reviews the advances in materials, specifically ceramics and polymers for sensor applications. Ceramic sensor materials may be classified according to the physical properties of the bulk or the properties of the grain boundary and according to the surface effects. Polymer dielectric and polyelectrolytes are the materials used for organic sensors. A comparison of the materials used in thin and thick film technologies will be given for both ceramic and organic materials. Thick-film based transducers and sensors, namely solid-state gas and humidity sensors and thermistors will be addressed. This paper will sketch the developments in materials and processing and the related performances for the various sensing devices.

## 1. Introduction

Sensors are needed in areas such as automotive, medical, aerospace, military, process and industrial controls, communications, and environmental applications. The need for gas sensors will increase as environmental pollution and worker safety regulations become more stringent. To function effectively, sensing devices are required to be small, inexpensive and to be sensitive, stable and reliable. Advances in technologies for thin and thick films and ceramics have made this possible. Because of the need for miniaturization and cost reduction, thick and thin film approaches have been the main techniques used for solid state sensors. For the thick film technology, screen printable materials have been investigated and in this area, both ceramic and polymer materials are very promising. The thick film processing of film with good porosity and reasonable thickness can be formed with one printing. Fabrication of sensors based on screen-printed thick-film multilayers faces significant technical problems although there have been remarkable improvements in the fabrication of components incorporating dozens of active layers [1]. In the case of chemical sensing, extensive use has been made of thick-film technology which allows the fabrication of low cost arrays of sensors [2].



*W. K. Jones et al. (eds.), MCM C/Mixed Technologies and Thick Film Sensors*, 239–257.

Gravure offset printing technique offers the possibility of printing narrow lines (50-100 μm) and the possibility of 3-D printing [3]. Vacuum evaporation, sputtering, and the use of lasers are the thin film fabrication techniques used for sensitive layers as well as electrodes.

## 2. Ceramic sensors

### 2.1. MATERIAL CLASSIFICATION

Ceramic sensor materials can be classified according to the bulk properties, to the properties of the grain boundary and according to surface effects [4]. Because most ceramic are polycrystalline, it is essential to determine which behavior dominates.

#### 2.1.1 *Bulk properties*

Bulk properties depend upon the crystallographic structure and the doping of the semiconductor oxides[4,5]. By the introduction of foreign atoms, a wide range of resistivities can be achieved. One example is the case of NiO semiconducting ceramic oxide doped with $Li^+$[4]. The ionic conduction via oxygen vacancies is another bulk property that can be used for oxygen sensing. For zirconia ($ZrO_2$)doped with Y or Ca, the development of large concentrations of oxygen vacancies makes the material an ionic conductor in which oxide ion moves from one vacancy to another. Creating lattice defects having energy levels in the band gap will modify the resistivity of a semiconductor material. The band gap for stoichiometric titanium oxide is 3.2 eV and $TiO_2$ has an electronic resistivity that depends on the partial pressure of oxygen $pO_2$. At low oxygen ambient partial pressure, oxygen ions escape from the solid, leaving oxygen vacancies in the material therefore the lattice defects are donors. As oxygen deficiency increases, the concentration of the electrons in the conduction band increases and the resistivity of the material decreases [4].

#### 2.1.2 *Grain boundaries*

For ceramic materials governed by grain boundary behavior, any impurity present in the material may accumulate at the grain boundary and may alter the nature of the barrier layer. This space-charge barrier layer is on the surface of every ceramic grain and forms a high resistivity skin around a low resistivity semiconductor grain core. Then, in the grain boundary region, it will take more energy to make the charge carrier mobile. $BaTiO_3$ and its solid solution are examples of ceramic materials governed by the grain boundary behavior rather than the bulk properties.

#### 2.1.3 *Surface effects*

The process whereby a semiconductor can change its resistivity through surface absorption is the phenomenon which makes the development of semiconductor gas sensors possible. Surface absorption is followed by ionic conduction. For humidity sensors, the key mechanism involves the ionic conduction of the absorbed water film. Typically, the surface of the material has to be porous and the pores of the material must be relatively small and uniform in shape to yield a maximum amount of surface in a given volume. Materials such as $ZnCr_2O_4$ meet these criteria. The amount of $OH^-$ absorbed causes a change in the resistivity which serves as an indication of humidity levels [10].

## 2.2 APPLICATIONS

### 2.2.1 *Gas sensors*

Almost all current metal oxide n-or p-type semiconductors such as the oxides of Sn, Fe, Ti, In, Zn are used as gas sensors. The production of ceramic oxides involves precipitating the hydrous metal oxide, calcining, adding if needed a catalyst, grinding into a powder and forming the powder into a paste. The use of additives is an alternative approach to improve the performance of gas-sensitive materials. Gas responses are dependent on the catalytic properties of the surface and the electronic properties of the bulk. Most semiconductor gas sensors are prepared from tin dioxide $SnO_2$. Improvements of $SnO_2$ gas sensors can be obtained if care is exercised so that impurities from grinding and milling are not introduced at any stage of the fabrication [6]. We can switch from a p- to n-type behavior and this will change the manner in which the sensor responds to gases. For n-type semiconductors, in presence of oxidizing gas, the resistance increases while for p-type semiconductors, resistivity decreases (different for reducing gas) [5]. The bulk characteristics of $SnO_2$ are changed by doping with elements such as Al, Y, Bi, Sn [6]. Another way to improve the selectivity for different gases is to sense gases at different temperatures. By addition of $PtO_2$ to $SnO_2$, it was found that there is a decrease in the optimum temperature for detecting reducing gases. With addition of Pt, the main response peak for $H_2$ is at 100°C instead of 400-500°C [5]. An exhaustive table reviews the sensing mechanisms and the performance and reliability for gas sensors.

*Oxygen sensors.*

Oxygen sensors are playing an important role in improving combustion and controlling emissions for automotive applications. The electro-chemical sensor requirements are a high ionic conductivity, a low electronic conductivity, a high density, a wide $p(O_2)$ range of stability, a high catalytic activity, and a high resistance to thermal/shock. Y-$ZrO_2$ solid-electrolyte sensor generates voltage depending on $p(O_2)$ [7] and is presently used as well as $TiO_2$ for stoichiometric air/fuel control [4]. For semiconductor oxygen sensors, the variation of the resistance is dependent on the oxygen concentration $[O_2]$. The resistive sensor requirements are a high electronic conductivity, a strong dependence on $p(O_2)$, a high catalytic activity, a high porosity and small particle size, and a wide $p(O_2)$ range of stability. Materials investigated for this application are $TiO_2$, $SnO_2$, $Nb_2O_5$

and $CeO_2$ [4]. Oxygen sensors based on the resistance variation of screen printed film of fired barium titanate on an $Al_2O_3$ substrate operate at 600°-800°C and are able to detect $p(O_2)$ from 100% to 1 ppm with remarkable resolution and accuracy [1]. Thick film $YZrO_2$ is a sensor with excellent performance in terms of sensitivity, selectivity and stability for $p(O_2)$ values lower than 10% [10].

*Combustion gas sensors*
The key materials selected for this application are ceramic metal oxides and bismuth molybdate [4]. For combustion gas sensors, the predominante sensing mechanism is due to electrons moving through the grain boundaries surrounding a semiconductor grain core. Bismuth molybdate as a porous state or a thin film coated on quartz, functions at higher temperatures (300°C), the sensing mechanism being via electronic conduction [4]. The advantages of ceramics are the low cost of production, the adequate sensitivity of the semiconductor (100 ppb) and the small size (mm range). The disadvantages are in the lack of selectivity for particular gases, the lack of stability and the ease of contamination by the various processing atmospheres, and the poor reproductivity of ceramics. A good selectivity of combustion gas sensor is achieved by the addition of catalysts to speed up the response and promote the selection of gas by temperature control and by specific surface additives and filters [7,16].

### 2.2.2 *Humidity sensors*

Ceramic humidity sensors can be fabricated from a variety of semiconducting oxides, mainly stannates, perovskites, and apatites [8]. The detection of moisture involves the chemical and physical adsorption of water. Materials for humidity sensors are governed by the surface effects properties [4, 9]. Semiconducting oxides such as $SnO_2$, ZnO, $Fe_2O_3$ and $Ag_2O$ that contain interconnected porosity exhibit a decrease of resistivity when exposed to humid atmospheres at normal ambient temperatures. Devices based on porous, anodic $Al_2O_3$ are capacitive and require complex circuitry and frequent calibration after exposure to wet ambients (RH > 50%). Spinel solid solution of $MgCr_2O_4$-$TiO_2$ with porous solid $RuO_2$ electrodes, $SiO_2$-ZnO, $ZnCr_2O_4$-$LiZnVO_4$, and $K^+$-doped $ZrO_2$ exhibit a decrease in resistivity ($10^7$ to $10^4$ Ω.cm) of several decades between 10 to 90% RH. Most recently, efforts have led to the development of ceramic sensors without heaters, capable of stable operation near saturation. This has been achieved by introducing a second phase which forms a thin coating on the grains, effectively isolating the porous substrate material from condensate at high humidity and from contaminants. As an example, $SiO_2$-ZnO tested at 60°C and above 99% RH exhibits very stable response [10].

### 2.2.3 *Thermistors*

Thermistors exhibit a resistance change in accordance to the absorbed energy. NTC (Negative Temperature Coefficient) and PTC (Positive Temperature Coefficient) can be easily prepared using commercial pastes and are widely used as discrete sensors or in hybrid circuits for the temperature compensation of transducers or other chemical and physical quantities [1]. The electrical characteristics of thick film NTC and PTC resistors are strongly affected by the geometrical proportions of the sensing elements as well as by the firing conditions. Materials for PTC sensors are ceramics dominated by the grain boundary behavior. As an example, the ferroelectric titanate polycrystalline ceramics $BaTiO_3$ and its solid solutions with strontium or lead titanate are used as PTC sensors [11, 12]. Additives may include $SiO_2$, $Al_2O_3$, and excess of $TiO_2$ to promote and control doping and grain growth by means of liquid phase sintering. Adding transition metal ions has the effect of enhancing the resistivity ratio. PTC sensors are used in high stress environments for a variety of applications such as voltage regulators, compensating elements in electronic circuits, and temperature sensing. Materials for NTC sensors are usually semiconducting ceramic oxides composed of transition metals such as doped monoxides and complex spinels, and also of ceramic rare-earth oxides [9], and are governed by the bulk properties. NTC thermistors are needed to measure the temperature of intake air, water and fuels levels (engine efficiency) for automotive applications and for cryogenic applications. Usually, the temperature range of operation of NTC thick film thermistor is from [-100 to 500 °C] when based on ceramic oxides [9]. High-temperature materials have been developed for sensing at higher operating range, [300 - 1000 °C], [1, 9]. For these higher temperatures, refractory or noble metals in ceramic multilayer structure are valuable solutions. Exhaust gas temperature sensors are made from high temperature oxides such as $Al_2O_3$-$Cr_2O_3$, $ZrO_2$ ($T > 350$ °C) and spinel solid solution in the oxide system Ni-Mn-Co-Cu [9,12]. A close tolerance of the resistance values can be achieved by laser trimming for thick film thermistors but, the resistivity drifts and the control of the temperature coefficients are still of concern.

## 3. Polymeric Sensors

### 3.1. MATERIAL CLASSIFICATION

#### 3.1.1 *Polymer Dielectric*

One objective of present microelectronic technology is to minimize high-temperature processing and wet-etching chemistry. Polymeric sensors offer a solution to this problem. There are two main types of polymeric sensors: (i) capacitive sensors which use thin-film cellulose esters or polyimide resins as humidity-dependent dielectrics and (ii) resistive sensors which use polyelectrolytes that exhibit humidity-dependent ac resistivity [10]. Polymeric sensors have been used longer than ceramic sensors. Obviously, polymers are less robust than ceramics, and exhibit a greater sensitivity to both thermal and chemical changes so the applications are limited to lower temperatures. They are more resistant to contamination than porous ceramic. The latter is an important property if the sensor mechanism is a bulk rather than a surface phenomenon. This is the case of negative temperature coefficient thermistors NTC where resistivity decreases when temperature increases. Cellulose esters and polyimide resins are the two main class of polymer dielectrics. The thin film preparation for cellulose esters is easier. Polyimide resins such as Kapton and Pyralin are more expensive but exhibit a superior thermal and chemical stability.

### 3.1.2. *Polyelectrolytes*

The ion-exchange resins are cross-linked polymers with hydrophilic ionic species appended to the polymeric backbone. The ionic appendages are in the sulfonate group or the quaternary ammonium group, and the amide group. Examples are polystyrene sulfonate, Nafion, polymethacrylates and quaternized polyvinyl pyridine [10]. Polyelectrolytes are used as humidity sensors, the moisture absorption capacity depends on the cross-link density and the conductivity depends on the nature and concentration of ionic appendages.

## 3.2. APPLICATIONS

### 3.2.1. *Gas sensors*

The conductivity, permittivity or mass of certain polymers change slightly when they are inserted in gas atmospheres. An important class of sensing devices with a high specificity for certain gases is the metal phthalocyanines [13]. Devices based on metal phthalocyanines semiconductors are attractive because of the response at near ambient temperature and the sensitivity to toxic gases. The significant aspect of the phthalocyanine response based sensors is that changing the central metal atom of the material and/or the operating temperature gives rise to a significant change in the conduction characteristics from exposure to reducing and oxidizing gases [5]. A research group from the University of South Hampton, England, has compared thick and thin film gas sensitive organic semiconductors [13]. The thick phthalocyanine film is made into a screen printable ink of thickness of 15-20 μm. Low pressure vapor deposition allows the fabrication of thin films up to 1 μm in thickness. Sensors produced by the two methods exhibit distinct morphological differences which significantly affects their respective sensitivities. The conduction mechanism is strongly dependant on the porosity of the sensing film and seems to be diffusion limited. As an example, for lead-phthalocyanine, the conductance increases with [$NO_2$] and the response is faster and more stable [21].

### 2.2 *Humidity Sensors*

Humidity sensors are needed for air-conditioning control and auto-defogger systems as well as for medical applications because human health is affected by extreme relative humidity [8]. Polymeric sensors can be lower in cost than their ceramic counterparts, since high temperature processing and heat cleaning are not required. Polymers have a dielectric constant of 5, comparatively with water 80. Therefore, absorption of water is expected to increase the dielectric constant significantly [10]. Applications are limited to low temperatures. Common polymer dielectric are cellulose acetates and polyimides such as Kapton or Pyralin. Cellulosic materials exhibit hysteresis related to film thickness. Polyimide resins possess a superior thermal and chemical stability. For the polyelectrolytes, the resistance and capacitance strongly depend on %RH. These materials are an excellent choice for midhumidity range (20% to 80% RH). A protective coating of melamine formaldehyde resins or porous Teflon is usually applied over the sensor film providing protection against degradation [10].

## 4. Conclusions

This article describes the wide variety of sensing materials using thin and thick film technologies. Improvements of ceramic sensors performance are in the control of the contamination of semiconducting metal oxides during the fabrication process. Also, the use of catalyst and microstructure engineering allows the temperature of operation to be modified [5]. By a suitable choice of materials, and a better understanding of the sensing mechanisms, a significant degree of control may be exercised over selectivity between different gases, temperatures, with reliability and stability.

*My acknowledgments to Ana Velasquez and Prof. P. E. Schmidt.*

| Gas Sensors | | | | |
|---|---|---|---|---|
| **Materials** | **Gas** | **Reliability/Performance** | **Sensing Mechanisms** | **Ref.** |
| $SnO_2$<br>+Pt<br><br>dopants (oxides of Sb, Ta, Nb, V, Fe, Al, Y, Cr, Bi and In) | $O_2$<br>$CH_4$/CO/$C_2H_4OH$<br>$H_2$ | Good reproducibility.<br>Response time within less than 30s.<br>Slow recovery time (mn)<br>Maximum sensitivity within [300-500°C] range.<br>T [140-180°C] for sensing CO | Oxide semiconductor n- or p-type.<br>Operating T critical for selectivity.<br>Variation of the resistance with gas concentration.<br>Sb improves CO selectivity | [3,6,17, 33,39, 40,41, 52,57] |
| $SnO_2$ + Pd | $CH_4$/CO/$C_2H_5OH$ | $\Delta$R/R = 2/1.1/8.3 % at 1000 ppm<br>Operating T [400 °C]<br>Higher sensitibity in TFT | Catalytic combustion type is humidity independent | [14,25, 33,49, 59,71] |
| $SnO_2$ + Pd, Pt | $C_2H_5OH$/CO/$H_2$ | $\Delta$R/R = -60/-2/+45% at 1000 ppm<br>Operating T [250 °C]<br><br>$\Delta$R/R = -90/-5/-50% at 1000 ppm<br>Operating T [400 °C] | | [14,33, 71] |
| $SnO_2$ + $Al_2O_3$ | $H_2S$ | T cycling inproves recovery and response time | | [55] |
| $SnO_2ThO_2$ + hydrophobic silica | CO | $\Delta$R/R$\Delta$T = 3.3%/°C<br>$\Delta$R/R = 90 (500ppm)% at 1000 ppm<br>$\Delta$R/R = 2.2% per %RH selective with respect to $H_2$<br>Operating T [100-200 °C] | | [14,33, 71] |
| $SnO_2$ + $La_2O_3$ | $CO_2$ | Operating T [400°C]<br>High sensitivity, quick response | Variation in capacitance with gas concentration | [33] |

| Gas Sensors | | | | |
|---|---|---|---|---|
| **Materials** | **Gas** | **Reliability/Performance** | **Sensing Mechanisms** | **Ref.** |
| $BaTiO_3$ | $O_2$<br>$CO_2$ | Good resolution , good accuracy<br>Operating T [800 °C] | Variation of capacitance with gas concentration | [1,4,7,33] |
| $CeO_2$ | $O_2$ | Operation T [600-800 °C] | | [1] |
| $ZrO_2$ doped, YSZ<br>Y doped | $O_2$ | Response time [50-200 ms]<br>Function T [300-900 °C]. Good sensitivity and stability from $pO_2 < 10\%$ | Solid electrolyte<br>$V = f(pO_2)$<br>Limiting current type sensor | [1,4,7]<br>[15,33,<br>50,54,59] |
| Metal oxides<br>Ti, Fe, In, Zn | Combustion gas | Sensitivity 100 ppb<br>Lack of stability and selectivity | Variation of resistivity | [4,7] |
| $TiO_2$<br><br><br>$TiO_2$ with Al,Ga,Sc,In dopants<br>$TiNb_2O_7$ | $O_2$<br><br>$H_2$<br>$NO_x$<br><br>$CH_4$ | Response time [50-200 ms]<br>Function T [300-900 °C]<br>Operating T [25-300 °C]<br>Sensitivity, selectivity, fast response<br><br>Operating T > 500°C | n-type semiconductor<br>Resistivity = $f(O_2)$ | [4,53,59]<br><br><br>[33]<br><br>[5] |
| $Bi_6Fe_4Nb_6O_{30}$ | Reducing and oxidizing gases | | Semiconductor variation of resistivity | [5] |
| $CaZr_{0.9}In_{0.1}O_{3-x}$<br>$Ca_{0.4}Sr_{0.6}CoO_3$ | $H_2/H_2O$ | Operating T [700 °C] | Solid electrolyte<br>Proton conduction | [33] |
| $ABO_3$<br>A= Sm, La, Gd, Sr<br>B= Co, Mn, Fe, Cr,Ti | $CH_3OH/CH_4$<br>$CO/O_2$ | Operating T [400-1067°C] | TFT, absorption and oxidation<br>Catalytic combustion<br>Catalytic oxidation mechanism | [33,34,<br>38,42] |
| ZnO + Pd | $CH_4/CO/C_2H_5OH$ | $\Delta R/R$ = 1/2.2/17 % at 1000 ppm<br>Operating T [400 °C] | Resistivity variation | [14,44,<br>71] |

| Gas Sensors | | | | |
|---|---|---|---|---|
| **Materials** | **Gas** | **Reliability/Performance** | **Sensing Mechanisms** | **Ref.** |
| $Al_2O_3$ | CO | Operating T = 600°C | TFT, V = f(CO) | [1,10,14,34] |
| $Bi(MoO_4)_3$ | Combustion gas breathalyzer (alcohol) | Working T [RT, 300 °C]<br>Insufficiently gas selective | Electronic conduction<br>Lack of sensitivity to water vapor | [4] |
| $SrFe^{3+}{}_{0.7}\ Fe^{4+}{}_{0.3}\ O_{2.65}$ | Ethanol | $\Delta R/R$ = 0.1 % at 1000 ppm<br>Operating T [470 °C] | | [14,71] |
| $SrCe_{0.95}Yb_{0.05}O_3$ | $H_2$ | Operating T [400-1000 °C] | Protonic conduction | [33] |
| $(Mn_xZn_{1-x}Fe_2O_4)/RuO_2$ /Carbon black/binders | Ammonia, acetic acid, ethanol gas, methanol gas. | High sensitivity, high stability | Specific variation of resistance with time for each gas.<br>Dependance T due to ferromagnetic and paramagnetic properties of the ferrite | [20] |
| $WO_3$<br>$WO_3$ + Pd | $NO_x$<br>$CH_4/CO/C_2H_5OH$ | Low concentration of NO and $NO_2$<br>$\Delta R/R$ = 1.5/9/25 % at 1000 ppm<br>Operating T [400 °C] | | [33]<br>[14,71] |
| Nasicon $Na_4Zr_2(SiO_4)_3$ | $CO_2$<br><br>$H_2$ | Better conductivity at 300°C for the thin film than pressed and sintered bulk Nasicon.<br>Working T [200°C]<br>Detection 100ppm $H_2$ to 100% in air.<br>Response time = 1s. | Solid electrolyte.<br>Variation of ionic conduction for small quantity of gas (ppm)<br>TFT | [19,22,42]<br><br>[1,15] |
| Metal phtalocyanines Co, Cu, Ni, Pb, Zn | CO<br>$NO_2$<br>$NH_3$<br>HCl | Fast response, stability, better sensitivity than thin film counterpart. High sensitivity (25 ppb), reversibility, fast response times at room T. FET device structure for comparable sensitivity but lower noise. | n- or p-type semiconductor<br>Variation of conductance with sub-ppm concentrations of gases<br>Conduction mechanism diffusion limited<br>Water vapor enhances device sensitivity | [1,5,13,21,60,62,63] |

| **Humidity Sensors** | | | |
|---|---|---|---|
| **Materials** | **Reliability/Performances** | **Sensing Mechanisms** | **Ref.** |
| glass ceramic + $Al_2O_3$ | Recalibration necessary above 50% RH.<br>$\Delta C/C\Delta T < 1\%/°C$<br>$(\Delta C/C)\%/RH\% = 2.5$ (50Hz)<br>Range T [+5 to +35 °C] | Capacitance variation with %RH | [1,8,10, 14,28, 71] |
| Semiconducting perovskites $ABO_3$<br>A= Ca, Ba<br>B= Sn, Ti | Humidity range [25-90 %RH] at 1MHz<br>300% change in capacitance | Capacitance variation with %RH | [8,31, 46,47] |
| $BaTiO_3$ + glass<br><br>$K_2O$ - $BaSrTiO_3$ | $\rho = 9.0 \times 10^6\ \Omega\cdot cm$<br>2500 R(75%RH)/R (@ dew pt.)<br>Range T [+5 to +60 °C]<br>Range T [25 to 85 °C] for 65-95 %RH | Capacitance and resistance variation with %RH | [1,14, 35,45, 69,71] |
| $RuO_2$ + glass | $\rho = 9.0 \times 10^6\ \Omega\cdot cm$<br>3500 R(75%RH)/R (@ dew pt.)<br>Range T [+5 to +60 °C] | Resistance variation with %RH | [15,71] |
| ZnO - $Cr_2O_3$ | Suitable for aggresive environment | Resistance variation with %RH | [4,8] |
| $(Na,K)_2\ O\cdot Al_2O_3\cdot nSiO_2$<br>Porous zeolite | | Conduction of Alkaline metals (NaK) | [4,8] |
| $(Ca(PO_4)_6\ (OH)_2$<br>Apatites | | Conduction of alkaline metals.<br>Substitution of OH/Ca or $H^+$ in P site | [4,8] |
| $Zn_2Al(OH)_6ClnH_2O$<br>hydrotalcite | $\sigma = 1.3 \times 10^{-5}\ \Omega^{-1}cm^{-1}$ at 294°K and 50 %RH<br>$\sigma$ x 50 for %RH [10 to 90%] | Ionic conduction.<br>Resistivity variation with %RH.<br>TFT | [8,29, 48] |

**Humidity Sensors**

| Materials | Reliability/Performances | Sensing Mechanisms | Ref. |
|---|---|---|---|
| $ZrCr_2O_4$ - $LiZnVO_4$ | $\rho = 6.2 \times 10^7$ Ω·cm<br>$\Delta R/R\% = 9.3$ with %RH<br>Range T [-10 to +40 °C] | Resistance variation with %RH | [1,4,8, 10,14, 71] |
| $ZrO_2$ - MgO | High stability for T [400-600°C] for long durations. | n type semiconductor<br>Resistance decreases with increase of moisture and reducing gas. | [8] |
| $ZrO_2$ stabilized, $K^+$ doped YSZ | T [500°C] | Polyelectrolyte sensor | [10,33] |
| Nasicon | | Ionic conductivity with no P | [19] |
| $MnWO_4 + V_2O_5$ | $\rho = 6.2 \times 10^7$ Ω·cm<br>$\Delta R/R\% = 8.8$ with %RH<br>Range T [-10 to +40 °C] | Resistance variation with %RH | [1,14, 71] |
| $NiWO_4$ | $\rho = 8.0 \times 10^4$ Ω·cm<br>$\Delta R/R\% = 8.2$ with %RH<br>Range T [-10 to +40 °C] | Resistance variation with %RH | [1,14, 71] |
| $MgCr_2O_4$ + glass<br><br>$MgCr_2O_4$ - $TiO_2$<br>$MgAl_2O_4$, $MgFe_2O_4$ | $\rho = 8.3 \times 10^7$ Ω·cm<br>$\Delta R/R\% = 8.3$ with %RH<br>Range T [-10 to +40 °C]<br>Range [1 to 100 %RH] for up to 150°C | Resistance variation with %RH, ionic conduction<br><br><br>p-type semiconductor | [1,8,10, 14,23, 31,43, 71] |
| $SnO_2$ | Operating T [300-400°C].<br>Sensitivity to reducing gas. | n type, dependance of conductivity with T | [8,39] |
| $CoAl_2O_4$ | $\Delta R/R\% = 3.1$ with %RH<br>Range T [-10 to +40 °C] | Resistance variation with %RH | [1,14, 71] |

**Humidity Sensors**

| Materials | Reliability/Performances | Sensing Mechanisms | Ref. |
|---|---|---|---|
| Nafion film | Excellent chemical stability. Melamine formaldehyde resins or porous Teflon as protective coating. Good performance [20-80 %RH]. | Polyelectrolyte sensor, variation of the electrical conductivity with %RH. | [10] |
| $Nb_2O_5$ | [30 to 90 %RH] | Variation of electrical resistance and capacitance with %RH | [26,27] |
| Poly(styrene) sulfonate | Excellent chemical stability.<br>Good performance [20-80 %RH]. Protective coating. | Polyelectrolyte sensor<br>Variation of the electrical conductivity with %RH. | [10] |
| Poly(methyl methacrylate) PMMA | 25 mg/g at 90% RH, linear response of the sensor<br>Fast response time: 30 s for 20-80 %RH change | Variation of capacitance with %RH. | [64,67] |
| Polyethylene-poly-(2-hydroxy-3-methacryloxypropyl trimethyl-ammonium chloride) HMPTAC | Response time increases with grafting % | Variation of resistivity with %RH. Electrical properties of the sensor depend on the metal of graph polymerization. | [64,66] |
| Quaternized poly(4-vinyl pyridine) | Excellent chemical stability. Protective coating.<br>Good performance [20 to 80 %RH] | Polyelectrolyte sensor<br>Variation of the resistivity with %RH | [10,65, 68] |
| Cellulose Acetate Butyrate CAB | Exhibit low hysteresis, very stable | Variation of the resistivity with %RH | [10,67] |
| Kapton | Excellent characteristics except stability at high T. | Variation of capacitance with %RH. | [10,70] |
| Pyralin | Superior thermal and chemical stability | Variation of capacitance with %RH. | [10] |
| Acetylene-terminated polyimide resins | Less hygroscopic compared with other polyimides. | Variation of capacitance with %RH. | [70] |
| Metal phtalocyanines<br>Co, Cu, Ni, Pb, Zn | Water vapor enhances device sensitivity, reduces response and recovery times (TFT and thin film) | TFT and thin film technologies | [1,5,13, 21,60, 63] |

| Thermistors | | | |
|---|---|---|---|
| **Materials** | **Reliability/Performances** | **Sensing Mechanisms** | **Ref.** |
| $Ba^{2+}_{(1-x)} M^{3+}_{x} Ti^{4+}_{(1-x)} O_3$<br>$M = Y^{3+}$<br>$M = Sb^{3+}$<br>$M = La^{3+}$<br>$M = Nb^{5+}$<br>$M = Ta^{5+}$ | Good response time<br>Composition, sintering temperature, atmosphere and annealing conditions strongly influence the PTC anomaly.<br><br>Addition of Pb for 50% of Ba, increases $T_c$= 320°C | PTC, properties of the grain boundary.<br>Crystallographic change with T.<br>Variation of resistance with T. | [4,11] |
| $BaTiO_3 + SrTiO_3$ | Addition of Sr for 70% of Ba, decreases $T_c$=-100°C<br>$\Delta C/C$ = 65%<br>Range T [5 to 100°C] | PTC<br>Variation of capacitance with T. | [11, 14,71] |
| $V_2O_5$ + Ag | R(75°C)/R(60°C) = 0.5-3<br>Sharp $\Delta R$ at 68°C | Variation of capacitance with T.<br>TFT, increased conductivity with firing T | [14, 36,71] |
| Transition metal oxides (Ti, Mn, Fe, Ni, Co, Cu, etc) | High sensitivity in a narrow operating range. | NTC, bulk properties dependent. | [9] |
| $(FeTi)_2O_3$ hematite<br>$(Li,Ni)_2O_3$<br>$Al_2O_3$ - $Cr_2O_3$ | Good response time | NTC, for cryogenics, exhaust gas temperature sensors<br>Properties of the grain boundary | [4]<br>[7,9] |
| $ZrO_2$<br>$ZrO_2$-$RuO_2$-$RhO_2$<br><br>$ZrO_2$-$RuO_2$ | T > 350°C<br>$\alpha$ = -3000 ppm/°C<br>Range T [20 to 120°C]<br>$\alpha$ = 1270 ppm/°C<br>Range T [20 to 120°C] | NTC<br><br><br>PTC | [14, 71] |
| Ru/Pyrochlore/glass | Range T [Liquid Nitrogen to 250°C]<br>$\alpha$ = 2600 - 3300 ppm/°C | PTC | [75, 76] |
| Complex spinels $AB_2O_4$<br>$(Ni,Mn)_3 O_4$, $(Ni,Mn,Fe,Co)_3 O_4$ | High sensitivity in a narrow operating range. | NTC, bulk properties dependent. Interfacial aspects important to control the reproductivity of the device. | [9] |

| Thermistors | | | |
|---|---|---|---|
| **Materials** | **Reliability/Performances** | **Sensing Mechanisms** | **Ref.** |
| Mn-Co-Ru spinel type | $\rho = 10^4$-$10^6$ Ω·cm<br>B= 2500 - 5000K<br>Range T [-50 to 200°C] | | [14, 71] |
| Mn-Ru spinel type | r = $10^3$ Ω/□<br>B= 2000K<br>Range T [-50 to 200°C] | | |
| $Co_2RuO_4/RuO_2$/glass | Range T [-55 to 150°C]<br>B = [1600 to 2600 °C]<br>r = 1 KΩ/□ - 1 MΩ/□ | | [74] |
| Alloy or noble metal pastes on ceramic substrates: | ~ 20µV/°C<br>Range T [-200 to +500°C] | Dependance of thermocouple EMF on temperature | [1,14, 18,71] |
| PdPtAu | | TFT | |
| AuPd | | | |
| Au + PtAu | | | [14, 71] |
| Au | r = 10 mΩ/□<br>α = 2700-3400 ppm/°C<br>Range T [0 to 700°C] | | |
| Au | r = 2-3 mΩ/□<br>α = 2840 ppm/°C<br>Range T [0 to 700°C] | | |
| Pt | r = 80-90 mΩ/□<br>α = 3700-3900 ppm/°C<br>Range T [0 to 200°C] | | |
| Pd, Ag | r = 25-35 mΩ/□<br>α = 275 ppm/°C<br>Range T [0 to 700°C] | | [14, 71] |
| Ni | r = 90-120 mΩ/□<br>α = 5290 ppm/°C | | [12, 14,71] |
| Au-Ni, Cr-Ni, Ag-Ni, Ag-Cr | Range T [0 to 200°C] | | |

**References**


1- Prudenziati M., and Morten B. (1992) The State of the Art in Thick-film Sensors, *Microelectronics Journal* **23**, 133 .
2- Atkinson, J.K., and Cranny, A.W.J., (1991) The detection of gases using statistical pattern matching to learned responses of gas sensor arrays. *Sensors: Technology, Systems and Applications,* K.T.V. Grattan & Adam Hilger, 69-74
3- Ylinampa, A., Lantto, V., Leppavuori, S. (1993) Some differences between Au and Pt electrodes in SnO2 thick-film gas sensors. *Sensors and Actuators. Part B, Chemical* **14**, 602-604.
4- Ketron, L. (1989) Ceramic Sensors, *Bulletin of the American Ceramic Society* **68**, 860.
5- Moseley, P.T. (1991) New trends and future prospects of thick- and thin-film gas sensors, *Sensors and Actuators. Part B, Chemical* **3**, 167-174.
6- Stein S.J., Huang C., Grunstein, T., and Sykora G. (1993) Thick film sensors for methane detection, *ISHM 93 Proceedings, Spring Meeting.* 1-6.
7- Sheppard, L.M. (1992) Automotive Sensors Improve Driving Performance, *American Ceramic Society Bulletin* **71**, 905 .
8- Fagan, James G., Amarakoon, V.R.W. (1993) Reliability and reproducibility of ceramic sensors: Part III humidity sensors. *American Ceramic Society Bulletin* **72,** 119-129.
9- Fagan, J.G., and Amarakoon, V.R.W. (1993) Feature: Reliability and Reproducibility of Ceramic Sensors: Part I, NTC Thermistors, *American Ceramic Society Bulletin* **72**, 70 .
10- Kulwicki, B.M. (1991) Humidity Sensors, *Journal of the American Ceramic Society* **74,** 697.
11- Fagen, J.G., Amarakoon, V.R.W. (1993) Reliability and Reproductivity of Ceramic Sensors: Part II, PTC Thermistors, *American Ceramic Society Bulletin* **72,** 69-76.
12- Smetana, W., Nicolics, J. (1993) Application of integrated thick-film thermocouples for a laser power detector. *Sensors and Actuators. Part A, Physical* **37/38**, 565-570.
13- Cranny, A.W.J., Atkinson, J.K., and Burr, P.M. (1991) A comparison of thick- and thin-film gas-sensitive organic semiconductor compounds, *Sensors and Actuators. Part B, Chemical* **4**, 169.
14- Khazan, A.D. (1994) *Transducers and Their Elements*, PTR Prentice Hall, New Jersey.
15- Chu, W.F., Leonhard, V., and Erdmann, H. (1991) Thick-film chemical sensors, *Sensors and Actuators. Part B, Chemical* **4**, 321.
16- Dell'Acqua R. (1989) Thick film sensors' future, *Automotive Engineering* **97**, 41-45.
17- Son, Y.M. and Kwon, H.W. (1991) The Effects of Residual Chlorides on Gas Sensing Characteristics of Tin Oxide Gas Sensors, *J. Electrochem. Soc.* **138**, 2185-2189.
18- Hrovat, M., Zgonik, M., and Belavic, D. (1992) Thick-film materials for heat flux sensors, *Journal of Materials Science Letters* **11**, 89.
19- Ivanov, D., Currie, J., Bouchard, H., Lecours, A., Andrian, J., Yelon, A. (1993) Evaluation of the Ionic Conductivity of Sputtered Silicate-Limite Nasicon Thin Films for Electrochemical Sensors, *Proceedings: The 7th International Conference on Solid-State Sensors and Actuators*, 382-385.
20- Seki, K., Hori, M., Osada, H. (1992) Characteristics of Gas Sensors Using Magnetic Semiconductor Thick Film. *IEICE Transactions on Electronics* **75**, 1291-1293.
21- Belghachi, A. and Collins, R.A. (1988) Humidity response of phthalocyanine gas sensors, *Analytical Chemistry: Letter to the Editor*, 1647-1651.
22- Leonhard, V., Fischer, D., and Erdmann, H. (1993) Comparison of thin- and thick-film $CO_2$ sensors. *Sensors and Actuators. Part B, Chemical* **14**, 530-531.
23- Cantalini, C. and Pelino, M. (1992) Microstructure and Humidity-Sensitive Characteristics of $\alpha$-$Fe_2O_3$ Ceramic Sensor, *J. Am. Ceram. Soc.* **75**, 546-551.
24- Dutronc, P., Carbonne, B., Menil, F. (1992) Influence of the nature of the screen-printed electrode metal on the transport and detection properties of semiconductor gas sensors. *Sensors and Actuators. Part B, Chemical* **6**, 279-284.
25- Tsai, P.P., Chen, I.-C., Tzeng, M.H. (1993) Tin oxide (SnO2) alcohol sensor from metal organic decomposed (MOD) thick film. *Sensors and Actuators. Part B, Chemical* **14**, 610-612.

26- Tsuiki, H., Shimanouchi, T., Takahashi, N., Kakuta, N., and Ueno, A. (1990) Preparation of Niobium Oxide Films as Humidity Sensor, *Chemistry Letters*, 1729-1732.
27- Kurioka, N., Watanabe, D., Haneda, M., Shimanouchi, T, Mizushima, T. Katura, N., Ueno, A., Hanaoka, T. and Sugi, Y. (1993) Preparation of Niobium Oxide Films as a Humidity Sensor, *Catalysis Today* **16**, 495-501.
28- Chen, Z., Jin, M-C., Zhen, C. and Chen G-H. (1991) Properties of Modified Anodic-Spark-Deposited Alumina Porous Ceramic Films as Humidity Sensors, *Journal of the American Ceramic Society* **74**, 1325-1330.
29- Moneyron, J.E., de Roy A., Besse, J.P. (1991) Protonic conductivity of hydrotalcite-type compound thick films: Application to a humidity sensor, *Solid State Ionics* **46**, 175-181.
30- Noguchi, H., Uchida, Y., Nomura, A., Mori, S-I. (1989) A highly reliable humidity sensor using ionene polymers, *Journal of Materials Science Letters* **8**, 1278-1280.
31- Nitta, T., Hayakawa, S. (1980) Ceramic Humidity Sensors, *IEEE Transactions on Components, Hybrids, and Manufacturing Technology* **CHMT-3**, 237-243.
32- Hayes, M.B., Bhutta, I.A., Elshabini-Riad, A. and Riad, S. (1993) Application of Multilayer Circuit Technologies to Sensors, *ISHM '93 Proceedings*, 13-19.
33- Azad, A.M., Akbar, S.A., Mhaisalkar, S.G., Birkefeld, L.D., and Goto, K.S. (1992) Solid-State Gas Sensors: A Review, *Journal Electrochemical Society* **139**, 3690-3704.
34- Lalauze, R., Visconte, E., Montanaro, L., (1993) A new type of mixed potential sensor using a thick film of beta-alumina. *Sensors and Actuators. Part B, Chemical* **13**, 241-243.
35- Yeh, Y.C., Tseng, T.Y. (1989) Electrical Properties of $K_2O$ - doped $Ba_{0.5}Sr_{0.5}TiO_3$ ceramic humidity sensor, *IEEE Transaction on Components, Hybrids, and Manufacturing Technology* **12**, 259-266.
36- Umadevi, P., Nagendra, C.L., Thutupalli, G.K.M. (1993) Structural, Electrical and Infrared Optical Properties of Vanadium Pentoxide (V2O5) Thick-film Thermistors. *Sensors and Actuators. Part A, Physical* **39**, 59-69.
37- Huyberects, G., Honore, M., Roggen, J. (1993) Characterization of strontium containing tin dioxide based thick film humidity sensors. *Sensors and Actuators. Part B, Chemical* **16**, 281-287.
38- Schonauer, U. (1991) Response times of resistive thick-film oxygen sensors. *Sensors and actuators. Part B, Chemical* **4**, 431-436.
39- Schmitte, F.-J., Wiegleb, G. (1991) Conductivity behavior of thick-film tin dioxide gas sensors. *Sensors and Actuators. Part B, Chemical* **4**, 473-477.
40- Shimizu, Y., Nakamura, Y., Egashira, M. (1993) Effects of diffusivity of hydrogen and oxygen through pores of thick film SnO2-based sensors on their sensing properties. *Sensors and Actuators. Part B, Chemical* **13**, 128-131.
41- Ionescu, R., Vasilescu, V., Vancu, A. (1992) Conduction-concentration relationship in chemoresistive thick-film SnO2 gas sensor. *Sensors and Actuators. Part B, Chemical* **8**, 151-154.
42- Chu, W.F., Fischer, D., Erdmann, H. (1992) Thin and thick film electrochemical CO2 sensors. *Solid State Ionics* **53-56**, 80-84.
43- Gusmano, G., Montesperelli, G., Nunziante, P. (1993) Microstructure and Electrical Properties of MgAl2O4 and MgFe2O4 Spinel Porous Compacts for Use in Humidity Sensors, *British Ceramic Transactions* **92**, 104-108.
44- Pizzini, S., Butta, N., Narducci, D., Palladino, M. (1989) Thick Film ZnO Resistive Gas Sensors. Analysis of Their Kinetic Behavior. *Journal of the Electrochemical Society* **136**, 1948.
45- Yeh, Y.C. and Tseng, T.Y. (1989) Analysis of the d.c. and a.c. properties of $K_2O$ - doped porous $Ba_{0.5}Sr_{0.5}TiO_3$ ceramic humidity sensor, *Journal of Materials Science* **24**, 2739-2745.
46- Yeh, Y.C., Tseng, T.Y. and Chang, D.A. (1990) Electrical Properties of $TiO_2$-$K_2Ti_6O_{13}$ Porous Ceramic Humidity Sensor, *Journal of American Ceramic Society* **73**, 1992-1998.
47- Chen, S.N., Ramakrishnan, E.S., Huang, R.S. and Grannemann, W.W. (1984) A New Thin-Film Humidity Microsensor, *IEEE Electron Device Letters* **EDL-5**, 452-453.
48- Moneyron, J.E., De Roy A., and Besse, J.P. (1991) Realization of a Humidity Sensor, *Sensors and Actuators B* **4**, 189-194.

49- Eastwood, P.G., Claypole, T.C., Watson, J., and Coles, G.S. (1993) The behaviour of tin dioxide sensors in exhaust environments at low and intermediate temperatures, *Meas. Sci. Technol.* **4**, 524-533.

50- Usui, T, Asada, A, Ishibashi, K. and Nakazawa M. (1991) Humidity-Sensing Characteristics in Wet Air of a Gas Polarographic Oxygen Sensor Using a Zirconia Electrolyte, *Journal of Electrochemical Society* **138**, 585-588.

51- Harkoma, A., Torvela, H., Romppainen P. and Leppavuori, S. (1988) Detection of CO Levels by Semiconductor Gas Sensors in Combustion Gases Containing NO, *Combust. Sci. and Tech.* **62**, 21-29.

52- Van Geloven, P., Honore, M., Roggen, J., Leppavuori, S. and Rantala, T. (1991) The Influence of Relative Humidity on the Response of Tin Oxide Gas Sensors to Carbon Monoxide, *Sensors and Actuators B* **4**, 185-188.

53- Takami, A. (1988) Development of Titania Heated Exhaust-Gas Oxygen Sensor, *Ceramic Bulletin* **67**, 1956-1960.

54- Usui, T., Asada, A., Nakazawa, M., and Osanai, H. (1989) Gas Polarographic Oxygen Sensor Using an Oxygen/Zirconia Electrolyte, *J. Electrochem. Soc.* **136**, 534-542.

55- Lantto, V. and Romppainen, P. (1988) Response of Some $SnO_2$ Gas Sensors to $H_2S$ after Quick Cooling, *Journal of Electrochemical Society* **135**, 2550-2556.

56- Hardwood, J.A., Moseley, P.T., Peat, R., and Reynolds, C.A. (1991) The Use of Low Power Carbon Monoxide Sensors to Provide Early Warning of Fire, *Fire Safety Journal* **17**, 431-443.

57- Romppainen, P. and Lantto, V. (1988) The effect of microstructure on the height of potential energy barriers in porous tin dioxide gas sensors, *J. Appl. Phys.* **63**, 5159-5165.

58- Sanjines, R., Coluzza, C., Rosenfeld, D., Gozzo, F., Almeras, Ph., Levy, F., and Margaritondo, G. (1993) Photoemission spectromicroscopy: A new insight in the chemistry of $SnO_x$ films for gas sensors, *J. Appl. Phys.* **73**, 3997-4003.

59- Ippommatsu, M., Ohnishi, H, Sasaki, H., and Matsumoto, T. (1991) Study on the sensing mechanism of tin oxide flammable gas sensors using the Hall effect, *J. Appl. Phys.* **69**, 8368-8374.

60- Temofonte, T.A. and Schoch K.F. (1989) Phthalocyanine semiconductor sensors for room-temperature ppb level detection of toxic gases, *J. Appl. Phys.* **65**, 1350-1355.

61- Brina, R., Collins, G. E., Lee, P.A., and Armstrong, N.R. (1990) Chemiresistor Gas Sensors Based on Photoconductivity Changes in Phthalocyanine Thin Films: Enhancement of Response toward Ammonia by Photoelectrochemical Deposition with Metal Modifiers, *Anal. Chem.* **62**, 2357-2365.

62- Belghachi, A. and Collins, R.A. (1988) The effects of humidity on phthalocyanine $NO_2$ and $NH_3$ sensors, *Anal. Chem.*, 223-227.

63- Sakai, Y., Matuguchi, M., Sadaoka, Y., and Hirayama, K. (1993) A Humidity Sensor Composed of Interpenetrating Polymer Networks of Hydrophilic and Hydrophobic Methacrylate Polymers, *J. Electrochem. Soc.* **140**, 432-436.

64- Sakai, Y., Sadaoka, Y., Matsuguchi, M., Kanakura, Y., and Tamura, M. (1991) A Humidity Sensor Using Polytetrafluorethylene-Graft-Quaternized-Polyvinylpyridine, *J. Electrochem. Soc.* **138**, 2474 -2478.

65- Wang, K., Seiler, K., Haug, J.P., Lehmann, B., West, S., Hartman, K., and Simon, W. (1991) Hydration of Trifluoroacetophenones as the Basis for an Optical Humidity Sensor, *Anal. Chem.* **63**, 970-974.

66- Sakai, Y., Sadaoka, Y., Matsuguchi, M., Rao, V.L. (1989) Humidity sensor using microporous film of polyethylene-graft-poly-(2-hydroxy-3-methacryloxypropyl trimethyl-ammonium chloride), *Journal of Materials Science* **24**, 101-104.

67- Matsuguchi, M., Sadaoka, Y., and Sakai, Y. (1991) A Capacitive-Type Humidity Sensor Using Cross-Linked Poly(methyl methacrylate) Thin Films, *J. Electrochem. Soc.* **138**, 1862-1865.

68- Sakai, Y., Sadaoka, Y., Matsuguchi, M. (1989) A Humidity Sensor Using Polytetrafluorethylene-Graft-Quaternized-Polyvinylpyridine, *J. Electrochem. Soc.* **136**, 171-174.

69- Roubik, J.R., and Payne, D.A. (1993) Sol-Gel Derived Thin-Layer Capacitors Integrated on Silicon, *ISHM '93 Proceedings*, 147-152.

70- Matsuguchi, M., Sadaoka, K., Nosaka, K., Ishibashi, and Sakai, Y. (1993) Effect of Sorbed Water on the Dielectric Properties of Acetylene-Terminated Polyimide Resins and Their Application to a Humidity Sensor, *J. Electrochem. Soc.* **140**, 825-829.

71- Prudenziati, M., and Morten, B. (1986) *Sensors and Actuators* **10**, 65-82.

72- Mandelis, A. (1993) *Physics, Chemistry, and Technology of Solid State Gas Sensor Devices*, John Wiley & Sons, Inc., New York.

73- Neuman, M.R., Buck, R.P., Cosofret, V.V., Lindner, E., and Liu, C.C. (1994) Fabricating Biomedical Sensors with Thin-Film Technology. *IEEE Engineering in Medicine and Biology*, June/July, 409-419.

74-Hormadaly, J., June 16, (1992), U.S. Patent 5, 122, 302.

75-Hormadaly, J., March 6, (1990), U.S. Patent 4, 906, 406.

76-Hormadaly, J., October 9, (1990), U.S. Patent 4, 961, 999.

# PTC THICK FILM THERMISTORS

**J. HORMADALY, S.J. HOROWITZ, J.R. LARRY AND P. O'CALLAGHAN**
*DuPont Electronic Materials*
*Wilmington, DE 19880*

Electrical properties of commercially available and developmental PTC thermistor compositions are presented. The electrical properties of commercially available PTC compositions 5091, 5092, 5093 and their developmental extensions to 16 kohm/sq/mil, were measured on various conductors. Detailed process sensitivities including effects of peak firing temperature, thickness, conductor and firing profile were obtained for the commercially available series. The PTC series is compatible with Ag, Ag/Pd and Au conductors. Sensitivity to processing is a function of the resistance. Low end members 5091 and 5092 have high linear TCR and low process sensitivity, typical of thick film resistors. The 1 kohm/sq/mil member 5093 is more sensitive to processing.

## I. INTRODUCTION

Positive temperature coefficient (PTC) thermistors are usually (1) separated into two categories, switching PTC thermistors based on $BaTiO_3$ and elemental semiconductors based on silicon and germanium. PTC thermistors have been used for a variety of applications such as temperature sensing, self-regulating heaters, control elements in appliances, voltage regulators, compensating elements in electronic circuits in TV sets and telephone systems. Thick film (TF) PTC thermistors are a fairly new category of thermistors. They have a lower temperature coefficient of resistance (TCR) than elemental thermistors. However, they offer robust-ceramic, high tolerance elements with wide operating temperature range. These advantages coupled with the flexibility in design, thermal management and other advantages of thick film technology, makes them a unique group.

The 509x PTC thermistor series is based on Ru chemistry and displays unique properties. While most Ru based thick film compositions are typical TF resistors with low TCR in the -55C to 150C temperature range, the 509x compositions have high positive TCR of the order of 3000ppm/C. This high positive TCR is typical to metals with the advantage of higher and adjustable resistivity range. Resistances of 509x members are essentially linear with temperature over a fairly wide temperature range, spanning at least 300C.



*W. K. Jones et al. (eds.), MCM C/Mixed Technologies and Thick Film Sensors,* 259–265.

These properties can be realized with common thick film processings, including laser trimming. Trimmed thermistors have very good stabilities of <0.5% after 1000 hours storage at 150C, room temperature and 80%RH/90C. Because of their very good post laser trim stability, 509x compositions do not require overglazing, therefore their intrinsic response time is not altered.

The purpose of this poster is to report studies on the processing parameters of 5091, 5092 & 5093 PTC thermistor compositions. A second objective is to present initial electrical properties of developmental compositions which extend the resistance range of 509x series to ~16kohm/sq.

## II. RESULTS & DISCUSSIONS

Figure 1 presents the effects of peak firing temperature on the resistances of 509x series. Compositions were fired in the short Birox® profile ( 30 minutes long profile and 10 minutes at peak,which is the recommended profile for 509x series ) and terminated by 7484 Ag/Pd conductor. 5091 shows small decrease in resistance from 840C to 850C and very slight change between 850C to 860C. 5092 and 5093 have resistance that is monotonically increasing with peak temperature; 5092 exhibits small change while 5093 is more sensitive.

Figure 2 collects the TCR characteristics of 509x series as a function of peak firing temperature for processing in the recommended profile on 7484 Ag/Pd conductor. TCR increases with temperature for 5091, reaching ~3400ppm/C at 860. 5092 and 5093 have TCR which are monotonically decreasing with temperature. In all case the change in TCR is small and predictable.

TCR as a function of conductor termination are given in Figure 3. Three conductors were evaluated: 6160 Ag, 7484 Ag/Pd and 5715 Au at recommended short Birox® profile (30 minutes long and 10 minutes at 850C). Figure 3 shows that 509x series is compatible with a variety of conductors.

Dry thicknesses were varied between ~20 to ~30 microns. The relationship between resistance and dry thickness is given in Figure 4. All compositions were processed in the short Birox® profile and terminated by 7484 Ag/Pd conductor. Normalized resistance of 5091 does not change with dry thickness. 5092 has a small change of R with thickness. 5093 normalized resistance decreases with the dry thickness in a steep manner.

Table 1 collects the preliminary electrical properties of developmental compositions which are blendable with and extend the resistance range of series 509x.

**FIGURE 1**

**RESISTANCE AS A FUNCTION OF PEAK FIRING TEMPERAURE**

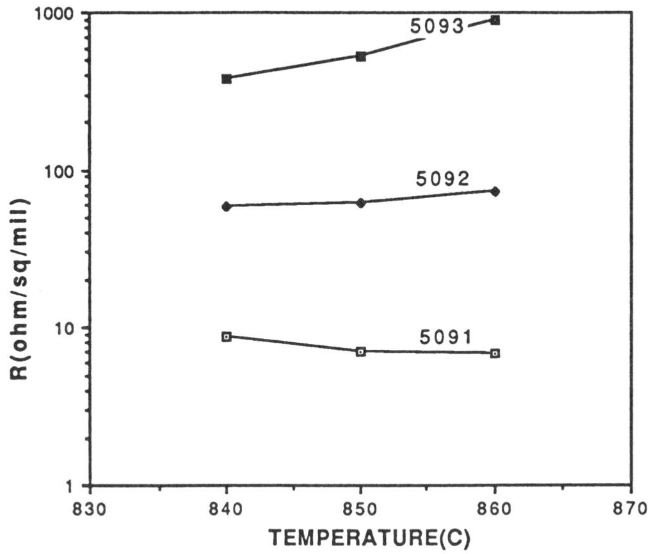

FIGURE 2
TCR AS A FUNCTION OF CONDUCTOR

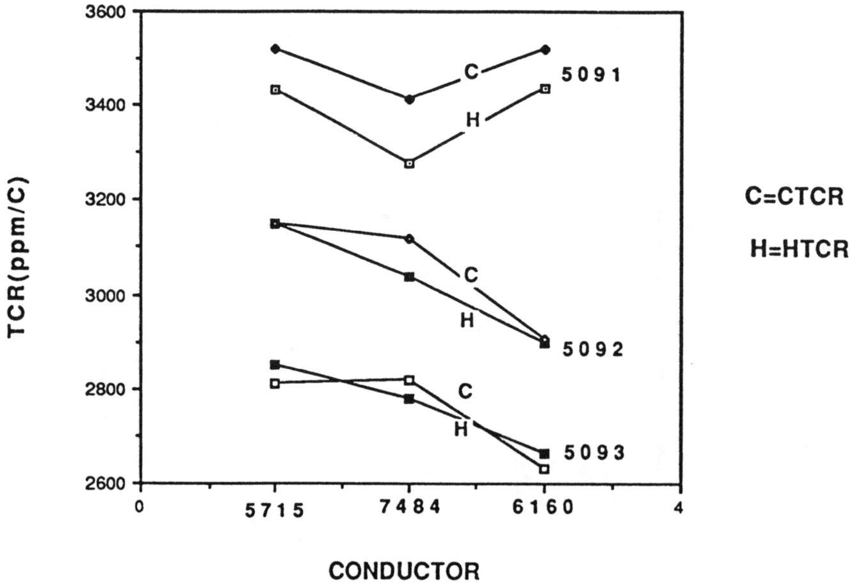

FIGURE 3
TCR AS A FUNCTION OF PEAK FIRING TEMP.

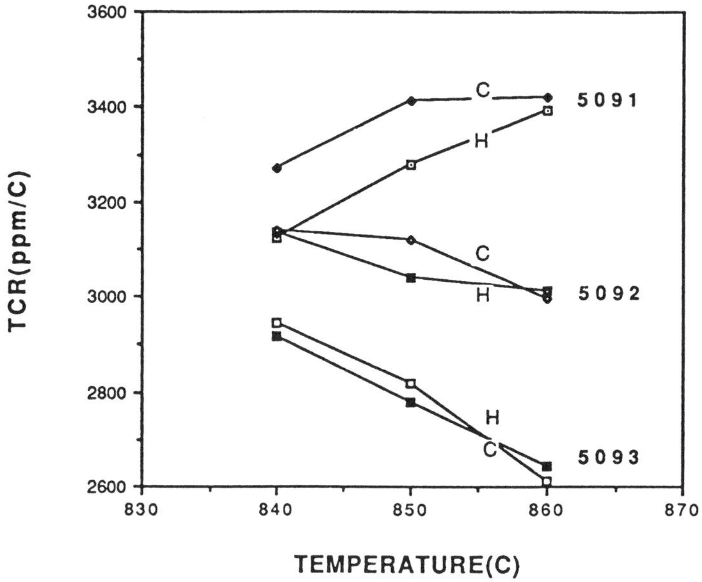

FIGURE 4
RESISTANCE AS A FUNCTION OF DRY THICKNESS

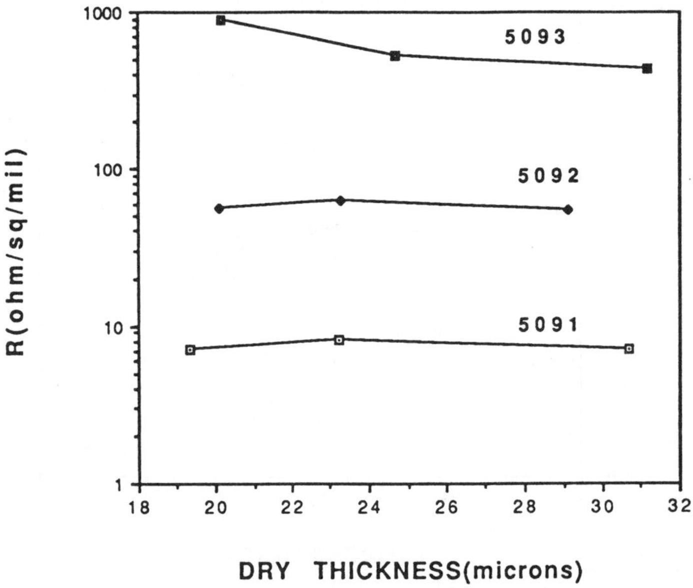

TABLE 1: Resistance and TCR of developmental compositions

| | A | B | C |
|---|---|---|---|
| R (kohm/sq/mil) | 4.12 | 6.18 | 12.97 |
| HTCR (ppm/C) | 2295 | 2018 | 1444 |
| CTCR (ppm/C) | 2277 | 1996 | 1467 |

## III. CONCLUSIONS

1. Process sensitivities of 509x series were measured.

2 5091 & 5092 are robust compositions, 5093 is more sensitive to processing.

3. Resistance range of 509x series was extended to ~16kohm/sq/mil.

## IV. REFERENCES

1. Macklen, E. D., 1979, *Thermistors*, Electrochemical Publications Limited, 29 Barns Street, Ayr, Scotland, 1979.

# EVALUATION OF SOME THICK FILM MATERIALS FOR TEMPERATURE, FORCE, AND HUMIDITY SENSORS

M. HROVAT, D. BELAVIČ*, J. HOLC
*Jožef Stefan Institute, Jamova 39, 61000 Ljubljana, Slovenia*
** Iskra HIPOT, Trubarjeva 7, 68310 Šentjernej, Slovenia*

## 1. Introduction

Thick film technology is a well known, mature and versatile technology for producing hybrid circuits. Also, for some applications it provides a relatively inexpensive and convenient way to produce sensors which can be integrated in to hybrid circuits. The aim of this paper is to present some results obtained with commercial thick film pastes for temperature and strain sensors (thick film conductors and resistors, respectively), and results on thick film humidity sensors based on $Nb_2O_5$ doped $TiO_2$.

## 2. Temperature sensors

Thermocouples were made with different combinations of thick film conductors based on Ag, Au, Au/Pd, Au/Pt, Ag/Pd, Ag/Pt, Ni and Cr, printed and fired on alumina substrates. Noble metal based pastes (Du Pont) were fired at 850°C, while air-firing Ni and Cr pastes (Electro Science Labs.) were fired at 750°C. Absolute EMFs (at 25°C) and nonlinearity (dependence of thermocouple EMF on temperature) are presented in Table I [1,2].

Table I: Absolute EMF and nonlinearity of thick film thermocouples

| Conductor combination | EMF (25 °C) [μV/K] | Nonlinearity [%/K] | Conductor combination | EMF (25 °C) [μV/K] | Nonlinearity [%/K] |
|---|---|---|---|---|---|
| Au/Pd-Au | 16.85 | 0.24 | Au/Pd-Ag/Pd | 8.29 | 0.35 |
| Ag/Pd-Au | 6.62 | 0.22 | Au/Pt-Ag/Pd | 2.38 | 0.18 |
| Au/Pd-Ag | 14.41 | 0.22 | Au/Pt-Au | 10.75 | 0.13 |
| Au/Pt-Ag | 9.06 | 0.24 | Au/Pt-Au/Pd | 6.61 | 0.08 |
| Au-Ag | 0.04 | ? | Ag/Pt-Ag | 1.49 | - 0.07 |
| Ni-Cr | 13.65 | - 0.22 | Ag/Pd-Ag/Pt | 5.48 | 0.13 |
| Ag/Pd-Ag | 6.15 | 0.21 | | | |

The prototype of a thick film heat flux sensor with 110 thermocouples, connected in series, was designed [1]. Heat flux sensors, realized in thick film technology have some good characteristics, for example low thermal resistivity and relatively high operating temperatures [3,4]. The thermal resistance layer is made of thick film multilayer dielectric pastes, printed and fired on $Al_2O_3$ substrates. Multilayer dielectrics are glass-like materials with higher heat resistivity than alumina substrates. The heat flow is perpendicular to the ceramic substrate. The temperatures on the top (cold junction) and underneath (hot



*W. K. Jones et al. (eds.), MCM C/Mixed Technologies and Thick Film Sensors*, 267–272.

junction) the dielectric layer are measured by thermocouples, made from thick film conductors. The principle of operation and a schematic representation of the thick film heat flux sensor are shown in Fig. 1 and 2. Thermocouples were made with a Pd/Au-Au combination with the highest measured EMF. The thermal resistance layers (thickness after firing around 70 μm) were made with dielectrics based either on crystallizable glass or a mixture of glass and ceramic. These two materials have heat conductivities of 0.64 W/mK and 1.83 W/mK, respectively. The voltage response of sensors made with dielectrics based on crystallizable glass and a mixture of glass and ceramic was 2.3 mV and 1.2 mV, respectively, when measured with a heat flux of 10 $kW/m^2$.

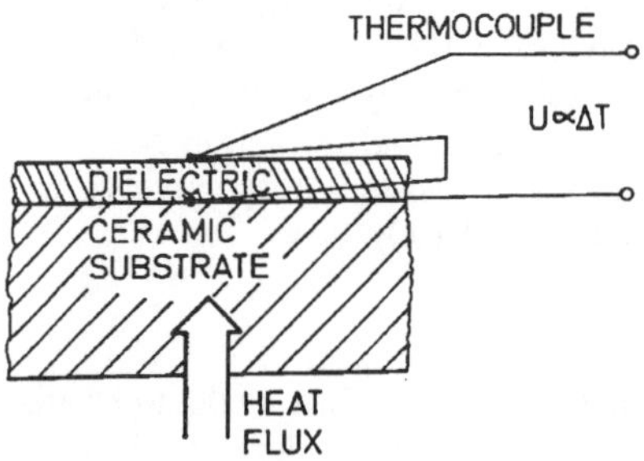


Fig. 1: The principle of operation of a thick film heat flux sensor.
Thermocouples measure the temperature difference generated by a heat flow through the dielectric layer.

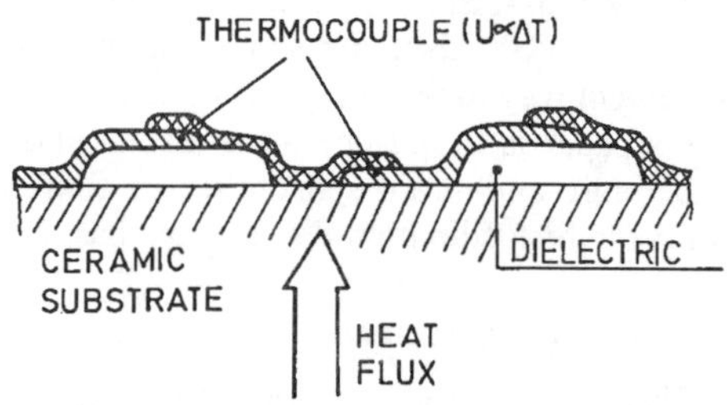


Fig. 2: A schematic representation of a thick film heat flux sensor. The heat flow is perpendicular to the ceramic substrate. Thermocouples made from thick film conductors measure the temperature on the top (cold junction) and under (hot junction) the dielectric layer [3].

Also, a prototype of the sensing element for a true RMS meter was realised [2]. A thick film resistor (heater) was printed and fired on one side of a ceramic substrate. The power dissipated in the heating resistor, which is proportional to $I^2$ or $U^2$ causes a thermal gradient along the length of the substrate which is measured by 12 thick film thermocouples, connected in series. In the case of a Ni-Cr combination the voltage response was 38 mV for 1 W of power dissipated in the heating resistor (10 ohm, 320 mA). The transfer function K(T) of the RMS meter was found to be relatively independent of the ambient temperature, which means that the voltage response is usably linear.

## 3. Strain sensors

The gauge factors (GF) of thick film materials are higher than those of thin metal films and lower than those of semiconductors. Due to their stability, the low temperature coefficient of resistivity (TCR) of thick film materials and the relatively low cost, strain gauges realized in thick film technology offer advantages in some applications over both metal film (low GF, low TCR) and semiconducting elements (high GF, high TCR) [5-8].

Some commercial thick film resistors with sheet resistivities from 1 kohm/ to 100 kohm/ , were evaluated for strain gauge applications. Resistor paste 3414 (Electro Science Labs.) is specially formulated for high gauge factors [9]. The conductive phase in the resistors was determined by X-ray powder diffraction analysis and EDS (Energy Dispersive X-ray Microanalysis). Nominal sheet resistivities, conductive phases, current noise indexes and longitudinal gauge factors ($GF_l$) are given in Table II [10].

Table II: Nominal sheet resistivities, conductive phases, noise indexes and longitudinal gauge factors of thick film resistors

| Thick film material | Nominal sheet resistivity [ohm/ ] | Conductive phase | Noise index [dB] | $GF_l$ |
|---|---|---|---|---|
| Du Pont HS 8031 | 1 k | $RuO_2$ | - 14 | 2.5 |
| Du Pont HS 8029 | 1 k | $Bi_2Ru_2O_7$ | - 14 | 8.5 |
| Du Pont HS 8041 | 10 k | $RuO_2$ | - 4.4 | 3.5 |
| Du Pont HS 8039 | 10 k | $Bi_2Ru_2O_7$ | - 14 | 11 |
| Heraeus R 8241 | 10 k | $RuO_2$ | - 6.0 | 16 |
| ESL 3414-A | 10 k | $Bi_2Ru_2O_7$ | + 8.3 | 16.5 |
| Du Pont HS 8049 | 100 k | $Bi_2Ru_2O_7$ | - 11 | 12 |

Data in the literature show that the gauge factors of thick film resistors based on ruthenates ($Bi_2Ru_2O_7$ or $Pb_2Ru_2O_{6.5}$) are usually higher than these of resistors based on $RuO_2$ [6]. However, as seen in Table II, a notable exception is the $RuO_2$ based paste 8241 with the second highest $GF_l$. This indicates that the values of GFs of thick film resistors depend not only on the composition of the conductive phase, but also on the microstructure of the material. TEM microstructures of two $RuO_2$ based 10 kohm/o resistors are shown in Figs. 3 (HS 8041, GF=3.5) and 4 (R 8241, GF=16) [11]. There seems to be a significant difference in the microstructures of $RuO_2$ based thick film resistor materials with low or high gauge factors. HS 8041 with a low GF has numerous small conductive grains close together in the glass phase, while the conductive grains in R 8241 resistor materials are larger and further apart.

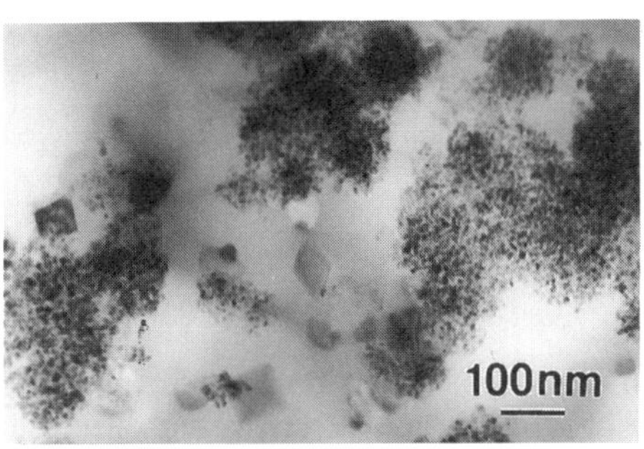


Fig. 3: TEM microstructure of $RuO_2$ based 10 kohm/ resistor HS 8041 with a low gauge factor (GF=3.5)

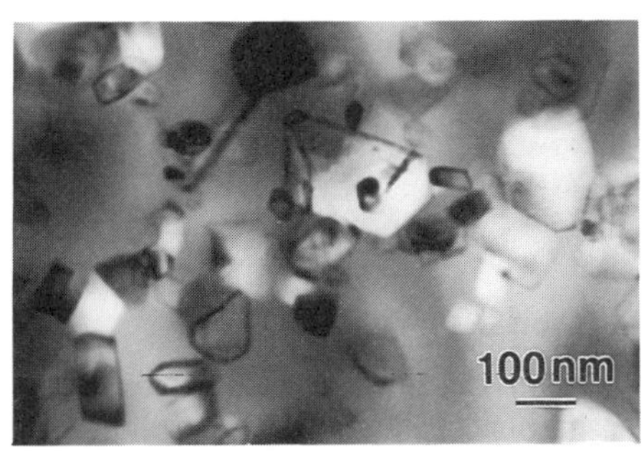


Fig. 4: TEM microstructure of $RuO_2$ based 10 kohm/ resistor R 8241 with a high gauge factor (GF=16)

Therefore, the connection between these two types of microstructures and the gauge factors can be tentatively interpreted as follows: a more homogeneous material with many small grains and short distances between the grains will have a low gauge factor, while a material with larger average distances between the grains will have a higher gauge factor. This is also in accordance with the fact that the gauge factors of thick film resistors increase with increasing sheet resistivity [6]. Higher sheet resistivities mean less conductive phase and therefore larger average distances between conducting sites.

When resistor pastes are fired on or under multilayer dielectrics instead of on alumina substrates, the interaction between the resistor and the multilayer material changes the electrical characteristics of the resistors [12]. The possibility of increasing the gauge factors of resistors in this way was investigated.

Three thick film resistor pastes (HS 8029, ESL 3414 and R 8241) were fired on and under multilayer dielectrics [13]. Gauge factors and current noise indexes are shown in Figs. 5 and 6, respectively. The highest gauge factors, over 30 and over 35, were obtained with 3414 resistors fired on 5704 (Du Pont) and 9117 (Heraeus) dielectrics, respectively. These values are close to the gauge factors obtained with specially formulated thick film resistor materials [14]. Higher values of GFs are presumably obtained when the conductive grains or clusters do not form continuous conductive paths through the resistor but are at least partially separated by a glass phase which diffuses from the dielectric into the resistor during firing. The increases in GFs are therefore accompanied by increased sheet resistivities and noise indexes.

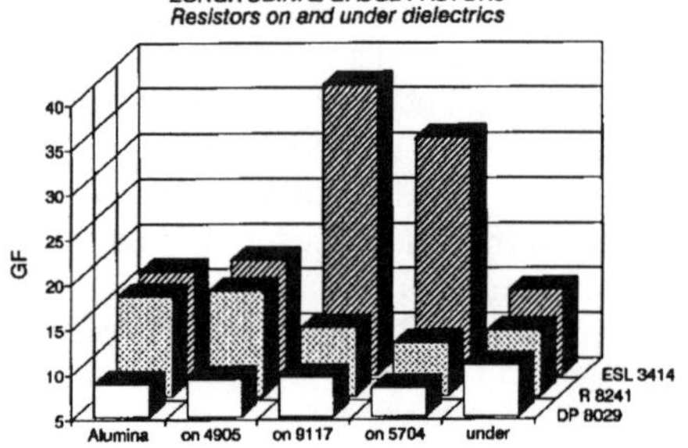


Fig. 5: Gauge factors of thick film resistors, fired on or under multilayer dielectrics

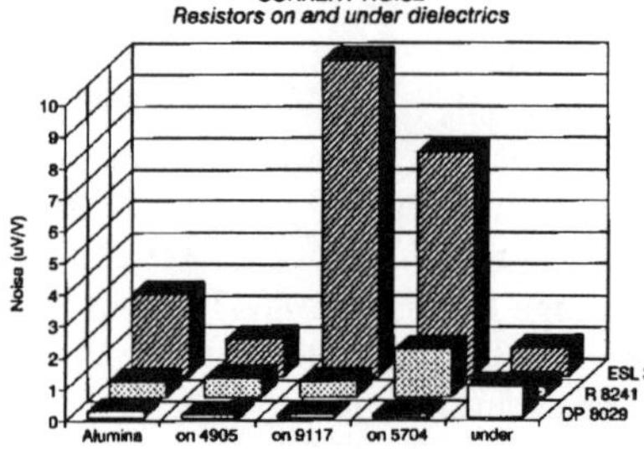


Fig. 6: Current noise indexes of thick film resistors, fired on or under multilayer dielectrics

## 4. Humidity sensors

Porous ceramic materials are considered to meet most of the requirements for humidity sensors such as high sensitivity, short response time, durability, selectivity and chemical and thermal stability [15]. The electrical properties of ceramic sensors, such as conductance and capacitance, are dependent on the relative humidity (RH) due to capillary condensation and physisorbtion of water [16]. Katayama [17] studied porous $Nb_2O_5$ doped titania as a humidity sensing material. The best results were obtained with fritted $RuO_2$ electrodes on the pellet type sensor. Because of the relatively thin layer of sensing material in thick film sensors, the possible reaction between the sensing material and the electrodes can influence the electrical characteristics of the sensing layer more than in the case of bulk form sensors [18-20].

Titania thick film pastes doped with 0.5 wt. $Nb_2O_5$ were prepared by mixing both powders with an organic vehicle. Pastes used for electrodes were based on fritted $RuO_2$ and Pt material and pure fritless platinum. First the lower electrode was printed and fired at 1000 °C on an alumina substrate. The sensing layer was fired at 1200 °C. Then the upper electrode was printed and fired at 850 °C and 1000 °C for $RuO_2$ and Pt based electrodes, respectively. AC electrical properties were measured between 5 and 90 % RH at frequencies from 150 Hz to 100 kHz.

The results obtained with different electrodes are summarized in Table III.

Table III: Summary of results obtained with different electrode pastes.

| Sample | Lower/upper electrode paste | Reaction with $TiO_2$ , layer | Conductance at 10 % RH, 1 kHz [S/cm] | Sensitivity: 0-50 % RH | Conductance at 90 % RH, 1 kHz [S/cm] | Sensitivity: 50-98 % RH |
|---|---|---|---|---|---|---|
| A | fritted Pt/fritted Pt | Yes | $\sim 1*10^{-10}$ | Low | $\sim 1*10^{-7}$ | Medium |
| B | fritted Pt/pure Pt | Yes | $\sim 1*10^{-9}$ | High | $\sim 1*10^{-5}$ | High |
| C | Pure Pt/pure Pt | No | $\sim 1*10^{-7}$ | High | $\sim 5*10^{-4}$ | High |
| D | pure Pt/fritted $RuO_2$ | Yes | $\sim 1*10^{-8}$ | Nil | $\sim 5*10^{-8}$ | Very low |

The glass fritted Pt electrode is more porous and contains a glass phase based on Si, Ba and B. Elements of the glass phase were detected by EDX analysis in the $TiO_2$ sensing layer of samples A and D. In sample B those elements were not detected but the electrical measurements indicate interactions between the fritted platinum paste and the $TiO_2$ fi lm.

The sample with fritted $RuO_2$ electrodes (sample D) had the highest initial conductance, but due to diffusion of glass phase into the $TiO_2$ layer its sensitivity was very low. The sensitivity of samples with fritted Pt electrodes (sample A) was also low within the range between 0 and 50 % RH. In both cases (A and D) this is due to the presence of glass phase in the sensing layer, which closes the pores with small diameter, which are "responsible" for sensitivity characteristics at low RH [16].

## 5. References


1. M. Hrovat, M. Zgonik, D. Belavič, S. Maček, Thick film materials for heat flux sensors, *J. Mater. Sci. Lett.*, 11, (2), (1992), 89-90
2. M. Hrovat, M. Zgonik, D. Belavič, A thick film sensing element for true RMS meter, *Active and Passive Elec. Comp.*, 14, (1), (1990), 39-45
3. A. C. van Dorth, F. van der Graff, G. K. Steenvoorden, Thick film flux sensor, *Sensors and Actuators*, 4, (1983), 323-331
4. T. Kwikkers, Two thick film thermal sensors, *Hybrid Circuits*, (16), (1988), 39-43
5. A. Cattaneo, R. Dell'Acqua, G. Dell'Orto, L. Pirozzi, C. Canali, A practical utilization of the piezoresistive effect in thick film resistors: a low cost pressure sensor, Proc. 1980 Int. Symp. on Microelectronics ISHM-80, New York, 1980, 221-227
6. M. Prudenziati, B. Morten, Piezoresistive properties of thick film resistors; an overview, Hybrid Circuits, (10), (1986), 20-23, 37
7. N. White, A. Cranny, Design and fabrication of thick film sensors, *Hybrid Circuits*, (12), (1987), 32-35
8. S. Satoh, Y. Takatsuji, F. Katoh, H. Hirata, Thick film pressure sensor using zirconia diaphragm, Proc. 1991 Int. Symp. on Microelectronics ISHM-91, Orlando, 1991, 148-152
9. S. Chitale, C. Huang, M. Stein, High gauge factor thick film resistors for strain gauges, *Hybrid Circuits Technol.*, 6, (5), (1989)
10. M. Hrovat, D. Belavič, J. Holc, S. Šoba, An evaluation of some commercial thick film resistor materials for strain gauges (To be published in J. Mater. Sci. Lett.)
11. M. Hrovat, G. Dražič, J. Holc, D. Belavič, The correlation between microstructure and gauge factors of thick film resistors (Submitted to J. Mater. Sci. Lett.)
12. M. Hrovat, D. Belavič, Thick film multilayer circuits with "buried" resistors - resistors under multilayer dielectric, Proc. 6 European Microelectronics Conf. ISHM-Europe 87., Bournemouth, 1987, 305-312
13. M. Hrovat, D. Belavič, J. Holc, S. Šoba, A possibility to increase the gauge factors of thick film resistors (To be published)
14. O. Abe, Y. Taketa, A new thick film strain sensor, Proc. 1988 Int. Symp. on Microelectronics ISHM-88, Seattle, 1988, 217-221
15. B.M. Kulwicki, Humidity sensors, *J. Am. Ceram. Soc.*, 74 (4), (1991), 697-708
16. N. Yamazoe, Y. Shimizu, Humidity sensors: principles and applications, *Sensors and Actuators*, 10, (1986), 379-398.
17. K. Katayama, K. Hasegawa, T. Takahashi, T. Akiba, H. Yanagida, Humidity sensitivity of $Nb_2O_5$-doped $TiO_2$ ceramics, *Sensors and Actuators* A, 24, (1990), 55-60.
18. J. Slunečko, J. Holc, M. Hrovat, S. Bernik: Interactions among thick film $TiO_2$ paste, alumina substrates and platinum electrodes, *J. Mat. Sci. Lett.* 11, 1992, 782-784
19. J. Slunečko, J. Holc, M. Hrovat, M. Čeh: Thick film humidity sensors based on $(Ba,Sr)TiO_3$ porous ceramic doped with MgO and CaO, *Sensors and Actuators* B, 7, 1992, 439-442
20. J. Slunečko, J. Holc, M. Hrovat, M. Čeh, S. Bernik: Evaluation of materials for thick film humidity sensors, Proc. 8th European Hybrid Microelectronics Conf. ISHM-Europe 91, Rottherdam, 1991, 236-242

# PARAMETERS AND TECHNOLOGY OF THICK FILM ELECTROLYTIC $SO_2$ SENSORS

L.J. GOLONKA
Institute of Electronic Technology, Technical University of Wrocław
Wybrzeże Wyspiańskiego 27, 50-370 Wrocław, POLAND

**Keywords/abstract:** gas sensor, thick film, solid electrolyte, sulphur dioxide

Short review of various microelectronics gas sensors is presented. Special attention is paid to the elements made by thick film technology. Principle of work, construction and the properties of solid state electrolytic sensors are presented. The basic characteristics of thick film $SO_2$ sensor are shown.

## 1. Introduction

There are various kinds of microelectronics sensors which can detect gases. The most important sensors are: conductive type gas sensors, electrochemical sensors and chemically sensitive field effect transistors. Unfortunately, only the electrochemical sensors are able to measure the sulphur oxides partial pressures.

Oxides of sulphur are produced during the combustion of fossil fuels, smelting of sulphide ores and in the manufacture of sulphuric acid. It is necessary to monitor the concentration of the gases escaping into the atmosphere, both for process control and reduction of environmental pollution. The sulphur oxides contribute to air and soil pollution in the form of the acid rain. The gas is very dangerous for human health. Its presence in air destroys plants as well as old historical sculptures and buildings. Conventionally the sulphur dioxides have been measured with analytical or optical methods, which are very expensive and not suitable for the continuous on-site monitoring. The solid electrolyte sensors have the advantage of working continuously, selectively and at the low cost.

The investigated thick film sensors were made on the alumina substrate. They worked in the potentiometric mode. The good agreement between the experimental value of the slope of electromotive force (E) against the logarithm of the sulphur dioxide partial pressure with the theoretical one was found. The response time was varied between a few seconds for increasing, and few minutes for decreasing the sulphur dioxide concentration.

## 2. Gas sensors

Gas sensors can detect toxic, explosive and flammable gases.

The following parameters characterize the toxic gases: Long Term Exposure Limit (LTEL) - the maximum allowable concentration of the toxic gas which does not



*W. K. Jones et al. (eds.), MCM C/Mixed Technologies and Thick Film Sensors*, 273–278.

cause any damage to human health during eight hours and Short Term Exposure Limit (STEL) - which is connected with short term exposition to the toxic gas.

The explosive gases are characterized by Lower Explosion Limit (LEL) and Upper Explosion Limit (UEL). The gas is explosive if the gas concentration is between the LEL and the UEL.

Gas concentration thresholds in air mixtures of the most important gas pollutants and toxic gases are presented in Table 1 [17].

TABLE 1. Gas concentration thresholds in air mixtures

| Gas | Formula | LTEL (ppm) | STEL (ppm) | LEL (vol %) | UEL (vol %) | Background conc. (ppm) |
|---|---|---|---|---|---|---|
| Carbon dioxide | $CO_2$ | 5000 | 15000 | - | - | 315 |
| Carbon oxide | CO | 50 | 300 | 12.5 | 74.2 | 0.1-2 |
| Chlorine | $Cl_2$ | 1 | 3 | - | - | |
| Hydrogen sulphide | $H_2S$ | 10 | 15 | 4.3 | 45.5 | 0.001-.01 |
| Nitrogen dioxide | $NO_2$ | 3 | 5 | - | - | 0.05-0.5 |
| Ozone | $O_3$ | 0.1 | 0.3 | - | - | |
| Sulphur dioxide | $SO_2$ | 2 | 5 | - | - | 0.05-0.5 |
| Methane | $CH_4$ | | | 5 | 15 | 1-2 |
| Ammonia | $NH_3$ | 25 | 35 | 15.5 | 27 | |

The gas sensors are made by semiconductor, thick film, thin film or ceramic technologies. The most important parameters which the sensors should exhibit are: stability, reproducibility, sensitivity, selectivity, low cost, short response time, low power consumption and small size.

The gas sensors usually are divided into the following groups :

- conductive type gas sensors,
- electrochemical sensors,
- chemically sensitive field effect transistors,
- others.

Conductive type gas sensors. Oxide semiconductors are the most widely used as a gas sensors. They detect gases by a change in the electrical resistance of the element. The density of the surface depends upon the adsorption and the desorption gas processes as well as chemical reaction between various gases. These effects influence the electrical conductivity of the sensor. The resistance of n-type semiconductor oxides ($SnO_2$, ZnO) is decreasing in reduction and increasing in oxidising atmosphere.

There are a lot of the thick film semiconductor sensors. Most of them is based on $SnO_2$ and various additives which are added to increase their selectivity [5,10,16]. Tin dioxide thick film doped with Nb, V, Ti and Mo are used to detect propane [18], with $ThO_2$ to measure CO [19], with Sb to monitor CO and $H_2$ [22], with $ZrO_2$ to find $H_2S$ [11] concentration. The semiconductor gas sensors work at elevated temperature (150÷ 500 $^oC$). To obtain this temperature the special thick film heater is used [2,5]. To increase the sensor selectivity the sensor array is constructed. The sensors in the array are made from various materials. They contain different dopants and work at specially determined temperatures.

Electrochemical gas sensors with solid electrolyte can detect $O_2$, $Cl_2$, $SO_2$, $CO_2$, CO, $H_2$ [12]. They work at various modes. Generally, the electromotive force of solid electrolyte gas sensor depends upon the gas concentrations on both side of the electrolyte film. This kind of sensor is described in more details in the Chapter 3.
Chemically sensitive field effect transistors. Field effect transistor (FET) is used as a gas sensor. The conductivity between the source and the drain depends upon the adsorbed gases. The sensor can detect hydrogen, carbon monoxide, ammonia [17].
Other sensors: catalytic gas sensor, optical based on the absorption of infra red light, surface acoustic wave sensors [17]

## 3. Solid state electrolyte gas sensors

The sensors exhibit several important advantages [12,20,21]:

- the generated electrical voltage and currents are easily measurable,
- the partial gas pressure is directly converted into the electrical signal,
- the potentials are independent upon the sensor dimension,
- the signal is selective for the transferred component of the electrolyte,
- the voltage is related logarithmically to the partial pressure and covers a wide range of gas pressures with constant accuracy,
- no calibration is usually required,
- good mechanical properties - all the components are solid and their sensitivity to mechanical parameters (vibration, gravity, acceleration) is very small,
- the chemical inertness can be very high, ensuring a long life time and prevents any contamination of the analysed gas,
- stability over wide temperature intervals,
- unattended operation, no maintenance, low energy consumption, low cost.

The disadvantages of the sensors are the long response time and difficulties in measuring the small variations of the gas partial pressures with a high accuracy on a logarithmic scale. The response time of the sensors becomes excessively long at low temperatures.

The sensors work at the potentiometric or amperometric modes.

The potentiometric gas sensor consists of the solid electrolyte, and two electrodes separated by the electrolyte. Ions diffusion takes place in the electrolyte at high temperatures due to the difference of the chemical potentials between the electrolyte and the reference electrode as a result of the reaction at the gas electrode - electrolyte interface. They are based on the Nernst law. It can be written as follows [12]:

$$E = E^0 + kT \log a \qquad \text{with } k = R/nF$$

where E is the measured voltage between the measuring electrode and the reference electrode, $E^0$ and k are the constants, n is the number of electrons involved in the sensing electrode reaction, and $a$ is the chemical activity of the analysed species.

The work of the amperometric sensor is based on the Faraday law [12]:

$$J = I/zF$$

where J is the flux of pumped matter, I the pumping current, F the Faraday constant, and z the number of electrons carried by the ions.

The electrochemical gas sensor can detect various gases. The typical examples of solid ionic conductors used in the electrochemical sensors are presented in [6,12].

The solid state electrolytic sensor are mostly made as the ceramic ones. Recently new thick film microelectronics sensors have been investigated. They can detect $CO_2$, $O_2$, $H_2$ [3,4,13] and $SO_2$ [8,9].

The cells for monitoring sulphur oxides are generally composed of a sulphate based electrolyte, a porous platinum working electrode and a reference electrode formed by dissolved silver sulphate. The use of sulphate based electrolyte was first proposed by Gauthier et al. [7]. The recent development of sulphur oxides sensors has been reviewed by Worrell et al. [21] and presented in the papers [1,12,15]. Solid state galvanic sensors for sulphur oxides have used $K_2SO_4$, $Na_2SO_4$ or $LiSO_4$. The alkali metal sulphates are good ionic conductors at the higher temperature.

## 4. Thick film solid electrolyte $SO_2$ sensor

Solid electrochemical sensor has been prepared by thick film screen printing method in the galvanic chain:

$$Ag \,/\, Ag_2SO_4 \text{ in } K_2SO_4 \,/\, K_2SO_4 \,/\, SO_2, O_2 \text{ (air), Pt}$$

The sensor has the typical sandwich structure [9]. It consists of the following films: electrode, reference electrode, solid electrolyte and Pt electrode. The heating element (Pt) was made on the reverse side of the 96 % alumina substrate. The sensor temperature was determined by the heater resistance measurements. The sensor chip dimension was 5 x 5 $mm^2$. Attention has to be paid to the compatibility of the single films with each other. To overcome a problem of the weak adhesion and formation of micro cracks, a small amount of glass was added to solid electrolyte and reference electrode inks. The addition of glass improved the mechanical properties of the sensor with only slight shifting the characteristic of electromotive force versus partial pressure of sulphur oxide [9].

Sulphates are attractive electrolytes for $SO_x$ sensors because the $SO_4^{2-}$ ion can equilibrate with either $SO_2$ or $SO_3$ in the gas mixture and establish the cell reaction. The potentiometric response of a sulphate - electrolyte sensor is directly related to the pressure or concentration of $SO_2$ or $SO_3$ in the gas.

The second major component of a solid electrolyte sensor is the reference electrode which establishes a constant $SO_3$ concentration. The most successful solid reference electrode for a sulphate sensor is the silver-silver sulphate mixture [7]. The silver-oxygen-sulphur system is unique because silver and $Ag_2SO_4$ can coexist as the equilibrium phases in contrast to the usual situation where an oxide phase separates the metal from the sulphate.

When silver-$Ag_2SO_4$ mixture is used, the half cell reaction at the reference electrode - electrolyte interface occurs:

$$2\,Ag + SO_4^{2-} = Ag_2SO_4 + 2\,e^-$$

The half cell reaction at the electrolyte - gas interface is the equilibration of the $SO_4^{2-}$ ion in the sulphate electrolyte with the $SO_3$ in the gas. With a catalyst like platinum the cell potential is established by the equilibrium pressure of $SO_3$ which can be related to the inlet $SO_2$ pressures. In the case when the air contains small amounts of $SO_2$ or $SO_3$ the oxygen pressure is known and is essentially constant. The half cell reaction at the gas electrode - electrolyte interface is the equilibration of the $SO_3$ in the gas with the $SO_4^{2-}$ ion in the electrolyte:

$$SO_3\,(g) + 1/2\,O_2\,(g) + 2\,e^- = SO_4^{2-}$$

The cell reaction is obtained by combining the half cell reactions for the gas electrode and the reference electrode, respectively:

$$2\ Ag + SO_3\ (g) + 1/2\ O_2\ (g) = Ag_2SO_4$$

In air and at constant silver ion activity the solid electrolyte sensor e.m.f. (E) depends only on the $SO_2$ partial pressure:

$$E = const. + RT/2F\ ln\ (p_{SO2})$$

where: R - gas constant 8.32 [J/mol K],
T - temperature of the cell [K],
F - Faraday's constant 96500 [C],
$p_{SO2}$ - $SO_2$ partial pressure.

The static properties of the sensor were measured in the chamber with the constant sulphur dioxide concentration. The response time was investigated in the gas installation with the mass flow controllers. The dependence of electromotive force against sulphur dioxide concentration at various sensor temperature is presented in [9]. The highest range of linearity was obtained for temperature 770$^o$C. The theoretical slope of the line was calculated for this temperature. The obtained theoretical value 103 mV/decade is very close to the experimental one.

The response time was determined for the sensors working at temperature 720$^o$C at the sulphur dioxide concentration range from 0 to 1000 ppm, 2000 ppm or 5000 ppm. The response time depends on the sensor temperature, the level of sulphur dioxide concentration and the direction of sulphur dioxide partial pressure changes. The time is shorter for higher sensor temperature or lower sulphur dioxide concentration. Increasing or lowering sulphur dioxide partial pressure in the gas mixture gives the response time shorter (15-30 sec.) or longer (12-15 min.), respectively. We hope, that future work on the sensor construction and the composition will give the further decrease of that time.

The reproducibility of sensor response was good [9].

## 5. Conclusions

The following conclusions can be drawn from our experiments:

- Solid electrolyte sulphur dioxide sensor can be made by thick film technology,
- Small addition of glass to the electrolyte ink improves the mechanical properties of the sensor films,
- The linearity of the sensor characteristic E = f (log $p_{SO2}$) depends on the sensor temperature. The highest range of linearity was obtained for 770$^o$C (30 ÷ $10^4$ ppm), The slope of the line is in a very good agreement with the theoretical value calculated on the base of Nernst equation,
- The response time depends on the sensor temperature (higher temperature - shorter time), the level of sulphur dioxide concentration and the direction of sulphur dioxide concentration change.

This work was sponsored by the Polish Scientific Research Committee (KBN), Grant No 8 0127 91 01.

## 6. References


1. Azad, A.M., Akbar, S.A., Mhaisalkar S.G., Birkefeld, L.D., and Goto, K.S. (1992) Solid-state gas sensors: a review, *J. Electrochem. Soc.*, **139**, 3690-3704.
2. Baumbach, P., Stein, M.A., Tait, R., and Whitmarsh, J. (1990) Materials, substrates and designs for manufacturing heaters using thick film technology, *Proc. of the 1990 International Symposium on Microelectronics ISHM USA*, 699-705.
3. Chu, W.F., Erdmann, H., Ilgenstein, and Leonhard, V. (1990) Solid state chemical sensors in thick film technique, *Proc. Micro System Technologies, Berlin Sept. 1990*, 1-6.
4. Chu, W.F., Fischer, D., Erdmann, H., Ilgenstein, M., Köppen, H., and Leonhard, V. (1992) Thin and thick film electrochemical $CO_2$ sensors, *Solid State Ionics*, **53-56**, 80-84.
5. Dziedzic, A., Golonka, L.J., Licznerski, B.W., and Hielscher G. (1994) Heaters for gas sensors from thick conductive or resistive films, *Sensors and Actuators B*, **19**, 535-539.
6. Fouletier, J. (1982/83) Gas analysis with potentiometric sensors. A review, *Sensors and Actuators*, **3**, 295-314.
7. Gauthier, M., and Chamberland, A. (1977) Solid state detectors for the potentiometric determination of gaseous oxides, 1. Measurements in air, *J. Electrochem. Soc.*, **124**, 1579-1583.
8. Golonka, L.J., Paściak, G., and Barycka, I. (1992) Thick film $SO_2$ sensor - preliminary data, *Proc. of 16th Conference ISHM Poland Chapter, Cracow September 1992*, 75-78.
9. Golonka, L.J., Paściak, G., and Barycka, I. (1993) Thick film $SO_2$ sensor, *Proc. 9th European Hybrid Microelecronic Conference ISHM, Nice June 1993*, 289-245.
10. Golonka, L.J., Kozłowski, J., and Licznerski, B.W. (1994) The influence of the electrode material on sensitivity of the $SnO_2$ thick film gas sensor, *Sensors and Actuators B*, **19**, 453-456.
11. Kanefusa, S., Nitta, M., and Haradome, M. (1988) $H_2S$ gas detection by $ZrO_2$ - doped $SnO_2$, *IEEE Trans. on Electron Devices*, **ED-35**, 65-69.
12. Kleitz, M., Siebert, E., Fabry, P., and Fouletier, J. (1991) Solid-state electrochemical sensors, in W. Göpel, *Sensors* **2**, VCH, Weinheim-New York- Cambridge.
13. Leonhard, V., Fischer, D., Erdmann, H., Ilgenstein, M., and Köppen, H. (1993) Comparison of thin and thick film $CO_2$ sensors, *Sensors and Actuators B* **13-14**, 530-531
14. Makovos, E.B., Montague, F.W., Dudik, L., and Lin, C.C. (1993) A calorimetric combustible gas detector employing platinum film heaters, *Sensors and Actuators B*, **12**, 91-94.
15. Mari, C.M., Beghi, M., Pizzini, S., and Faltemier J. (1990) Electrochemical solid-state sensor for $SO_2$ determination in air, *Sensors and Actuators B*, **2**, 51-55.
16. Martinelli, G., and Carotta, M.C. (1993) Influence of additives on the sensing properties of screen printed $SnO_2$ gas sensors, *Sensors and Actuators B* **15-16**, 363-366.
17. Moseley, P.T., Norris, J.O.W., and Williams, D.E. (1991) *Techniques and mechanisms in gas sensing*, IOP Publishing Ltd., Bristol
18. Nitta, M., Kanefusa, S., and Haradome, M. (1978) Propane gas detector using $SnO_2$ doped with Nb, V, Ti and Mo, *Journal Electrochem. Soc.*, **125**, 1676-1674.
19. Nitta, M., and Haradome, M. (1979) CO gas detection by $ThO_2$ doped $SnO_2$, *J. Electron. Materials*, **8**, 571-580.
20. Weppner, W. (1987) Solid-state electrochemical gas sensors, *Sensors and Actuators* **12,** 107-119.
21. Worrell, W.L. (1988) Solid-electrolyte sensors for $SO_2$ and/or $SO_3$, in T. Seiyama, *Chemical Sensor Technology* **1**, Elsevier, Amsterdam-Oxford-New York-Tokyo.
22. Yannopouls, L.N. (1987) Antimony - doped stannic oxide based thick film sensors, *Sensors and Actuators*, **12**, 77-89.

# SENSORS: A GREAT CHANCE FOR MICROELECTRONIC TECHNOLOGIES

R. DELL'ACQUA
*MiTeCo*
*Corso Cairoli 96, 27100 Pavia, Italy*

## Abstract

Electronic control systems, as well as the human beings, consist of a brain and of several sensors and actuators for outside world interfacing.
While system brains have been improved at full speed in the last decades, leading to the introduction of very powerful and inexpensive microprocessors, the same did not happen either for sensors and for actuators which still are considered as the weak points of the system.
It is possible to state that, with a few exceptions, the today sensors are rather expensive, not enough accurate, and sometimes unreliable even if many attempts have been done in the past to improve this situation.
In fact, when 15 years ago automotive and medical equipment market began by asking high volumes of low cost pressure sensors, it became clear that only by using the microcircuit implementation technologies the target would have been achieved.
Since then, a lot of work has been carried out leading to very effective products and promising solutions based on the technologies of silicon integration and micromachining, of thick and thin films, and of ceramics.
During this presentation the advantages and drawbacks of the above mentioned technologies, when used for sensors, will be briefly described mainly through several examples of either already existing and future devices.
In particular, the most interesting approaches for the development and manufacturing of pressure, acceleration, and gas sensors will be discussed.
These sensors are already or will be shortly needed in very high volumes by several market areas; but a continuous improvement of their cost/performance ratio will be required. By now, it is clear that this improvement will be



*W. K. Jones et al. (eds.), MCM C/Mixed Technologies and Thick Film Sensors,* 279–297.

strictly related to the proper use of the most advanced microelectronics packaging technologies.

## Introduction

During the past few decades, the high volume price of microprocessors has dropped below the cost of sensors that tie the processor to the analog world of pressure, temperature, acceleration, flow, and other physical and chemical variables. As a result, real-time control applications for microprocessors, particularly in high-volume consumer applications, were hampered by the lack of inexpensive, digitally-compatible sensors.
However, in the last years a growing trend in fabricating either micromachined sensors on silicon wafers and thick-film sensors on ceramics is appeared.
IC production techniques are leading to silicon sensor chips that integrate moving structures with resistive Wheatstone bridges (or sometimes with variable capacitors), laser trimmable resistor networks for calibrations, and even signal-processing circuits.
Thick (and sometimes thin) film technology, taking advantage of special materials developed on pourpose, provides extremely rugged and reliable physical and chemical sensors at very reasonable prices.
Even if they seem to be in competition each other, the above mentioned technology should be considered as complementary; their choice depends on the type of application, environmental conditions, size, performances, and so on.

## 1. Silicon Sensors

Micromachined silicon has several advantages for use in sensors. In particular, silicon has no mechanical hysteresis and is highly sensitive to mechanical stress. Its modulus of elasticity is the same as steel, and tensile yield strength is higher than steel. Silicon is also as hard as quartz, yet less dense than aluminium. Perhaps most important, silicon sensors having tightly controlled submicron geometries can be built and packaged with the same mature processes, equipment, and ultrapure materials used for producing high-volumes of integrated circuits.
A number of innovative fabrication techniques have recently

been developed specifically for micromechanical structures , and they fall into two categories: bulk micromachining and surface micromachining.

The first involves sculpturing the silicon substrate by means of chemical etchants, and the second etching layers of thin films deposited upon the substrate.

Alkaline solutions attack different facets, or planes, of the silicon crystal at very different rates. Indeed, one facet can hardly be etched at all. Exposing an area of silicon wafer with a specific crystal orientation to these solutions creates cavities with precisely angled walls that are aligned with the non-etching crystallographic planes. (Fig. 1).

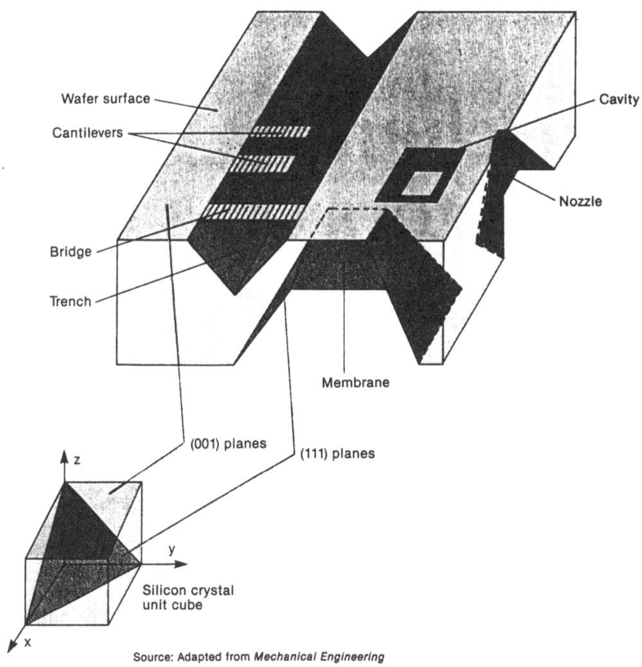


Fig. 1 Silicon bulk micromachining

The common microelectronic thin-film materials silicon dioxide or silicon nitride can serve to mask the portions of the wafer that are not to be etched. To further control geometry, microstructure designers can stop the etch either with a layer of silicon heavily doped with boron or by applying a passivation voltage to one side of a p-n junction. Bulk micromachining, a proven high-volume production process, is routinely used to fabricate microstructure with critical dimensions that are precisely determined by the crystal structure of the silicon wafer, by etch-stop layer thickness, or by the lithografic masking pattern.

To obtain complex structures, the ability to bond silicon to glass and silicon to silicon is an important adjunct to bulk micromachining.

Surface micromachining is based on the deposition and etching structural and sacrificial films. After deposition of the films, the sacrificial material is etched away, leaving a completely assembled micromechanical structure. Usually polycristalline silicon is used as structural material while silicon dioxide, silicon nitride or others are used as the sacrificial materials . The films are grown on silicon wafers by chemical vapor deposition (CVD). The process is useful for making a wide range of micromechanical elements, including beams, bearing and linkages.

Multiple depositions of structural and sacrificial films, each individually patterned, can build surprisingly complex micromechanical structures. Still there is a limit to the number of layers since each one increases surface roughness, gradually degrading the photolithographic process.

## 1.1. SILICON PRESSURE SENSORS

Mostly based on bulk micromachining, piezoresistive silicon pressure sensors are produced in very large volumes for several applications with an already well proved technology. Capacitive and piezoresistive pressure sensors are available, but the piezoresistive devices are more popular due to the lower cost.

The sensor itself, when a bipolar technology is used, is built in the following way (Fig. 2):

- an epitaxial layer is first deposited on the surface of the wafer (substrate). The typical thickness of this layer is 15 microns and depends on the required sensitivity of the pressure sensor.
- four piezoresistors are implanted into the epitaxial layer, forming a bridge whose output is proportional to the stress of the membrane and thus proportional to the pressure.
- the silicon wafer is etched from the back side after that a protective nitride layer is laid to control the exact position of the holes. The electrochemical etching is automatically stopped when the epitaxial layer is reached,

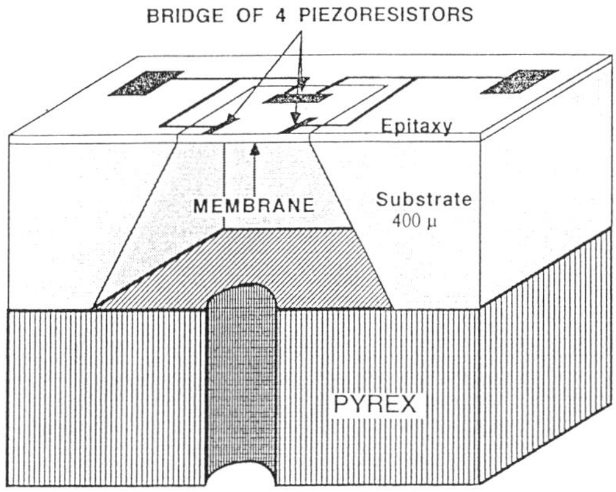


Fig. 2 Silicon pressure sensor configuration

thus giving an excellent control over the thickness of the epi-membrane. Silicon etching with a KOH solution is anisotropic and results in well defined edge angles of 54,74° given by the crystalline structure of the silicon.

- being the sensitive part of the sensor fragile and difficult to be handled, the silicon wafer is generally bonded on the back side to a glass (Pyrex) wafer. The process, called anodic bonding, requires a high voltage of 1500 V between the two parts to be bonded and a temperature of 400°C. An alternative to silicon-pyrex anodic bonding is given by the silicon to silicon fusion, a high temperature process which fuses silicon wafers together at the atomic level without a "glue" layer or an applied electric fileld.

In most cases silicon sensors are unusable without signal processing circuitry and adequate packaging. Usually the bridge output signal must be amplified and several adjustments and thermal compensations are needed to obtain the proper output characteristics. Moreover the die attach is a crucial step since mechanical stresses can be induced into the chip. This hinder dramatically their practical and industrial use. To overcome the problems two types of solutions are generally considered:

- the hybryd solution which involves mounting of the sensor and the signal processing implemented on a ceramic base.
- the monolithic solution which combines the sensor and electronics on the same silicon chip.

Due to the difficulties found in putting together the micromachining and the IC manufacturing processes, most of the today silicon sensors are of the hybrid style. On the other hand the market shows also a strong demand for entirely monolithic solutions which can lead to:

- remarkable miniaturization
- reduced costs for high volumes
- better performances due to the fact that the sensor and the relevant circuitry are on the same substrate.

In fact, the monolithic pressure sensor has still several limitations when operating in tough environmental conditions and high costs mainly due to the rather low yields.

However, in applications where the small size and the need for an amplified signal are mandatory the monolithic sensors have become a must.

In fig. 3 is shown a pressure/temperature integrated sensor produced by Ascom Microelectronics (Switzerland) and used to build a catheter for medical applications. On the chip (1 x 5 mm in size) a pressure and a temperature sensor are located together with a voltage reference and the amplifier. The thermal compensation is provided via software by the computer to which the output of the sensor is directly connected.

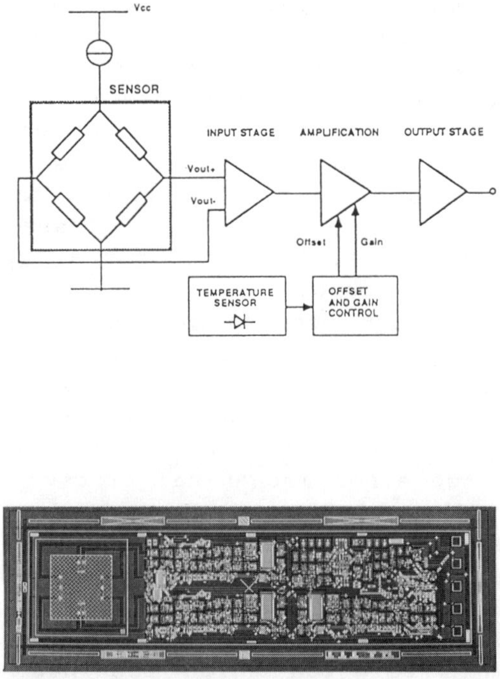


Fig. 3. Block diagram and layout of a monolithic pressure and temperature sensor (Ascom Microelectronics)

## 1.2. SILICON ACCELEROMETERS

Even if the feasibility of silicon accelerometers have already been demonstrated several years ago, their mass production is starting just now under the push of automotive industry which needs them for active suspensions control and air-bag deployment. These applications require only intermediate performances but very high reliability and low cost. Both bulk and surface micromachining seem to be suitable for high volume production of silicon accelerometers.

### 1.2.1. *Bulk-micromachined accelerometers*

A typical design incorporates a bulk-micromachined silicon mass (called the proof mass) suspended by silicon beams (fig. 4). Ion-implanted piezoresistors on the suspension beams, sense the motion of the proof mass produced by acceleration. Even here, as in the case of pressure sensors, a temperature compensation, mainly due to the implanted piezoresistors, is needed. This compensation is usually performed by mounting the sensor chip on a ceramic board, by connecting it with a proper circuitry, and by actively or passively trim suitable thick film resistors.

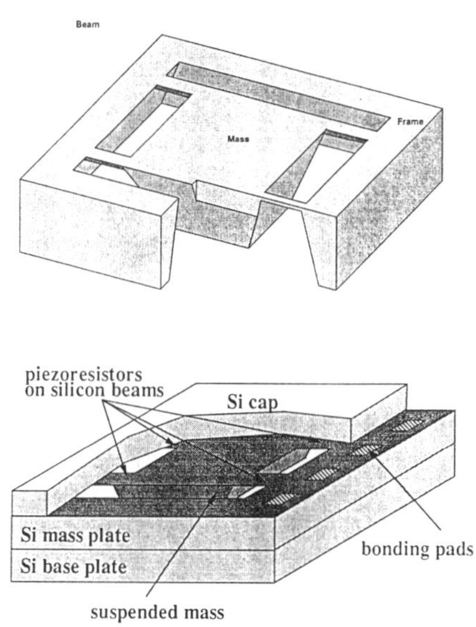


Fig. 4. Structure of a piezoresistive accelerometer

For high precision applications such as inertial navigation high-precision silicon accelerometers are needed. In this case, the preferred means for detecting movement of the proof mass is a change in capacitance. In some design, capacitor plates on top and bottom capping wafers (the two wafers that enclose the proof mass) also apply a restoring electrostatic force to the mass to null its displacement, offering improved reliability and dynamic range over "open loop" devices (fig.5 ).

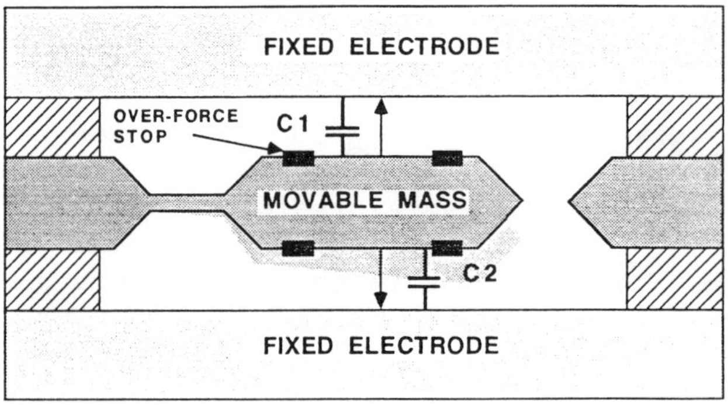


Fig. 5. Structure of a capacitive accelerometer

### 1.2.2. *Surface micromachined accelerometers*

A fully integrated surface machined silicon accelerometer developed for air-bag deployment, with a range of 50-g, has been recently presented by Analog Devices Inc (fig. 6).

A 3x3 mm, 3-microns minimum feature size BiCMOS chip contains a micromachined, polysilicon sensing element and complete circuitry, including a self test function.

Unlike bulk micromachining, where substrate silicon comprises the sensing element, surface micromachining utilizes deposited films, such as polysilicon, silicon nitride, and nichel.

The simple fixed-beam spring design used in the accelerometer necessitates tight control of the intrinsic stress present in the polysilicon film. The center member and movable capacitor plates (or beam fingers) are suspended by four springs or tethers (fig. 7). Fixed capacitor plates are interspersed between the beam fingers (fig. 8). When subjected to an acceleration, the proof mass (center member and attached fingers) move while the fixed plates remain stationary. The separation between the plates and beam fingers therefore changes. Each set of fingers and fixed

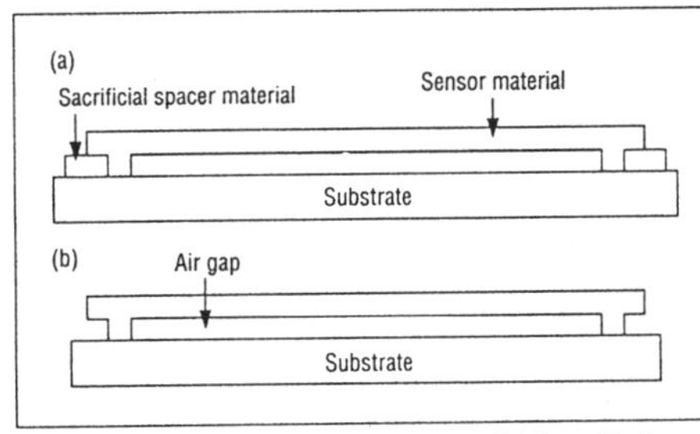


Surface micromachining (a) with sacrificial layer (b) after sacrificial layer removal.

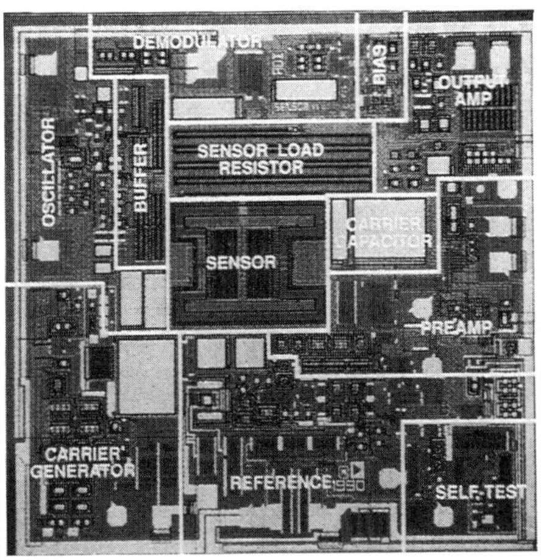


Fig. 6. The Analog Devices accelerometer

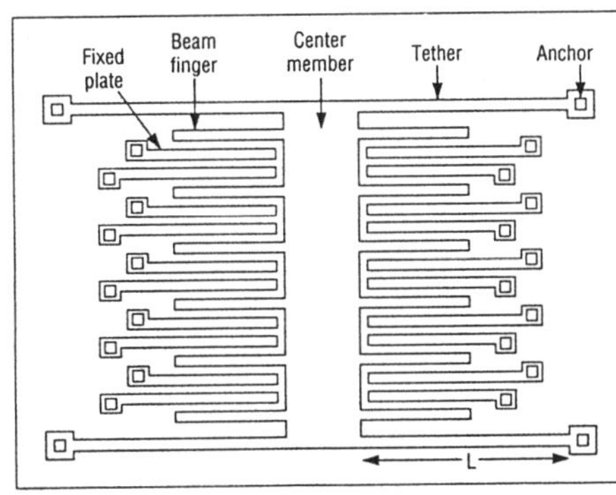


Fig. 7. Simplified view of sensor geometry

Fig. 8. Magnified view of tether and fixed plate anchors on polysilicon

plates comprises a parallel plate capacitor with the air gap as the dielectric. Therefore, the capacitance changes as this gap varies.
The manufacturing process of the sensor and of the relevant circuitry that also includes thin film resistors for functional trimming is rather complex (it needs more than 25 masking steps), but it leads to a device which could represent a turning-point in low-cost accelerometers manufacturing.

## 2. THICK FILM SENSORS

Thick film sensors are based on the outstanding properties of several different materials developed on pourpose and deposited on suitable substrates by screen printing techniques. The intrinsic semplicity of manufacturing process, the increasing availability of new sensing materials, the possibility of using many kinds of substrates (including cofired multilayer ceramics) have already lead to the implementation of temperature (even very high), pressure, force, acceleration, displacement, oxygen, gas, humidity sensors, and so on. The achiveable integration is the one typical of the thick film hybrid circuits as well as the size.
The most attractive properties of the thick film sensors are the very high flexibility, the low development cost and time, the easy handling, the low investments cost and the exceptional environmental behaviour. When needed, the signal conditioning electronics can be obtained on the sensor substrate by adding to the passive thick film sensor network the suitable semiconductor dices to complete the circuitry. This can be done in a very short time by using standard ICs thus greatly reducing the time to market. Of course, incidental design changes can be quickly performed and the functional trimming of the whole circuit can easily tailor the sensor parameters to the customer requirements.
Generally speaking the cost of thick film sensors is higher than that of the equivalent semiconductor sensors at least for very high volumes but very competitive for medium and low volumes.
An other interesting feature which gives some bonus to thick film sensors is the environmental behaviour. Usually they are less affected by temperature and by poisoning media thus offering a better global reliability.

To emphasize and possibly to better explain the above mentioned concepts a few examples are given below.

## 2.1. CERAMIC THICK FILM PRESSURE SENSORS

The classical structures of a thick film capacitive and of a piezoresistive pressure sensors are shown in fig. 9.

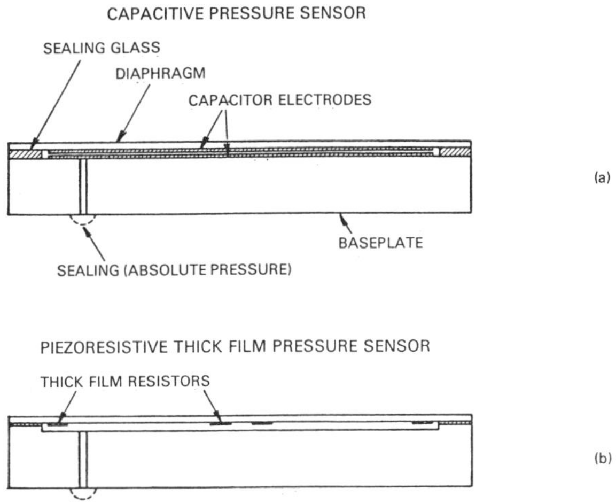


Fig. 9. Structures of thick film pressure sensors (a) capacitive, (b) piezoresistive

In both cases a thin (how thin depends on the pressure to be measured) ceramic diaphragm is bonded by using a suitable screen printed glass to a robust baseplate. In the case of the capacitive device, two metal layers are previously printed on the diaphragm and on the baseplate to form a capacitor; in the case of the piezoresistive sensor, four thick film resistors, connected into a Wheatstone bridge configuration are deposited in the points of diaphragm maximum deflection, thus maximizing the bridge unbalancing when deformated.

The electrical contacts between the diaphragm and the signal conditioning electronics, usually located on the back of the baseplate, are obtained by means of metallized through holes (Fig. 10).

Either the capacitive and the piezoresistive thick film pressure sensors have been produced in very large volumes in the last 15 years fulfilling the requirements of a wide variety of applications like automotive, medical, industrial, aerospace.

Fig. 10. Cross section of a ceramic piezoresistive pressure sensor

The overall accuracy, usually better than that of silicon, the linearity, the insensitivity to most of the dirty media have been their strong points.
The size of the ceramic element is much larger than that of silicon chips and usually ranges from 15 mm up to 40 mm in diameter. If this fact makes the ceramic sensors unusable for certain applications, on the other side it allows a much easier handling of the sensing element. It should also be noted that in most cases the dimension of the finished sensor depends mainly on the packaging and on the signal conditioning electronics size.
Unfortunately the cost of the above described structure is much higher (five to ten times) than that of the silicon pressure sensor chip. On the other side processing and assembly of thick film sensors as well as their packaging are slightly less expensive than those of silicon devices.
In fact, even if the gauge factor of thick film resistors is lower than that of silicon (15 versus 50) , the better temperature coefficients of resistance (50 versus 1500) and of gauge factor 200 versus 2000) allow a simpler signal conditioning electronics and easier funtional adjustments.
Looking at the silicon sensor manufacturing cost breakdown

shown in fig. 11, it clearly appears that even with a much higher cost of the sensing element (here indicated as processed die) it is possible for thick film sensors to achieve a still acceptable price.

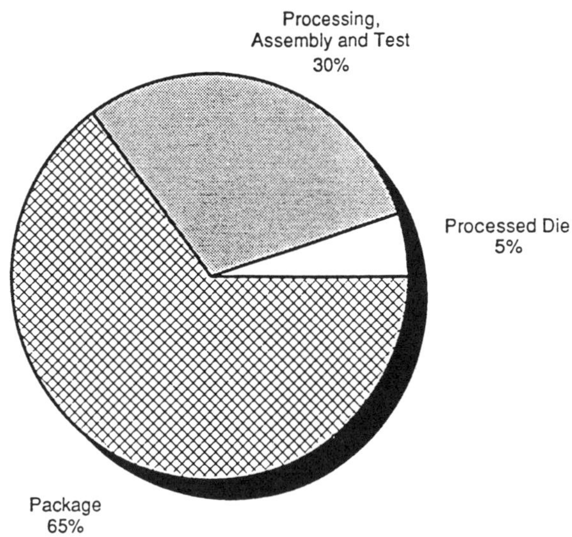


Fig. 11. Micromachined silicon pressure sensor manufacturing cost breakdown

*2.1.1. Trends in piezoresistive pressure sensors*

There are several applications where silicon sensors are not economically and sometimes technically convenient. This is the reason why the thick film piezoresistive pressure sensor technology is still in evolution.

This evolution is based on the improvement of the thick film resistive compositions in terms of gauge factor, noise, and long term stability and on the use of different types of substrate.

Partially stabilized Zirconia, for example, with a flexural strength three times higher than that of alumina, can broaden either towards the low and the high pressure the pressure range already covered.

High precision hot pressed ceramic parts with a built in cavity/diaphragm will greatly simplify the sensor structure.

High and low temperature green tape multilayers can be also

used to reduce the costs.
Most likely the next thick film pressure sensor generation, shown in fig. 12, will considerably reduce the price gap with the silicon sensor still mantaining the today outstanding performances.

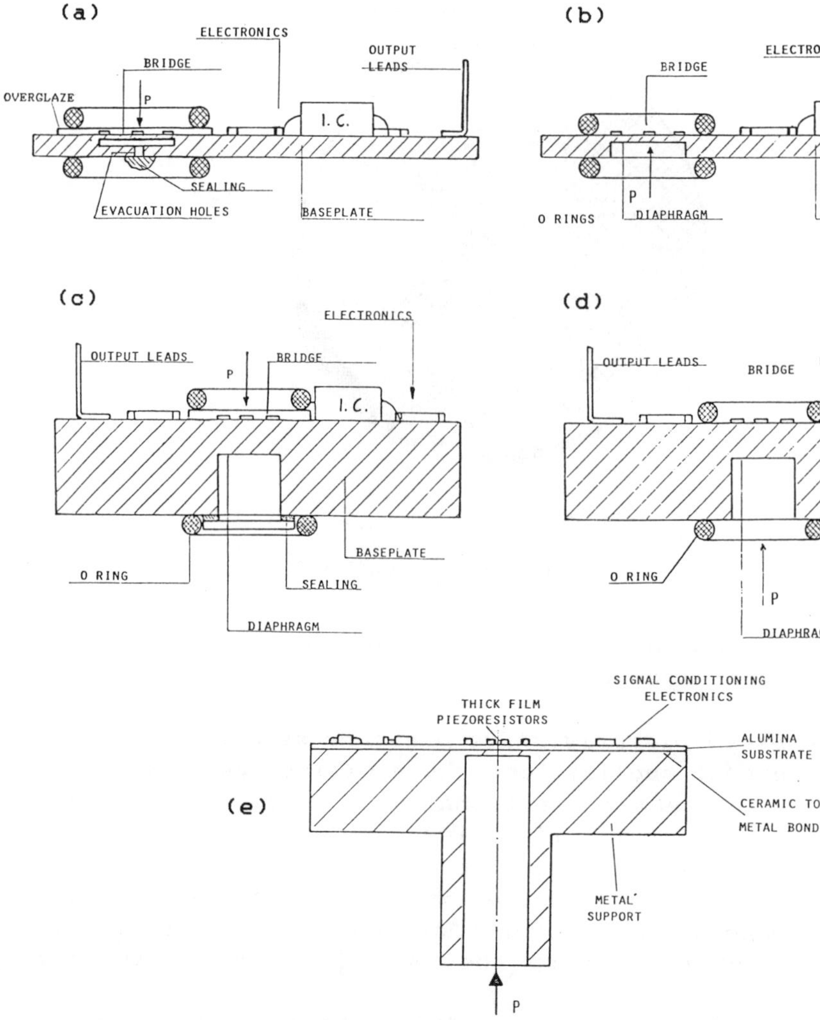


Fig. 12. Possible configurations of thick film piezoresistive sensors: (a) Low pressure absolute, (b) low pressure gage, (c) high pressure absolute, (d) high pressure gage, and (e) very high pressure (metal/ceramic diaphragm)

## 2.2. CERAMIC THICK FILM GAS SENSOR

A very heavy research and development activity has been carried out in the past ten years on gas sensors since the already existing ones were expensive and sometimes unreliable. One of the most interesting works in this area is tied to the exhaust gas sensor used on cars to control the air/fuel ratio. The classical structure was, in this case, a thimble-shaped zirconia solid electrolite cell, obeying the Nerst law. As the requirements became more stringent, planar solutions based either on a multilayer ceramic structure and on a full thick film structure have been developed, reproducing and improving the Nerst cell, that will be soon on the market.
Since a lot of literature exist on this matter, it is more useful to look at an other very interesting device.

### *2.2.1. Thick film methane sensors.*

Gas sensing properties of semiconductor oxides films, like ZnO and $SnO_2$, are well known since a long time and in fact $SnO_2$ sensors have been available for more than twenty years. The basic mechanism that gives these oxides their sensitivity to reducing gases is well known: the reduction of their surface makes electrons available for electrical conduction and causes the electrical resistance of oxide to decrease.
Gas sensors generally consist of a mixture of $SnO_2$ and electrically inert oxides as $Al_2O_3$ and $SiO_2$ with some noble metals acting as catalyst. Usually the properties of gas sensors depend very strongly on composition and preparation conditions.
Characteristic features for a sensor to detect fuel gas leakage are:

- reliable detection of the alarm threshold concentration, independence from the history of the sensor and in particular of previous exposures to other gases;
- selectivity versus the main interfering gases possibly present in the same environment (a false alarm produces disregarding the true danger signal);
- high reliability, in terms of working life.

To fulfil these requirements, not always fully satisfied by the today available sensors (one of which produced by Figaro, Japan, is shown in fig.13), ENIRICERCHE S.p.A., Italy, has chosen as manufacturing technique the thick film on ceramic approach which also provides high reproducibility and low production costs.

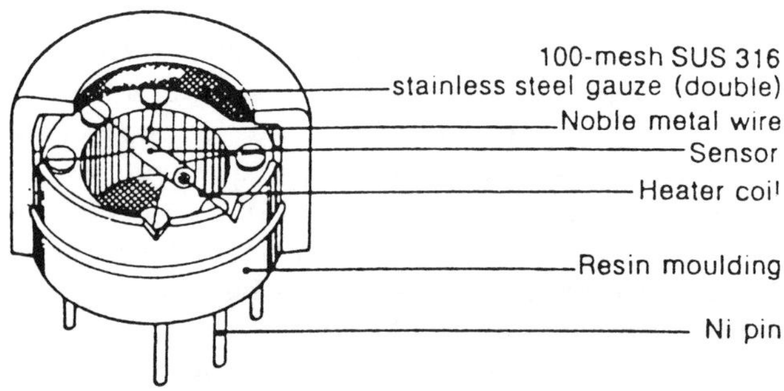


Fig. 13. Structure and configuration of the FIGARO gas sensor. The sensing material is deposited on a ceramic cylinder and heated by a coil.

Screen printing can be used both for sensing and for heating elements, with possibility of array integration of smart sensors.

The material for the sensing element is a blend of tin and aluminium oxides with dispersed a catalyst which is screen printed on a ceramic substrate. The heater is a resistive platinum stripe printed on the other side of the ceramics. The material chosen for the sensing element exhibits high sensitivity to methane at an operating temperature around 500 °C (Fig. 14).

Fig. 14 displays sensitivity versus operating temperature curves for the following gases: $CH_4$, CO, $NH_3$, $C_2H_5OH$, and $CH_3COCH_3$. The concentrations for the interfering gases are the highest expected in a home environment. As shown, the sensitivity to methane is highest at high sensor temperature while the highest sensitivity to interferent gases occurs at low temperature. An improvement of the selectivity against high concentration of interfering gases is obtained through comparison of measurements taken at different temperatures, e.g. 300 and 500 °C.

The comparison is easily done taking advantage of a simple arrangement integrating two sensing elements on the same substrate: the pattern of the heater has been designed so that the two elements are heated at two different temperatures (fig. 15).

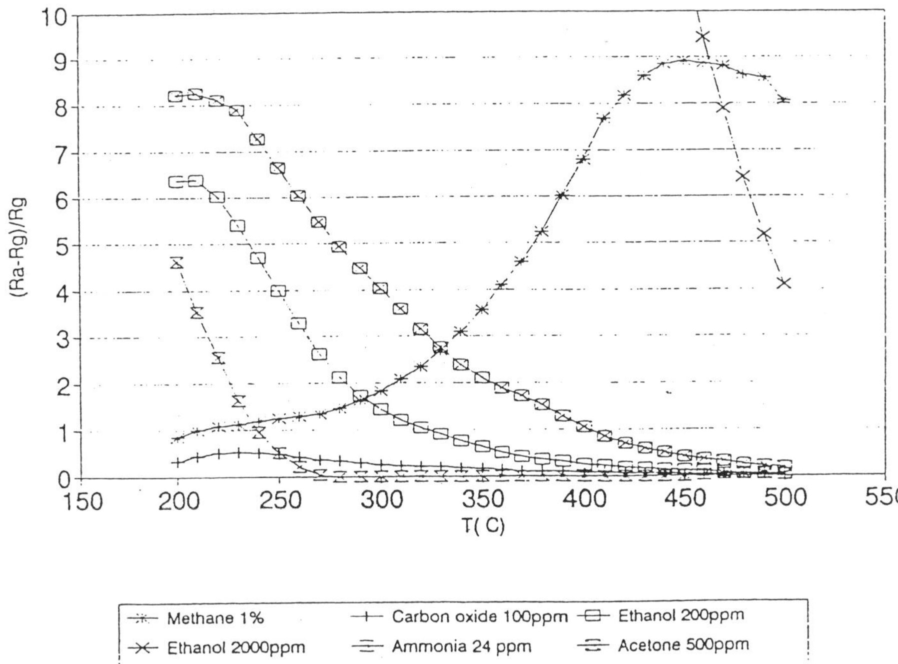


Fig. 14. Sensitivity vs temperature to methane and interfering gases of the ENIRICERCHE gas sensor. The sensitivity is given as (Ra-Rg)/Rg where Ra is the resistance of the sensing element in air and Rg its resistance in an atmosphere with the selected methane concentration.

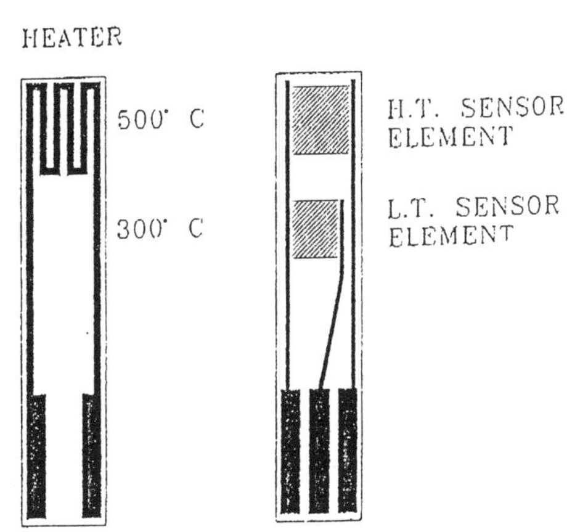


Fig. 15. Layout of the ENIRICERCHE methane thick film sensor

## CONCLUSIONS

The increasing demand for real time electronic control systems drives the growth of the sensor market and the increase of their features.
Very large volumes are already demanded in the automotive, industrial, medical, safety, and home appliances field. Since the volumes already exist, silicon and thick/thin film technologies can show at last their potentiality. The need for higher accuracy, reliability, and lower cost can be fulfilled by their proper and clever use.
The few examples given in the paper show that these two technologies have their own advantages and drawbacks; they should be very well understod and evaluated by sensor designers.
Silicon micromachining and the integration of signal conditioning electronics on the same chip offer incredible chances for several applications and when very high volumes are involved.
Silicon micromachined devices mounted on ceramic hybrids can represent the fastest approach to the market with still very reasonable costs.
Thick film technology seems particularly suitable and economical for very hard working condition sensors.

## REFERENCES


1. Dell'Acqua,R. (1987) Non Conventional Applications of Thick Film Technology, Hybrid Circuits.
2. McCarty,L.H. (1988) Tiny Accelerometer Weighs Just One Gram, Design News.
3. Leonard,M. (1989) IC Fabrication Techniques Sculpt Silicon Sensors, Electronic Design, 37-40.
4. Allen,H.V. Terry,S.C. and Knutti,J.W. (1989) Understanding Silicon Accelerometer, Sensors, 17-31.
5. Bianchi,V. Dell'Acqua,R. Geslot,F. (1990) The Role of Sensors in Future Automotive Applications, SAE Paper 901132, Convergence 90, Detroit.
6. Howe,R.T. Muller,R.S. Gabriel,K.J. and Trimmer,W.S.N. (1990) Silicon Micromechanics: Sensors and Actuators on a Chip, IEEE Spectrum, 29-36
7. Rahali,F. Ansermet,S. Ardalan,J. Otter,D. (1993) Low-Cost Integrated Silicon Sensors for Industrial Application, Proccedings of European ISHM Conference, Nice.

8. Core,T.A. Tsang,W.K. Sherman,S.J. (1993) Fabrication Technology for an Integrated Surface-Micromachined Sensor, Solid State Technology, 39-47.
9. De Angelis,L. Mercuri,R. Minnaia,N. Modica,L Guiducci,M. Occhio, L. (1993) Natural Gas Sensors for Home Application, Proceedings of 1992 International Gas Research Conference, Orlando.

# HIGH SENSITIVITY THERMOMETERS FOR MILLIKELVIN TEMPERATURE RANGE

I.BAŤKO[1] and K.FLACHBART[1]
M.SOMORA[2] and D.VANICKÝ[2]
*Institute of Experimental Physics, 043 53 Košice, Slovakia*
*Technical University Košice, 043 89 Košice, Slovakia*

## 1. Introduction

Commercial $RuO_2$ - based thick film resistors (TFRs) were successfully introduced as low temperature thermometers in the last years [1-8]. Their low magnetoresistance [1-5], possibility to predict their low temperature *R-T* behavior from room temperature resistance [5] and possibility to lower their heat capacity simply by shaving its alumina substrat without significant altering of the electrical properties [4,7] predict them as practical, easy-to-use and low cost thermometers, mainly in temperature range below 1K. Moreover, our earlier studies shown, that thermometers for milikelvin temperatures with high (comparable to germanium thermometers) sensitivities, sufficient stability and suitable resistance ranges may be prepared based on $RuO_2$ thick films, if they are designed for this purpose and if suitable types of pastes for their preparation are used [8]. The all mentioned facts were motivation for us to try to develop improved $RuO_2$ - based thermometers as alternative thermometers (in milikelvin temperature region) to germanium ones. During our research we found some technical limitations that may be interesting from both technical and physical point of wiew. The aim of this paper is a short description of them.

## 2. Experimental

The experiments were carried out in an Oxford TLE-200 dilution refrigerator. The ac resistance measurements of TFRs were made using VS-3 resistance bridge (Instruments for Technology, Finland). The temperature measurement and control were performed by a calibrated germanium thermometer (GR-200A-30, Lake Shore Cryotronics, USA) and Model 1000 Potenciometric Conductance Bridge/Controller (Biomagnetic Technologies, inc., USA), respectively.



*W. K. Jones et al. (eds.), MCM C/Mixed Technologies and Thick Film Sensors,* 299–305.

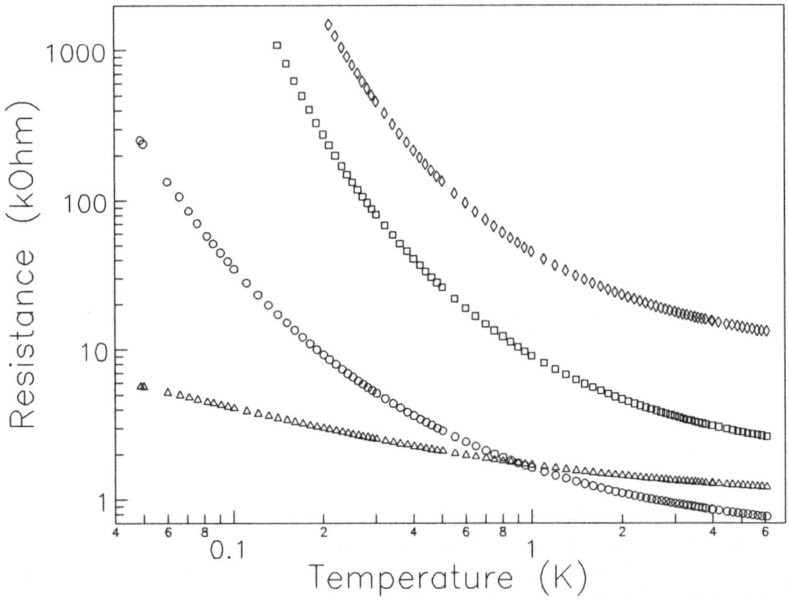


*Figure 1* Temperature dependence of the resistance of the 1000 Ohm unit (1kOhm/sq. paste triangles), 600 Ohm unit (10 kOhm/sq. paste circles), 2k Ohm unit (100 kOhm/sq. paste squares), 10 kOhm unit (1Mohm/sq. paste diamonds).

## 2.1. DESIGN

Based on experiences and results of previous experiments [8] we prepared samples with a new design. The aim of the design was to obtain samples with suitable resistances in the low temperature range.

The usually used dimensions of thick film resistors alter from 1mm to cca 10 mm , and the ratio between the lenght and the width alters from 1/5 to 5/1. These basic rules guarantee the reproducibility and the defined TKR of pastes. On the other hand the respectation of these rules leads to very high resistance of resistors at very low temperatures. The resistance increases to values, unmeasurable with standard equipment. The sensitivity of the $RuO_2$ pastes was investigated on pastes with higher sheet resistance only (10 -100 kOhm). Their temperature coefficient of resistivity is negative and increases with decreasing temperature.

Therefore, the design of our samples should enable to prepare resistors with sufficiently low absolute resistance by use of pastes with rather high sheet resistivity.

The additional task for the design of thermometers was to enable the direct acces of the active layer (the resistive one) to the cooling surface for better cooling and also, if possible to enable the cooling from the back side of the sample.

Both of these tasks were solved by an interdigital structure of the conductive electrodes and with the overcontacting of them to the back side conductive electrodes. These backside electrodes cover almost the entire area of the backside of the termometer. They enable the connection with the measurement wires by soldering. These wires cool at the same time through the backside electrodes the sample from its backside.

Such a structure can be easily positioned and well connected with the cooling surface by its flat frontside.

To achieve smooth surface for the die attaching to the cooling surface, and therefore for excellent cooling, it was necessary to use a complementary screen printing, when the resitive layer is printed only on the surface between the fingers. The more simple version, when the resistive paste was printed onto the entire surface of the frontside at once, created small unequalities in the surface and the cooling was unsufficient, or better to say, the self heating of the samples was mentioned.

## 2.2. $RuO_2$ PASTES

Pastes with different sheet resistivities, based on "homogenous" $RuO_2$ powder from the producer TESLA Lanskroun (Czech Republic) were chosen for experiments.

## 2.3. SCREEN PRINTING AND FIRING PROCESS

Conductive patterns were screen printed, dried and fired by a standard thick film process from the conductive pastes, based on Ag (Producer Tesla Lanskroun, CR). Thick film resistors were printed by a standard screen printing techniques, from both groups of pastes on 0.635 mm thick 96% $Al_2O_3$ substrates. After deposition films were dried for 15 min at 150 $^oC$ and then fired for 1 h in the standard thermal profile, recommended by producers of pastes. (The paek temperature - 850 $^oC$, the time at peak temp. - 10 min.)

## 3. Results and disscusion

As it was reported before [8], the change to common conditions of thick film preparation was connected with the fact, that the "low-resistance" 600Ω and 2kΩ units were printed from "high-resistance" pastes 10kΩ/sq and 100kΩ/sq, respectively. (Commercially available 600Ω and 2kΩ resistances are usually printed from paste 1kΩ/sq). This decision was based on the experience, that resistors prepared from 10kΩ/sq and 100kΩ/sq pastes show significantly higher sensitivities than those printed from the 1kΩ/sq one (see Fig. 1) [8]. On the other hand, the sensitivity of 10kΩ unit printed from an 1MΩ/sq paste is slightly worse than this of 2kΩ unit that seems to be surprising. However, this behavior is not property of the 1MΩ paste itself but it is a direct consequence of used printing geometry. It is a well-known rule in hybrid circuit fabrication, that there exists some low-end dimension $D_L$ characterizing recommended length limitation of using $RuO_2$ pastes. It requires to prepare resistors not "shorter" then $D_L$ , where $D_L$ is usually ≈ 1mm. In our case when we lowered the resistance, we ignored this criterion for each paste besides 1kΩ/sq and the most important "reduction" of the resistor length was done for 1MΩ/sq unit. (Resistors in which we ignored the above mentioned rule we will assign as "short resistors"). Due to this practically the whole $RuO_2$ resistor area is in immediate vicinity of Ag-Pd contact pads so that as can be expected it is strongly influenced by conduction pads as a consequence of diffusion of silver from contacts to $RuO_2$ resistor medium during the technological process of preparation. This diffusion can be assumed as doping of $RuO_2$ paste that alters its physical properties. Taking into account all above mentioned facts it is clear, that if we put some "upper limit" of resistor dimensions and resistances some competition between high sensitivity and low resistance will observed.

As it was reported in previous papers [1,3,5,7], the electrical conductivity at lowest temperatures can be adequately described in terms of variable-range hopping (VRH) between localized states in the vicinity of the Fermi level (useful reviews can be found in [9, 10]. In the most general case when the density of localized states varies as $|E-E_F|^m$ $(m>0)$ and a system is three dimensional, the resistivity is given by

$$\rho(T) = \rho_0 \exp[(T_0/T)^x] \tag{1}$$

where the exponent $x = (m+1)/(m+4)$ and $T_0$ is a characteristic temperature. In the special case when the density of localized states near the Fermi level is constant or depends on energy slowly $(m=0;\ x=1/4)$, the temperature dependence of the resistivity follows the Mott's law

$$\rho(T) = \rho_0 \exp[(T_0/T)^{1/4}]. \tag{2}$$

When the parabolic "Coulomb gap" $(m=2;\ x=1/2)$ is created as a consequence of the Coulomb interaction, the resistivity is described by

$$\rho(T) = \rho_0 \exp[(T_0/T)^{1/2}]. \tag{3}$$

The last mentioned behavior is known as Shklowski-Efros law.

In Fig. 2 are reploted resistances of above mentioned 600Ω , 1kΩ, 2kΩ and 10kΩ units shown in Fig. 1 in coordinates $T^{-x}$ *versus* ln $R$ where values of $x$ are indicated in Fig. 2. As can be seen in this figure, eq. (1) provides satisfied approximation.

It is a good experience in low temperature resistance thermometry, that the resistance of thermometer in the useful temperature range is less than 100kΩ. According to this criterion based on our experiences we designed thermometers for the temperature range above 50mK and 300mK, using pastes 10kΩ/sq and 100kΩ/sq, respectively. *R-T* characteristics of these thermometers are shown in Fig. 3 and their sensitivities are shown in Fig. 4. For comparison the same characteristics of commercial GR-200A-30 and GR-200A-50 Lake Shore thermometers and sensitivities of similar $RuO_2$-based units reported in [8] (with higher distance between conduction pads) are given. As can be seen from Fig. 3 and Fig. 4, shorter

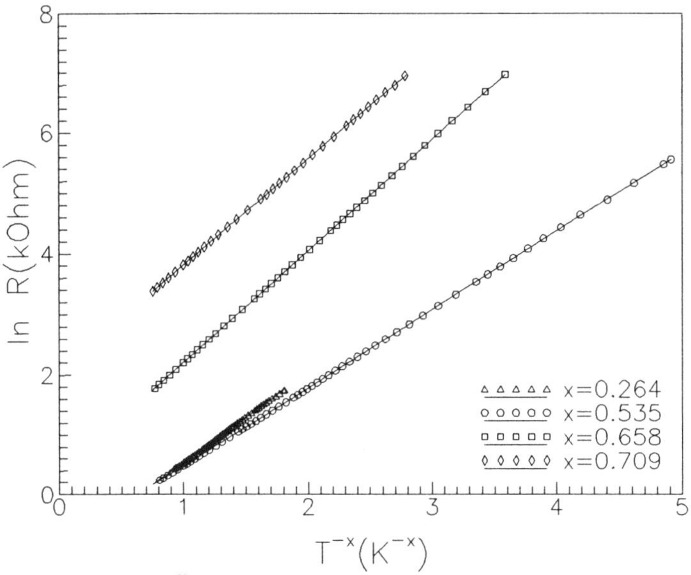


*Figure 2* ln R versus $T^{-x}$ for the 1000 Ohm unit (1kOhm/sq. paste triangles), 600 Ohm unit (10 kOhm/sq. paste circles), 2k Ohm unit (100 kOhm/sq. paste squares), 10k Ohm unit (1Mohm/sq. paste diamonds) in temperature interval between 50 mK and 1,6 K.

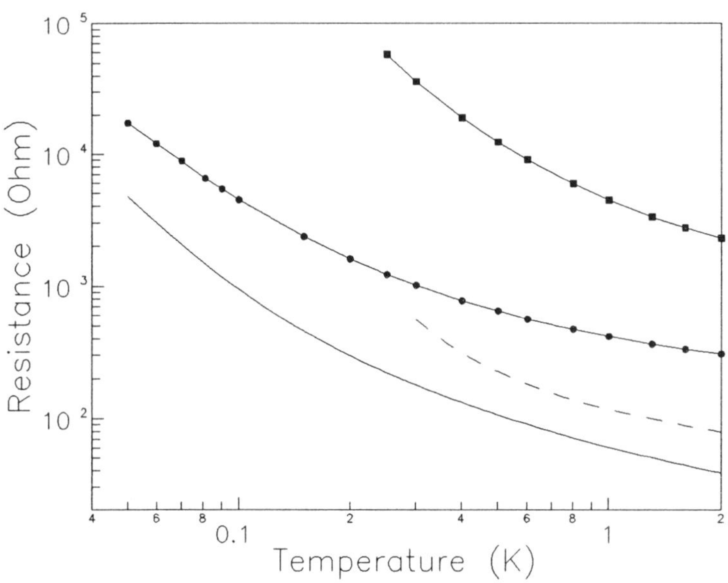


*Figure 3* Temperature dependence of the resistance of the $RuO_2$ - based thermometers designed for temperature range down to 50 mK (closed circles) and 300 mK (closed squares) and germanium thermometers GR-200A-30 (solid line) and GR-200A-50 (dashed line) [11].

distance between conduction pads provides resistances lower than 100kΩ, but sensitivities of thermometers are reduced too, in agreement with our above considerations.

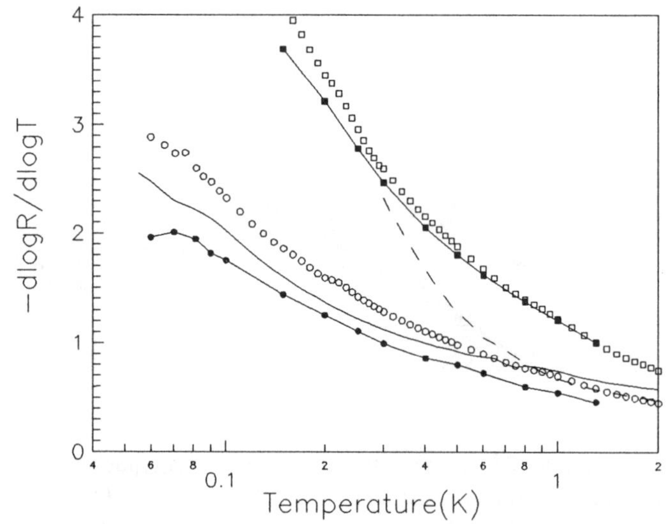


*Figure 4* Temperature dependence of hte negative logarithmic sensitivity of the $RuO_2$ - based thermometers designed for temperature range down to 50 mK (closed circles) and 300 mK (closed squares) and germanium thermometers GR-200A-30 (solid line) and GR-200A-50 (dashed line) (11). Open circles and open squares are sensitivities of the similar 600 Ohm unit (10 kOhm paste) and 2k Ohm unit (100 k Ohm paste) respectively (see Fig. 1 and [8]).

## 4. Conclusions

We conclude that (i) $RuO_2$ - based resistors can be prepared as alternative thermometers to germanium ones. (ii) Both *R-T* characteristics and sensitivities can be simply changed by the printing geometry (changing distance between conduction pads). On the other hand, (iii) low thermometers resistance and its sensitivity are competitive mechanisms which provide some technological limitations.

**References**


1. H.Doi, Y.Nahara and H.Nagano, *Proc LT-17 (Eds Eckern U. et al.)* North-Holland, Amsterdam, The Neterlands (1984) 405
2. W.A.Bosh, F.Mathu, H.C.Meijer and R.W.Willekers, *Cryogenics* (1986) ***26*** 3
3. Q.Li, C.H.Watson, R.G.Goodrich, D.G.Haase and H.Lukefahr, *Cryogenics* (1986) ***26*** 467
4. M.W.Meisel, G.R.Stewart and E.D.Adams, *Cryogenics (1989)* ***29*** 1168
5. R.W.Willekers, F.Mathu, H.C.Meijer and H.Postma, *Cryogenics* (1990) ***30***} 351
6. W.Schoepe, *Physica B* (1990) ***165/166*** 299
7. M.L.Siqueira, R.J.Viana and R.E.Rapp, *Cryogenics* (1991) ***31*** 796
8. I.Baťko, M.Somora, D.Vanický and K.Flachbart, *Cryogenics* (1992) ***32*** 1167
9. B.I. Shklovskii and A.L. Efros, {\it *Electronic Properties of doped Semiconductors*, Nauka, Moscow (1979) (*in russian*)
10. A.N. Ionov and I.S. Shlimak, in *Modern Problems in Condensed Matter Sciences*, ed. by V.M. Agranovich \& A.A. Maradudin, vol.20, p.397 (North - Holland, Amsterdam, 1991)
11. Certificats of calibration of germanium resistor thermometers, Model GR-200A-30, Sensor Serial No.: 23091, Calibration Report no.: 144301 and Model GR-200A-50, Serial No.:16565, Calibration LOW142-3, Lake Shore Cryotronics, Inc., USA

# THICK FILMS BASED ON GLASS AND POLYMERIC MATRICES, MECHANISM OF CONDUCTIVITY

I. KŘIVKA, J. KUBÁT*, R. KUŽEL, J. PROKEŠ
*Faculty of Mathematics and Physics, Charles University, Ke Karlovu 5, 121 16 Praha 2, Czech Republic*
**Department of Polymeric Materials, Chalmers University of Technology, S-412 96 Gothenburg, Sweden*

## Abstract

New types of two- and multi-component polymeric composites based on polyethylene or polystyrene were studied. Conductive parts were formed from metal powders (Pd, Ag, Ag-compounds). Additional dielectric components (glass particles, $BaTiO_3$ or piezosonic powder U5H-32) were used to modify mechanisms of electric charge transport. Investigation of the mixtures included blending curve measurements and evaluations, measurements of temperature dependence of DC resistivity and the complex impedance $|Z|$ at various frequencies and the study of the composite structure by means of SEM. Some relations between the electrical characteristics and the structure are mentioned.

## 1. Introduction

Microelectronics has spread into enormous number of applications. The need for new materials for specialized purposes is increasing. Glass composites has already found their place in hybrid circuit production. In recent years the attention is attracted to new systems based on non-traditional materials. Combinations of glass and various metal oxides can be replaced by systems based on polymers filled with different types of conductive components. Replacing glass matrix with polymeric one brings qualitatively new features in physical behaviour. Some results obtained by present authors on two-component systems based on polyethylene and polystyrene were published in [1] and [2].

New types of composite systems based on polymeric matrices (polystyrene, polyethylene) containing conductive (Pd, Ag, Ag-compounds) and dielectric components such as glass, $BaTiO_3$ or piezosonic powder were developed and investigated in bulk and as thick films.

The composition of the samples had a significant influence on their key electri-



*W. K. Jones et al. (eds.), MCM C/Mixed Technologies and Thick Film Sensors,* 307–314.

cal parameters such as the DC resistivity and its temperature dependence as well as the complex permittivity and its dependence on temperature and frequency.

## 2. Sample preparation

### 2.1. PRECURSOR MATERIALS

Three polymers were used for the composite preparation: polystyrene (PS) and high-impact polystyrene (HIPS) (both $\rho = 1.04\ \mathrm{g/cm^3}$) and low density polyethylene (LDPE) ($\rho = 0.915\ \mathrm{g/cm^3}$).

The following conductive solid ingredients (in the form of powder) were selected for filling the polymers: palladium (type Pd-213), silver (type Ag-023), compound powders (Ag-Cu — type Ag-816 and Ag-C — type Ag-839) and co-precipitated powder Ag-Pd (type Pd-630). The metal powders were produced by Shoei Chemical Inc., Japan. Perfect spherical particles of the same diameter ($\approx 0.30\ \mu$m) were found in the case of Pd and Ag-Pd powders by means of SEM. The particles of other powders were mostly spherical as well but their diameters were not uniform ($\approx 0.2$ to $2.5\ \mu$m).

Three dielectric ingredients were selected for filling the polymer: glass powder with a typical particle size of 1 to 4 $\mu$m, two types of $BaTiO_3$ with particle size of 1 to 2 $\mu$m (type A) and 0.2 to 0.3 $\mu$m (type B), and piezosonic powder (particle size of 1 to 2 $\mu$m) marked U5H-32 produced by Ultrasonic Powders Inc., U.S.A.

### 2.2. COMPOSITE PREPARATION

The mixture of a given polymer with conductive particles and/or dielectric filler was first mixed and homogenized at high temperature (from about 130 °C to 180 °C depending on the polymer). Then the compounds were compressed at high temperature and cooled to room temperature at a rate of $10\ ^\circ\mathrm{C\,h^{-1}}$. The prepared samples were pellets of diameter 12 mm and thickness 1 to 2 mm. The electrodes were applied by evaporating gold *in vacuo* or by painting the planar surfaces with a thin layer of a colloidal silver dispersion.

The PS and HIPS thick films of about 25 $\mu$m thickness were prepared by a screen-printing technique applying the polystyrene+powders composition dissolved in toluene on alumina substrates at room temperature. The alumina substrates were provided with Ag-Pd terminals fired at 860 °C.

## 3. Experiment

### 3.1. DC RESISTIVITY

Sheet resistance of thick films is strongly dependent on the composition. The relation between the conductive powder content and the sheet resistance is plotted

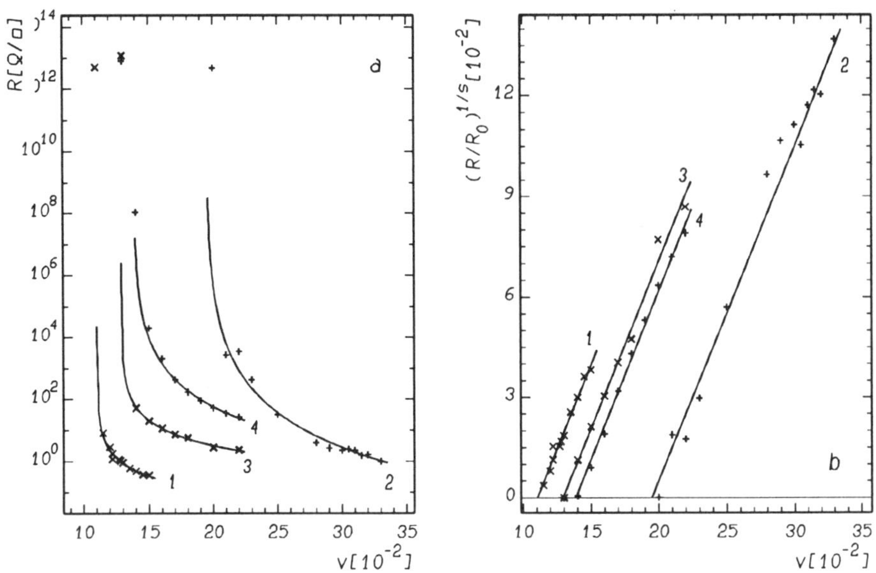


*Figure 1*: Dependence of the sheet resistivity $R$ (a) and $(R/R_0)^{1/s}$ (b) on the volume fraction $v$ of the conductive component for polystyrene thick films filled with silver (curve 1), palladium (curve 2), 50 % (vol.) piezosonic powder and silver (curve 3), and 50 % (vol.) piezosonic powder and palladium (curve 4).

in various scales (see Fig. 1) for four different systems: a polystyrene matrix filled with silver or palladium powders, and a polystyrene matrix containing 50 % (vol.) piezosonic powder and filled with silver or palladium powders. The dependence of the sheet resistance on the conductive powder content is quite evident.

The observed dependences are well described by means of the blending curve equation [3]

$$R = R_0(v - v_c)^s \ , \tag{1}$$

where $v$ is the volume fraction of the conductive phase, $v_c$ is the critical volume fraction (the threshold), $s$ is the critical exponent and $R_0$ is a material parameter. The most important of the parameters is the critical volume fraction $v_c$. From the mathematical point of view, it is the concentration where the resistivity rapidly goes to infinity. All the unknown parameters—$v_c$, $s$ and $R_0$—were determined using the least squares fit to a straight line

$$R^{1/s} = R_0^{1/s}(v - v_c) \ . \tag{2}$$

The best fit is that yielding the correlation coefficient closest to unity. The solid lines in Fig. 1 represent blending curves obtained from a fit to the measured values. Considering lines in Fig. 1b, fits are found to be very good and Eq. 1 appropriate.

*Figure 2*: Micrographs of composite fracture for two compositions: (a) Ag-filled polyethylene and (b) Pd-filled polystyrene.

The parameters gained with the described procedure are listed in Table 1. The results prove a strong influence of the matrix on the parameters—critical volume fraction $v_c$ varies from about 20 % (vol.) down to 11 % (vol.).

SEM investigation showed different character of conductive fillers: the distribution of palladium particles in a polymeric matrix was spatially homogeneous (see Fig. 2b) while silver powder tended to creating clusters and chains (see Fig. 2a). Consequently, critical volume fraction $v_c$ for PS/Pd systems was considerably higher than $v_c$ for PS/Ag ones (see Fig. 1, curves 1 and 2). The value $v_c$ could be changed by adding dielectric filler into the matrix. In samples with a spatially homogeneous distribution of the metal particles, as was the case with Pd, incorporation of the dielectric component caused a marked decrease of the perco-

*Table 1*: Parameters in the blending curve equation for different matrices.

| matrix | $R_0$ | $v_c$ | $s$ |
|---|---|---|---|
| PS + Ag | $4.2 \times 10^{-3}$ | 0.11 | −1.3 |
| HIPS + Pd | $3.5 \times 10^{-4}$ | 0.20 | −4.0 |
| PS + U5H + Ag | $5.5 \times 10^{-2}$ | 0.13 | −1.5 |
| PS + U5H + Pd | $3.5 \times 10^{-4}$ | 0.14 | −3.1 |

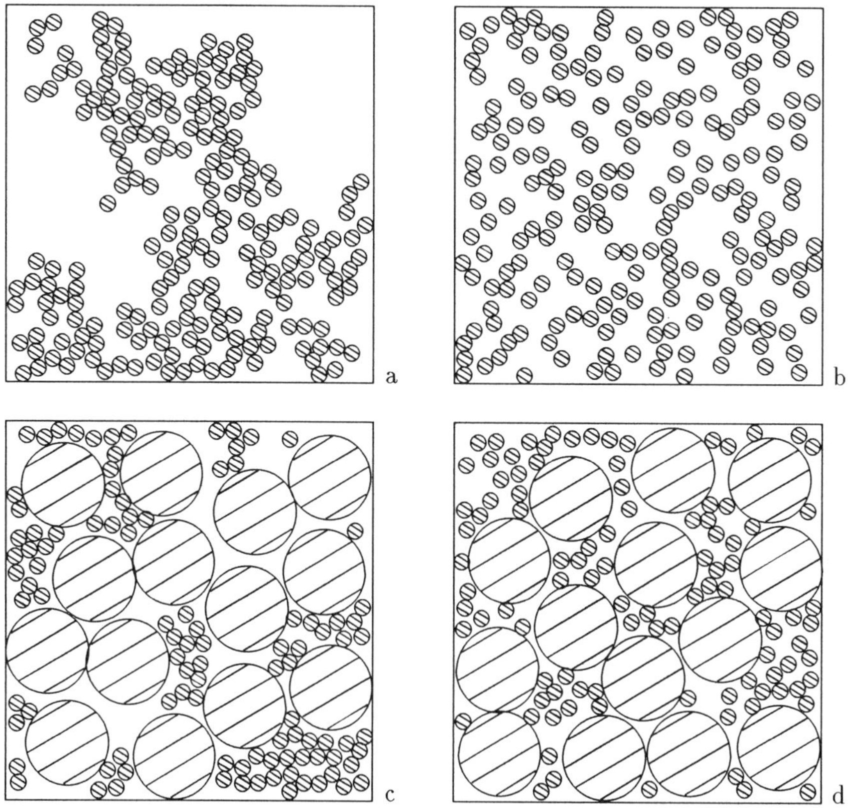


*Figure 3*: 2D model of composite structure: small circles—conductive particles (**C**), large circles—dielectric filler (**D**) and the area in between—a polymeric matrix. (a) 25% **C** with accumulation; (b) 25% **C** without accumulation; (c) 12.5% **C** and 50% **D** with accumulation; (d) 12.5% **C** and 50% **D** without accumulation.

lation threshold. The opposite was true when the particles were inhomogeneously distributed (Ag). A common value of the threshold at about 14 % (vol.) metal content was reached in both cases (see Fig. 1, curves 3 and 4).

2D computer model can be used to explain this behaviour. Figures 3a and 3b illustrate the situation in polymeric matrix filled with conductive filler only. Both the samples contain the same concentration of conductive particles. In Fig. 3a, particles were generated randomly, but many of them were not placed: the particle touching another one was always placed while free particles were placed with probability 1:100. In Fig. 3b, particles were placed randomly. It can be seen that a tendency to create clusters has a significant influence on the critical volume fraction $v_c$. In Fig. 3c and Fig. 3d, the conductive particles were placed (with the same rules as in Fig. 3a and Fig. 3b, respectively) into the area between randomly

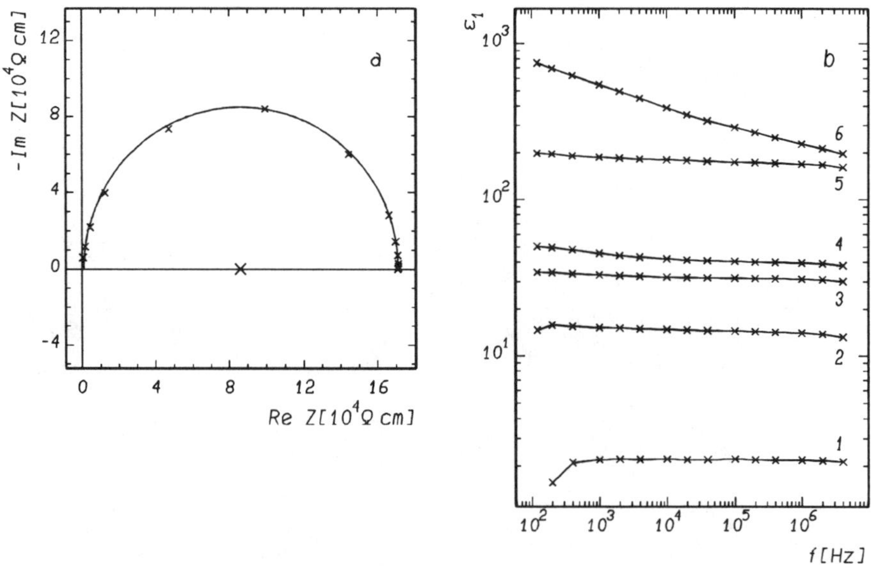


*Figure 4*: (a) Cole-Cole diagram of the composite consisting of 50 % (vol.) glass and 11 % (vol.) Pd-Ag powder in a polystyrene matrix. (The cross on the horizontal axis represents the centre of the semicircle.) (b) Real part of the complex permittivity $\varepsilon_1$ vs. frequency for pure polystyrene (curve 1), HIPS + 25 % (vol.) Pd (curve 2), PS + 50 % (vol.) $BaTiO_3$ (type A) (curve 3), PS + 50 % (vol.) $BaTiO_3$ (type B) (curve 4), PS + 50 % (vol.) $BaTiO_3$ (type A) + 12 % (vol.) Pd (curve 5), and PS + 50 % (vol.) $BaTiO_3$ (type B) + 12 % (vol.) Pd (curve 6).

placed dielectric ones. The dielectric particles we used had several times larger diameter than conductive ones and this is the reason why almost the same structures (often called segregated) arose. Of course, the critical volume fraction values $v_c$ are also close.

The thick film resistor pastes with the sheet resistance ranging from 1 $\Omega/\square$ to 1 $M\Omega/\square$ can be prepared and even values as high as $10^{12}$ $\Omega/\square$ can be reached near the critical volume fraction.

Characteristics measured on bulk samples resemble those of thick films. Conclusions very similar to those presented were obtained for polyethylene based composite matrices.

## 3.1. AC MEASUREMENTS

The volume fraction range is divided by the critical volume fraction value $v_c$ into two parts. As it could be seen, the composite exhibits resistive behaviour above the threshold. On the contrary, the capacitor made of composite below the threshold can be acceptably characterized by means of an equivalent circuit composed of

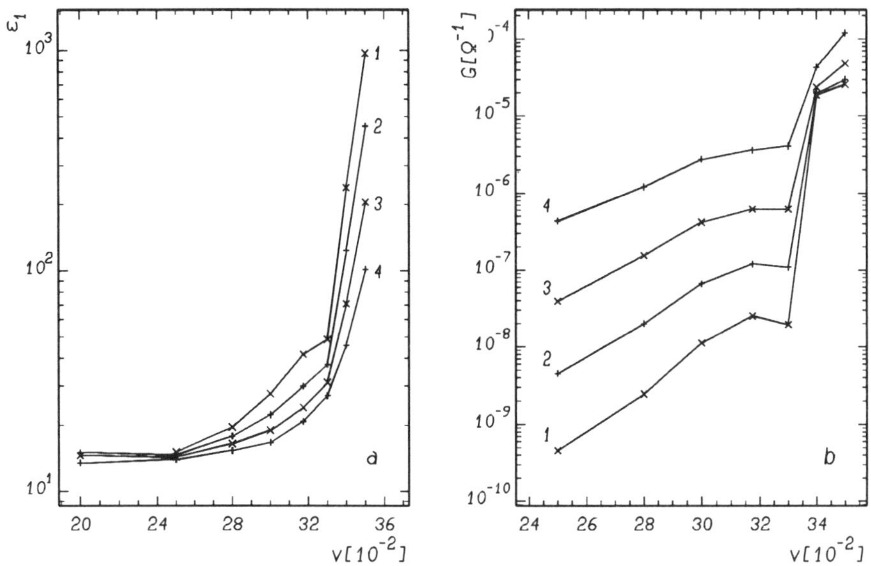


*Figure 5*: Dependence of real part of the complex permittivity $\varepsilon_1$ (a) and AC conductivity $G$ (b) on the volume fraction of the conductive filler for HIPS + Pd system measured at 1 kHz (curve 1), 10 kHz (curve 2), 100 kHz (curve 3) and 1 MHz (curve 4). Critical volume fraction $v_c$ of the system is about 0.32.

parallel ideal resistor and capacitor. It is demonstrated by the perfect semicircle in the Cole-Cole diagram (Fig. 4a) measured at some frequencies in range from 120 Hz to 10 MHz. The real part of the complex permittivity $\varepsilon_1$ of systems varies from several units to 1000 or more. The capacitance of the filled polymers is higher than that of the pure one (see Fig. 4b) disregarding type of the filler (conductive or dielectric), but the reasons are different:

- Filling a polymer with conductive powder creates an irregular series-parallel network of capacitors with very thin dielectric. In other words, the filling drastically decreases effective thickness of the dielectric (compare curves 1 and 2 in Fig. 4b). The concentration of the filler cannot exceed the threshold value $v_c$. In fact, the permittivity $\varepsilon_1$ really increases rapidly in the vicinity of $v_c$ (see Fig. 5a), but it is accompanied by rising conductivity $G$ of parallel resistor in the equivalent circuit (see Fig. 5b)—it means the system near $v_c$ is not a good capacitor. Moreover, the permittivity $\varepsilon_1$ is strongly dependent on frequency near $v_c$.

- Filling a polymer with Hi-K ceramic powder increases the permittivity of the dielectric (see curves 3 and 4 in Fig. 4b). The concentration of the filler can be 60 % (vol.) or more depending on size and shape of its particles.

The above described mechanisms can be combined (see curve 5 in Fig. 4b). The finer Hi-K filler is used, the thinner dielectric of the elemental capacitor can be achieved (see curve 6 in Fig. 4b; the permittivity $\varepsilon_1$ is dependent on frequency due to change in structure).

## 4. Conclusion

The technology of polymeric materials filled with different conductive powders provides a possibility to control both the DC and AC properties. The effect of additional insulating component in the system, especially powders with high dielectric constant, seems to be useful for future tuning electrical parameters. It is possible to think of a variety of applications using thick film resistors and bulk and thick film capacitors prepared at room temperature.

### References

1. Kužel, R., Kubát, J., Křivka, I., Prokeš, J., Stefan, O. and Klason, C. (1993) Heterogeneous systems based on precious metal powders and polymers, *Materials Science and Engineering B* **17**, 190–195.
2. Kubát J., Kužel, R., Křivka, I., Bengtsson, P., Prokeš, J. and Stefan, O. (1993) New conductive polymeric systems, *Synthetic Metals* **54**, 187–194.
3. Kirkpatrick, S. (1973) Percolation and Conduction, *Reviews of Modern Physics* **45**, 574–588.

# Authors Index

# Key Word Index